Lecture Notes in Mechanical Engineering

Lecture Notes in Mechanical Engineering (LNME) publishes the latest developments in Mechanical Engineering—quickly, informally and with high quality. Original research reported in proceedings and post-proceedings represents the core of LNME. Volumes published in LNME embrace all aspects, subfields and new challenges of mechanical engineering. Topics in the series include:

- Engineering Design
- Machinery and Machine Elements
- Mechanical Structures and Stress Analysis
- Automotive Engineering
- Engine Technology
- Aerospace Technology and Astronautics
- Nanotechnology and Microengineering
- Control, Robotics, Mechatronics
- MEMS
- Theoretical and Applied Mechanics
- Dynamical Systems, Control
- Fluid Mechanics
- Engineering Thermodynamics, Heat and Mass Transfer
- Manufacturing
- Precision Engineering, Instrumentation, Measurement
- Materials Engineering
- Tribology and Surface Technology

To submit a proposal or request further information, please contact the Springer Editor of your location:

China: Ms. Ella Zhang at ella.zhang@springer.com
India: Priya Vyas at priya.vyas@springer.com
Rest of Asia, Australia, New Zealand: Swati Meherishi at swati.meherishi@springer.com
All other countries: Dr. Leontina Di Cecco at Leontina.dicecco@springer.com

To submit a proposal for a monograph, please check our Springer Tracts in Mechanical Engineering at http://www.springer.com/series/11693 or contact Leontina.dicecco@springer.com

Indexed by SCOPUS. All books published in the series are submitted for consideration in Web of Science.

More information about this series at http://www.springer.com/series/11236

Anand Parey · Rajesh Kumar · Manpreet Singh
Editors

Recent Trends in Engineering Design

Select Proceedings of ICAST 2020

Springer

Editors
Anand Parey
Department of Mechanical Engineering
Indian Institute of Technology Indore
Indore, Madhya Pradesh, India

Rajesh Kumar
Department of Mechanical Engineering
Sant Longowal Institute of Engineering
and Technology
Longowal, Punjab, India

Manpreet Singh
Department of Mechanical Engineering
Lovely Professional University
Phagwara, Punjab, India

ISSN 2195-4356 ISSN 2195-4364 (electronic)
Lecture Notes in Mechanical Engineering
ISBN 978-981-16-1078-3 ISBN 978-981-16-1079-0 (eBook)
https://doi.org/10.1007/978-981-16-1079-0

This Springer imprint is published by the registered company Springer Nature Singapore Pte Ltd.
The registered company address is: 152 Beach Road, #21-01/04 Gateway East, Singapore 189721, Singapore

Contents

About the Editors

Dr. Anand Parey is a Professor in the Discipline of Mechanical engineering IIT Indore. He holds a Post-Doctoral Fellowship from the University of Alberta, Edmonton, Canada, from 2006 to 2007. He obtained his Ph.D. from IIT Delhi in Mechanical Engineering in 2006. He completed his M.Tech. from MANIT Bhopal in 2001 and B.E., MITS Gwalior, 1998. Dr. Anand Parey had worked as a Manager in Heavy Engineering Division, Larsen and Toubro Ltd. Mumbai, from 2008 to 2009. He had also contributed as a Manager—Technology in Global R&D Centre, Crompton Greaves Ltd., Mumbai, from April 2008 to August 2008. He was a Lecturer in Department of Mechanical Engineering, BITS Pilani Goa Campus, from August 2005 to November 2006. He is also one of the Reviewers of *Measurement* (Elsevier), Reviewer of *Journal of Sound and Vibration* (Elsevier). His focus is on condition monitoring, noise and vibration isolation and in signal processing of mechanical systems. He has published nearly 50 articles in reputed journals.

Dr. Rajesh Kumar is affiliated to the Department of Mechanical Engineering, Sant Longowal Institute of Engineering and Technology, Longowal. Dr. Rajesh Kumar is currently providing services as a Professor. He has authored numerous peer-reviewed scientific papers and presented works at many national and international conferences. He has guided nearly 10 Ph.D.s and 25 Master's thesis. He is also actively associated with different societies and academies. Dr. Rajesh Kumar academic career is decorated with several reputed funded proposals. He has authored nearly 60 research articles.

Dr. Manpreet Singh is working as a Professor/HOD in the School of Mechanical Engineering, Lovely Professional University, Phagwara. He obtained his B. E. Mechanical from College of Engineering and Technology, Bathinda (now Giani Zail Singh College of Engineering and Technology, Bathinda) in 1998; M.Tech. (Manufacturing and System Engineering) from Sant Longowal Institute of Engineering and Technology, Longowal (affiliated from Punjab Technical University, Jalandhar) in 2004; and Ph.D. from the Sant Longowal Institute of Engineering and Technology, Longowal (Deemed University) in 2013. He has published more than 40 research articles in international journals, conferences of high repute. He has guided 6 M.Tech. and 1 Ph.D. thesis and 5 candidates are pursuing Ph.D. with him presently. He has also published 33 patents. His academic life includes serving as a peer reviewer in journals and conducting a number of training programs. His areas of interest include design for vibration systems, modal analysis, condition monitoring, fault prognosis, reliability analysis and green buildings. He has organized three international conferences at various institutes as organizing secretary and convener.

Parametric Investigation of Turning Process Using DEMATEL Approach

Swastik Pradhan

1 Introduction

Titanium alloy material is mostly used in various industrial, marine, aerospace, automobile, and medical sectors. Mostly lightweight, durable, corrosion, and wear resistance components are made from various grades of titanium alloy. The ability to withstand high load compared to its own weight. This is due to the high strength to weight ratio properties of titanium alloy. But due to low thermal conductivity properties and high chemical reactivity, the machining is a challenging task. Most of the time, the chip tries to weld in the machined surface at high speed machining. Also due to this, the build of edge formed premature failure of the tool and irregularity in the surface deteriorates the machinability of the titanium alloy. Mostly serrated type of chips is formed during machining of titanium alloy due to the low strain and high strain conditions. Various types of experiments have been performed with various conditions with lots of advancement have been seen but till date machining of titanium alloy is a concerned factor. Various types of optimization methods had been used to know the better combination of the input variables that increases the performance of machinability. Among those models, DEMATEL method is one of the methods which defines the relationship between the output variables. This method shows a direction and identifies the effective variables through a hierarchal approach. Some of the research work done so far in different fields are by Das and Chakraborty who used the DEMATEL method to analyze the process parameter during green electric discharge machining process. The interrelationships of the process parameter were visualized by splitting into cause and effect groups. The result shows that the used optimization techniques improved the machining performance by providing a hazardous free environment [1]. Bhowmik et al. used integrated M-DEMATEL with TOPSIS based method to optimize the process parameter during abrasive water

S. Pradhan (✉)
School of Mechanical Engineering, Lovely Professional University, Phagwara, Punjab 144411, India

A. Parey et al. (eds.), *Recent Trends in Engineering Design*, Lecture Notes in Mechanical Engineering, https://doi.org/10.1007/978-981-16-1079-0_1

jet machining of sundi wood dust polymer composites. The result revealed that the proposed method can be used for different advanced machining processes [2]. Asad et al. used the grey-based DEMATEL method for modeling the flexibility of information technology supply chain management. By using this method, the flexibility can be improvised so that it can attract the attentions in the field of planning and design of supply chain [3]. Li e al. used the hybrid fuzzy DEMATEL method for the machine tool selection. It can be revealed that the proposed method enables the decision-makers to select the best alternative over multiple conflicting criteria [4]. Ocampo et al. studied the mapping strategy for sustainable food manufacturing by utilization the fuzzy-DEMATEL-ANP and TOPSIS methods [5]. Shen et al. used the cluster-weighted DEMATEL combined with the ANP method for the supplier selection of the food industry. It revealed that both quality and service are the important criteria [6]. Liu et al. evaluated the supplier selection by adopting the DEMATEL along with the game theory. The DEMATEL method was to make the result reasonable in context to the comprehensive weights [7]. Wang et al. used the DEMATEL method to improve the feed system of the CNC machining center [8]. From various research studies, it can be concluded that DEMATEL was applicable in almost all the field of science and technology fields along with successful development of the business model and strategy for companies and organizations. DEMATEL approach provides a better, authentic, and accurate selection of the process parameter by creating the cause and effects group, where the performance of the sector, machine, or systems can improve in a directional manner leading toward the success growth of the firms, industries, society, and environment. In this paper, the DEMATEL method is used to know the significant cutting variables during dry machining of titanium alloy by using the WM25CT cutting insert.

2 DEMATEL Method

In the years between 1972 and 1976, Battelle memorial Institute of Geneva organized the science and human affairs program where this DEMATEL method was put forward to provide the solution for the complicated intertwined group problems [9]. This method also shows a direction and identifies the contribution parameter in the workable solution through the hierarchical composition. The DEMATEL methodology's comprehensive steps are as follows:

Step 1 Formation of decision matrix

A decision matrix (D) is formed consisting of alternatives or input parameters used and output responses measured during turning operation. The number of experimental run condition made through the RSM approach indicates the total number of rows. The output responses are considered as number of columns in the decision matrix. The quantitative and qualitative criterion combinedly influences the performance of the turning operation. The decision matrix was made as per Eq. 1.

$$D = \begin{array}{c} \\ A_1 \\ A_2 \\ A_3 \\ . \\ . \\ . \\ A_m \end{array} \begin{array}{c} C_1\ C_2\ C_3\ \ldots\ C_n \\ \begin{bmatrix} X_{11} & X_{12} & X_{13} & \ldots & X_{1n} \\ X_{21} & X_{22} & X_{23} & \ldots & X_{23} \\ X_{31} & X_{32} & X_{33} & \ldots & X_{3n} \\ . & . & . & \ldots & . \\ . & . & . & \ldots & . \\ . & . & . & \ldots & . \\ X_{m1} & X_{m2} & X_{m3} & \ldots & X_{mn} \end{bmatrix} \end{array} \tag{1}$$

where $A_1, A_2, A_3, A_4, \ldots, A_m$ are the alternatives and $C_1, C_2, C_3, C_4, \ldots, C_n$ are the criteria as chosen by the decision-makers. By combination of the alternatives and criteria, the performance is measured $X_{m1}, X_{m2}, X_{m3}, X_{m4}, \ldots, X_{mn}$. The number of output parameters is denoted by "*m*" and the number of inputs parameters is denoted by "*n*".

Step 2 Normalized decision matrix calculation

Since all the output responses are measured through different instruments with different least counts. Mostly all output responses are having different measurement systems like decimal, fractional, and real number with different units. So, comparison between the output responses is not feasible. Hence normalization (N_{ij}) of all the output parameter is done using Eq. 2. Normalization converts all the output parameters to a value range from 0 to 1. Each normalize value has the equal weight of its respective measured value.

$$N_{ij} = \frac{Y_{ij}}{\sqrt{\sum_{i=1}^{m} Y_{ij}^2}} \tag{2}$$

Step 3 Determination of total relation matrix

The total relation matrix is formed by using the identity matrix and normalization direct relation matrix. As per the number of attributes, the identity matrix is formed. By using Eq. 3, the total relation matrix (T) is formed. The indirect effect of *i*th criterion over the *j*th criterion is represented by each element t_{ij} of the matrix (T).

$$\begin{aligned} T &= \left[t_{ij}\right]_{n\times n} \quad (i, j = 1, 2, \ldots, n) \\ T &= Y(1 - Y)^{-1}. \end{aligned} \tag{3}$$

Step 4 Rows and columns sums evaluation in *T*matrix

The sum of rows and column are evaluated in the T matrix. The Ri indicates sum of rows and C_j indicates the sum of column. The R_i and C_j are calculated by using Eqs. (4 and 5).

$$R_i = \left[\sum_{j=1}^{n} t_{ij}\right]_{n\times 1} \quad \text{where} \quad i = 1, 2, \ldots, n \tag{4}$$

$$C_j = \left[\sum_{i=1}^{n} t_{ij}\right]_{1\times n} \quad \text{where} \quad j = 1, 2, \ldots, n \tag{5}$$

Step 5 Finding the threshold value

After the row and columns summation is defined, the threshold value is to be found out. The threshold value will be the average of all the total relation matrix. Threshold value is denoted by alpha. The threshold value is calculated by using Eq. 6.

$$\alpha = \frac{\sum_{i=1}^{n}\sum_{j=1}^{n}\left[t_{ij}\right]}{N} \tag{6}$$

The total number of elements present in the T matrix is represented by N.

Step 6 Casual diagram formation

This diagram depicts the most important criteria and the influence level of the criteria shown in the causal diagram helps to recognize the driving variable. With the help of the threshold value and the value calculated in the matrix, the causal diagram is designed. "Prominence" is the vector $(R_k + C_k)$ in horizontal axis. The prominence is calculated by addition of R with C. Similarly, "relation" is the vector $(R_k + C_k)$ in the vector axis. Thus, various criteria are categorized into two groups: one is cause group and other one is effect group. $(R_k + C_k)$ indicates the significance level that the entire system is exerted by the ith criterion. $(R_k - C_k)$ indicates the net effect that ith criterion conveys on entire system. Hence when the $(R_k - C_k)$ value is positive, it indicates the criterion fall on the cause group and when $(R_k - C_k)$ value is negative, then the corresponding criterion fall on the effect group. This casual diagram is also known as cause and effect diagram. The cause and effect diagram are modeled by using the coordinates sets $((R_k + C_k), (R_k - C_k))$. This diagram indicates the most important criteria and the influence of that criteria helps to reveal the driving variables.

3 Experimental Details

The cutting insert used for machining the workpiece was modern titanium carbonitrides coated carbide tool. This type of tool has a very good adhesive nature toward bonding cemented carbides and can sustain from wear and toughness due to the coating composition of the cutting tool. The titanium Ti-6Al-4V alloy was considered as the workpiece material. The titanium was difficult to be machined due to the low thermal and high chemical reactive properties. So, machining of titanium alloy with effective combination of cutting variables was necessary.

The experiment was conducted as per the response surface method. The layout for carrying out the experiments are made by selecting the combination of cutting variables. According to the number of variables and their levels the layout was formed. The cutting variable considered were cutting speed, feed, and depth of cut.

Table 1: Experimental layout [10]

Sl No	Cutting speed	Feed	Depth of cut
Units	M/min	mm/rev	mm
1	1	1	−1
2	0	0	1
3	0	0	0
4	0	−1	0
5	0	0	0
6	0	1	0
7	1	0	0
8	1	1	1
9	0	0	0
10	0	0	−1
11	−1	1	1
12	−1	0	0
13	−1	−1	1
14	0	0	0
15	0	0	0
16	0	0	0
17	1	−1	1
18	1	−1	−1
19	−1	−1	−1
20	−1	1	−1

The levels of cutting speed considered was 65, 124, 160 m/min, for feed it was 0.12, 0.14, 0.16 mm/rev and finally for depth of cut it was 0.5, 0.75, 1.0 mm, respectively. The three level was selected as per the lower, medium, and higher value of the cutting range of the workpiece. These parameter levels are selected on the basis of literature review. The layout consists of total 20 combinations of the input variables. The central composite design technique was used to form the layout [10] as shown in Table 1.

4 Result and Discussion

The turning operation of titanium grade 5 alloy is with coated carbide insert. The layout was made by RSM methodology of central composite design approach. The modeling was done by DEMATEL method. The decision matrix was created which includes all the criteria such as cutting force, surface roughness, flank wear, and chip reduction coefficient. The criteria are measured as per the alternatives and the matrix was formed. The relationships value for the output responses assigned by the decision-makers are categorized into various factors as show in the Table 2.

Table 2: Assign the relationship for all responses

0	No influence
1	Low influence
2	Medium influence
3	High influence
4	Very high influence

Direct relationships matrix was formed by assigning the relationship value to the corresponding output responses. The normalization of the matrix was carried out to form a comparable and non-dimensional data. The normalization value was calculated by using Eq. 2. The normalized value was shown in Table 3. The normalize value changes the output variables to a comparable range so that all the values of the variables lies between 0 and 1. The identity matrix was made and multiplied by the normalization matrix so as to form the total relation matrix as shown in Table 4. The total relation matrix was calculated through Eq. 3 as shown in Table 5. The threshold value is calculated by using Eq. 6, i.e., the average of all the total relation matrix value. Threshold value is found to be 1.904472. The sum of rows and sum of column matrix is formed and denoted by Ri and Ci, respectively, as shown in Table 5. Finally, the total effect and net effect are summarized for making the cause and effect diagram. The total effect is calculated by adding the Ri with Ci (Ri + Ci) and the net effect is calculated by subtracting the Ci from Ri, i.e., (Ri − Ci). The total effect and the net effect are estimated and tabulated in Table 6. If the net effects value is positive, this indicates that the response influences all the other output variables that the other influencing it. Whereas it can be also noted that same output variable is mostly influenced by other variables if the net effect values turn to be negative. The causal graph of the output variables is portrayed in Fig. 1.

Table 3: Initial direct relation matrix for the output responses

Responses	CF	SR	FW	CRC
CF	0	2	4	2
SR	2	0	4	3
FW	4	4	0	2
CRC	2	3	3	0

Table 4: Normalization direct relation matrix for the output responses

Responses	CF	SR	FW	CRC
CF	0.00	0.20	0.40	0.20
SR	0.20	0.00	0.40	0.30
FW	0.40	0.40	0.00	0.20
CRC	0.20	0.30	0.30	0.00

Table 5: Total relation matrix

Responses	CF	SR	FW	CRC
CF	1.560976	1.853659	2.219512	1.512195
SR	1.869919	1.845528	2.398374	1.707317
FW	2.113821	2.260163	2.276423	1.756098
CRC	1.707317	1.902439	2.146341	1.341463

Table 6: Total and net effects of responses

Responses	Ri	Ci	Ri + Ci	Ri − Ci	Identity
CF	7.146341	7.252033	14.39837	−0.10569	Effect
SR	7.821138	7.861789	15.68293	−0.04065	Effect
FW	8.406504	9.04065	17.44715	−0.63415	Effect
CRC	7.097561	6.317073	13.41463	0.780488	Cause

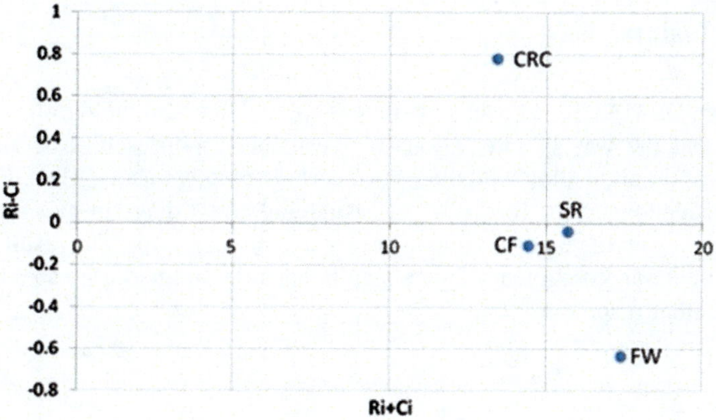

Fig. 1 Responses causal diagram

The positive value of the net effect constitutes on the cause group and the negative value of the net effect added to the effect group. In this, the CRC response falls in the category of cause group and the rest CF, SR, and FW fall on the effect group. It indicates that CRC is the main driving response for the remaining output responses. CRC is the most influencing responses among the other responses, hence it is considered to be the most significant responses among the others. Figure 2 shows the DEMATEL graph for the turning operation between the responses. Graph depicts the contextual interactions between the output responses. Bi-directional relationship exists between the cutting force, surface roughness, and flank wear, where chip reduction coefficient and flank wear have a unidirectional interaction.

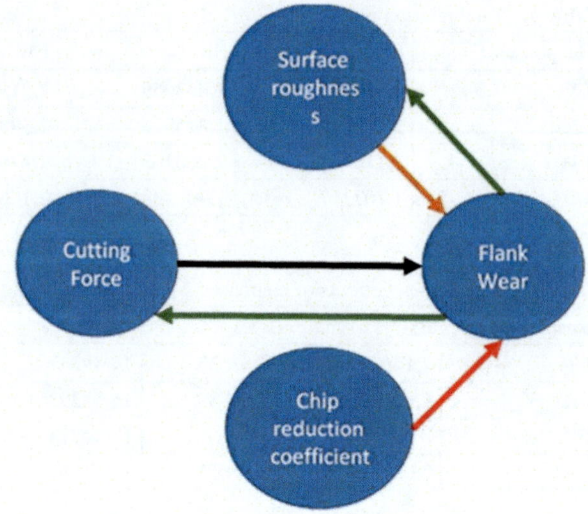

Fig. 2 DEMATEL interaction graph between the output responses

5 Conclusion

In this paper, DEMATEL, a novel technique is applied to reveal the significant output responses and the way of interactions between the responses formed during the machining of titanium alloy with the coated carbide cutting inserts. The DEMATEL approach leads toward the formation of casual and effect diagram by splitting and categorizes the output responses into the cause and effect groups. The result revealed that chip reduction coefficient plays a significant role and had a greater influence among the other responses.

References

1. Das PP, Chakraborty S (2019) Parametric analysis of a green electrical discharge machining process using DEMATEL and SIR methods. OPSEARCH 1–28
2. Bhowmik S, Gupta K (2019) Modeling and optimization of abrasive water jet machining process. Journal 29–44
3. Asad MM, Mohammadi V, Shirani M (2016) Modeling flexibility capabilities of IT-based supply chain, using a grey-based DEMATEL method. Procedia Econ Finan 36:220–231
4. Li H et al (2020) A novel hybrid MCDM model for machine tool selection using fuzzy DEMATEL, entropy weighting and later defuzzification VIKOR. Appl Soft Comput 106207
5. Ocampo L, Deiparine CB, Go AL (2020) Mapping strategy to best practices for sustainable food manufacturing using fuzzy DEMATEL-ANP-TOPSIS. Eng Manage J 1–21
6. Shen JL, Liu YM, Tzeng Y-L (2012) The cluster-weighted DEMATEL with ANP method for supplier selection in food industry. J Adv Comput Intell Intell Inf 16(5):567–575
7. Liu T, Deng Y, Chan F (2018) Evidential supplier selection based on DEMATEL and game theory. Int J Fuzzy Syst 20(4):1321–1333
8. Wang X, Zhang Y, Shen G (2016) An improved FMECA for feed system of CNC machining center based on ICR and DEMATEL method. Int J Adv Manuf Technol 83(1–4):43–54

9. Fontela E, Gabus A (1976) The DEMATEL observer, DEMATEL 1976 report. Battelle Geneva Research Center, Geneva
10. Maity K, Pradhan S (2017) Study of chip morphology, flank wear on different machinability conditions of titanium alloy (Ti-6Al-4V) using response surface methodology approach. Int J Mater Form Mach Process (IJMFMP) 4(1):19–37

Analytical Determination of Rheological Parameters of MR Fluids

Vivek Sharma, Gavendra Norkey, and Kalidasan Rathinam

1 Introduction

Magneto rheological (MR) fluid is used nowadays in various devices and equipment [1]. In the field of military operations where vibration mitigation is of utmost priority, the MR fluids play a significant role [2]. The fluids can also be used in the landing gear of the aircraft where safety of the passengers is of utmost priority [3]. One needs to ascertain the shear stress of MR fluids to determine their on-state behavior [4]. In this research, a novel technique of relative permeability is used to determine the yield stress of MR fluid. In this research, the fluid is made and is characterized. The fluid is put between the poles of the electromagnet and is subjected to high torque [5]. The following equations are used to determine the yield stress of MR fluid.

$$B_{\text{air gap}} = \mu_o H \tag{1}$$

While magnetic flux density for on-state MR fluid is given as

$$B_{\text{on}} = \mu_o \mu_r H \tag{2}$$

where μ_O is permeability of air and is equal to $4\pi \times 10^{-7}$ H/m. Using Eqs. (1) and (2)and , the relative permeability, μ_r, can be expressed mathematically as

V. Sharma · G. Norkey (✉) · K. Rathinam
Lovely Professional University, Phagwara, India
e-mail: gavendra.23303@lpu.co.in

V. Sharma
e-mail: vivek.23319@lpu.co.in

K. Rathinam
e-mail: kalidasan.22180@lpu.co.in

A. Parey et al. (eds.), *Recent Trends in Engineering Design*, Lecture Notes in Mechanical Engineering, https://doi.org/10.1007/978-981-16-1079-0_2

$$\mu_r = \frac{B_{\text{on}}}{B_{\text{air gap}}} \tag{3}$$

For the particle loading (ϕ) of a fluid as

$$\phi = \left[\frac{\tau_y}{C \times 271.7 \times \tanh(6.33 \times 10^{-6} H)}\right]^{\frac{1}{1.5239}}$$

$$\phi = \left[\frac{B_{\text{on}} - u_0 H}{1.91 \times \left(1 - \mathrm{e}^{-10.97 u_0 H}\right)}\right]^{\frac{1}{1.133}}$$

$$\left[\frac{\tau_y}{C \times 271.7 \times \tanh(6.33 \times 10^{-6} H)}\right]^{\frac{1}{1.5239}} = \left[\frac{B_{\text{on}} - u_0 H}{1.91 \times \left(1 - \mathrm{e}^{-10.97 u_0 H}\right)}\right]^{\frac{1}{1.133}}$$

The resultant yield shear stress can be given as

$$\tau_y = 271.7 \times C \tanh\left(6.33 \times 10^{-6} \frac{B_{\text{on}}}{12.56 \times \mu_r}\right)\left[\frac{B_{\text{on}}\left(1 - \frac{1}{\mu_r}\right)}{1.91\left(1 - \mathrm{e}^{\frac{-10.97 B_{\text{on}}}{\mu_r}}\right)}\right]^{1.345} \text{kPa}$$

2 Development of MR Fluid

The MR fluid mainly consists of the following components:

1. Carbonyl iron particles are used to make fluid.
2. Silicon oil is used to float the aforesaid particles.
3. Tetra methyl ammonium hydroxide is used to stop fluid from settling down under the effect of gravity.

3 Relative Permeability Approach for Fluid Characterization

In order to characterize the MR fluid, one needs an electromagnet equipped with a regulated power supply [6]. The current passes through the electromagnet, and thereby energizes the fluid. One needs to measure the on-state conductivity of the fluid and off-state conductivity of the fluid [7]. The ratio of the conductivities of the MR fluid can be used to find the shear stress of the MR fluid when activated above high magnetic fields of the range of 2 T [8]. However, other approaches can also be used to characterize the fluid which are, namely, experimental approach and

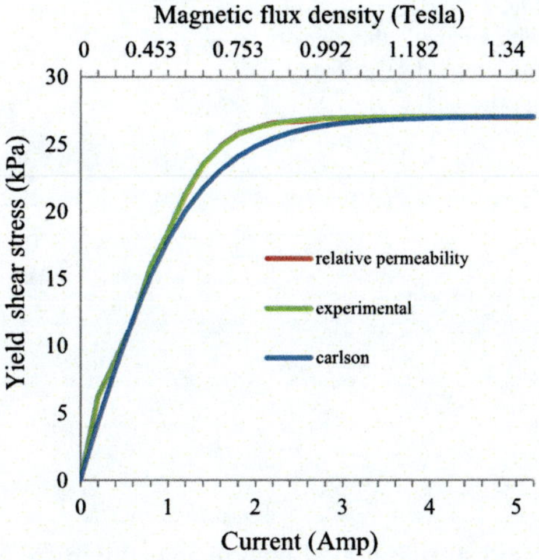

Fig. 1 Variation of yield shear stress with magnetic flux density for sample 1 by different approaches

Carlson approach [9]. The viscosity of the MR fluid can be measured using in-house designed and developed viscometer. Carbonyl iron particles of the particle density of 0.32 are used for formulating the MR fluid. These particles are mixed with oleic acid and silicon oil to make MR fluid.

4 Results and Discussions

The variation of the yield shear stress and viscosity with respect to magnetic field density for developed MR fluid samples is thus obtained using the three above different techniques, i.e., Carlson approach, relative permeability approach, and experimental approach. It can be observed from these tables and figures that the results obtained from the three approaches are matching quite well along with each other and are having a percentage error range from 2 to 7% (Figs. 1 and 2) (Tables 1, 2 and 3).

5 Conclusions

This paper paves away for any researcher to develop MR fluids at very less cost. The variation of the yield shear stress and viscosity with respect to magnetic field density for developed MR fluid samples is thus obtained using three above different techniques, i.e., Carlson approach, relative permeability approach, and experimental approach. It can be observed from these tables and figures that the results obtained from the three approaches are matching quite well along with each other and are having a percentage error range from 2 to 7%. It can be concluded that MR fluids

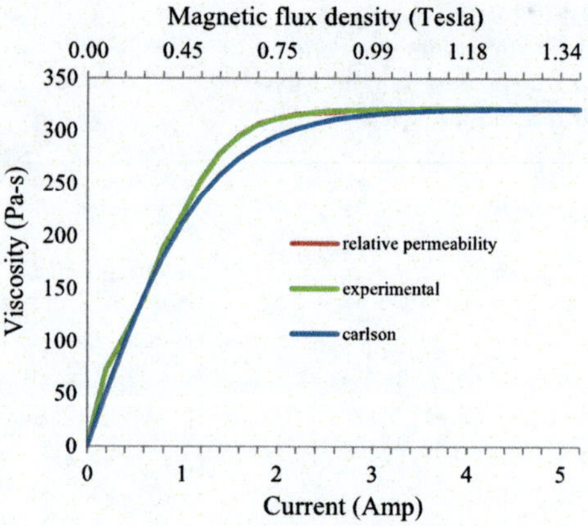

Fig. 2 Variation of viscosity with magnetic flux density for sample 1 by different approaches

yield stress can be found in an economical manner using a setup and can be used in various engineering ventures. This research can be used in determining the various parameters affecting fluid behavior. The present research work can help a researcher to compute the values of the yield stress in a novel approach known as relative permeability approach. Thus, with a minimalistic setup using a commercially available electromagnet equipped with gauss meter, one can use the values of the magnetic field density to calculate the yield stress and viscosity of the MR fluid.

Table 1 Viscosity and yield shear stress obtained from experimental approach

No	Current (A)	B_{on} Magnetic flux density (T)	Torque (Nm)	On-state yield shear stress (kPa)	On-state viscosity (Pa-s)
1	0	0	0	0	0
2	0.2	0.183	0.114	6.175	73.433
3	0.4	0.249	0.166	8.949	106.410
4	0.6	0.312	0.219	11.800	140.317
5	0.8	0.398	0.294	15.882	188.852
6	1.0	0.453	0.342	18.456	219.453
7	1.2	0.521	0.395	21.310	253.396
8	1.4	0.587	0.436	23.475	279.139
9	1.6	0.645	0.460	24.814	295.057
10	1.8	0.706	0.478	25.749	306.173
11	2.0	0.753	0.486	26.213	311.699
12	2.2	0.803	0.492	26.528	315.441
13	2.4	0.855	0.496	26.735	317.903
14	2.6	0.909	0.498	26.861	319.403
15	2.8	0.955	0.500	26.927	320.183
16	3.0	0.992	0.500	26.962	320.598
17	3.2	1.033	0.501	26.988	320.910
18	3.4	1.076	0.501	27.006	321.125
19	3.6	1.108	0.501	27.015	321.234
20	3.8	1.151	0.501	27.024	321.335
21	4.0	1.182	0.502	27.028	321.385
22	4.2	1.212	0.502	27.031	321.421
23	4.4	1.24	0.502	27.033	321.446
24	4.6	1.29	0.502	27.036	321.476
25	4.8	1.313	0.502	27.036	321.486
26	5	1.34	0.502	27.037	321.494
27	5.2	1.363	0.502	27.038	321.500
28	5.4	1.381	0.502	27.038	321.504
29	5.6	1.403	0.502	27.038	321.507

Table 2 Viscosity and yield shear stress obtained from relative permeability approach

No	B_{on} Magnetic flux density (Tesla)	$B_{air\ gap}$ magnetic flux density (T)	Relative Permeability of MR fluid	On-state yield shear stress (kPa)	On-state viscosity (Pa-s)
1	0	0	0	0	0
2	0.183	0.046	3.978	6.211	73.854
3	0.249	0.068	3.661	8.945	106.371
4	0.312	0.093	3.354	11.770	139.953
5	0.398	0.134	2.970	15.866	188.668
6	0.453	0.165	2.745	18.480	219.741
7	0.521	0.212	2.457	21.240	252.558
8	0.587	0.263	2.231	23.438	278.698
9	0.645	0.312	2.067	24.872	295.747
10	0.706	0.368	1.918	25.799	306.767
11	0.753	0.413	1.823	26.226	311.847
12	0.803	0.461	1.741	26.592	316.195
13	0.855	0.513	1.666	26.700	317.490
14	0.909	0.567	1.603	26.770	318.316
15	0.955	0.612	1.560	26.912	320.004
16	0.992	0.649	1.528	26.932	320.246
17	1.033	0.69	1.497	26.948	320.433
18	1.076	0.733	1.467	26.959	320.564
19	1.108	0.765	1.448	26.965	320.632
20	1.151	0.808	1.424	26.970	320.696
21	1.182	0.839	1.408	26.973	320.729
22	1.212	0.869	1.394	26.975	320.752
23	1.24	0.897	1.382	26.976	320.769
24	1.29	0.947	1.362	26.978	320.789
25	1.313	0.97	1.353	26.978	320.796
26	1.34	0.997	1.344	26.979	320.802
27	1.363	1.02	1.336	26.979	320.806
28	1.381	1.038	1.330	26.980	320.808
29	1.403	1.06	1.323	26.980	320.811

Table 3 Viscosity and yield shear stress obtained from Carlson approach

No	Magnetic field (Kamp/m)	B_{on} Magnetic flux density (T)	$B_{air\ gap}$ Magnetic flux density (T)	On-state Yield shear Stress (kPa)	On-state viscosity (Pa-s)
1	0	0.000	0	0	0
2	25	0.131	0.031	4.243	50.460
3	50	0.233	0.062	8.283	98.494
4	75	0.315	0.094	11.952	142.122
5	100	0.382	0.125	15.145	180.089
6	125	0.439	0.157	17.822	211.920
7	150	0.488	0.188	19.997	237.784
8	175	0.532	0.219	21.720	258.268
9	200	0.572	0.251	23.057	274.165
10	225	0.610	0.282	24.078	286.310
11	250	0.646	0.314	24.849	295.476
12	275	0.681	0.345	25.426	302.332
13	300	0.714	0.376	25.854	307.424
14	325	0.747	0.408	26.170	311.187
15	350	0.780	0.439	26.403	313.958
16	375	0.812	0.471	26.575	315.993
17	400	0.844	0.502	26.700	317.483
18	425	0.876	0.533	26.792	318.574
19	450	0.908	0.565	26.859	319.371
20	475	0.939	0.596	26.908	319.953
21	500	0.971	0.628	26.943	320.378
22	525	1.002	0.659	26.969	320.688
23	550	1.034	0.690	26.988	320.914
24	575	1.065	0.722	27.002	321.079

References

1. A Chaudhuri N Wereley R Radhakrishnan S Choi 2006 Rheological parameter estimation for a ferrous nanoparticle-based magnetorheological fluid using genetic algorithms J Intell Mater Syst Struct 17 261 269
2. Roszkowski A, Bogdan M, Skoczynski W, Marek B (2008) Testing viscosity of MR fluid in magnetic field. Meas Sci Rev 8
3. Spaggiari A, Dragoni E (2012) Effect of pressure on the flow properties of magnetorheological fluids. J Fluids Eng 134
4. A Dang L Ooi J Fales P Stroeve 2000 Yield stress measurements of magnetorheological fluids in tubes Ind Eng Chem Res 39 2269 2274
5. Shetty BG, Prasad P (2011) Rheological properties of a honge oil-based magnetorheological fluid used as carrier liquid. Defence Sci J 61:583–589

6. Shetty BG, Prasad P (2011) Study of magnetorheological fluid based flexible work holding fixture. J Manuf Eng 5:43–589
7. B Kumbhar S Patil S Sawant 2015 Synthesis and characterization of magneto-rheological (MR) fluids for MR brake application Eng Sci Technol Int J 18 432 438
8. C Sarkar H Hirani 2015 Synthesis and characterization of nano-particles based magnetorheological fluids for brake Tribol Online 10 282 294
9. (2011) Semi-active suspension system with M.R. Damper for car seat vibration—a review. Int J Curr Eng Technol

Design, Analysis, and Prototyping of USB-A Female Shield

Avinash Panchal, Nitish Sinha, Dattatray Chopade, and Ganesh Patil

1 Introduction

USB-A female is the “host” type of standard connector as shown in Fig. 1. This can be found in the computers, charging hubs, or any device having peripherals, devices, etc. plugged inside it. It is also quite useful to find extension cables with a female as shown in Fig. 2a. USB connectors have a wide use in today’s computer world. It has been researched for years by researchers and has been designed as a masterpiece. USB female connector shield failure Fig. 2b has put an attention on failure of USB. Anything that can wear will fail. If it is designed with less friction, then the chords would pop out easily. If it is designed tighter and with more friction, then it will wear faster due to there being more friction. USB ports themselves are rarely an issue. The biggest problems regarding USB device durability are USB cables going bad and strain relief. There are small wires inside the cables and they can break if flexed a lot over time. The solution to this is to use a much thicker wire with a stiffer jacket.

If the connector itself fails, there isn’t as much friction holding it in to the port, so it pops out. Port abuse is another issue. USB ports are fairly durable, but if one tries to handle it improperly like try to insert cables backwards, or attempt to connect incorrect cables to it, it can break the inside of the port. If the port is not inserted securely, there is a possibility that it can get twisted resulting in broken connection to the PCB. The wires are strained due to extreme flexing, and over time they do break internally.

USB-A has higher mechanical life (10,000 withdrawing and insertion endurance) [1]. But, as the most week points of the cables located in joints and terminations, a specific and precise control of electric field distribution at these points is important to maintain their reliability [2].

A. Panchal (✉) · N. Sinha · D. Chopade · G. Patil
Department of Mechanical Engineering, G H Raisoni Institute of Business Management, Jalgaon, India

A. Parey et al. (eds.), *Recent Trends in Engineering Design*, Lecture Notes in Mechanical Engineering, https://doi.org/10.1007/978-981-16-1079-0_3

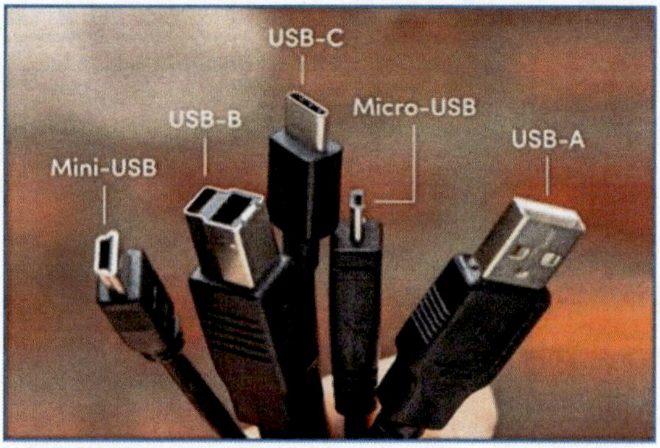

Fig. 1 Various types of USB connectors

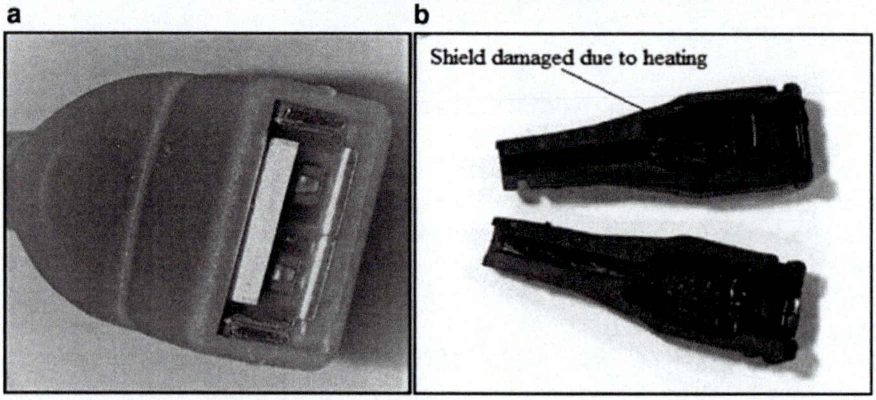

Fig. 2 **a** USB female socket and shield. **b** Common failure in USB

2 Design and Modeling Details

2.1 Design of USB Female Shield

3D Modeling and Drafting of the Plastic Product: In 3D CAD, i.e., FUSION 360 software product has been designed conceptually and checked for interference. The design assembly of USB female is shown in Fig. 3. The USB female CAD model is shown in Fig. 4.

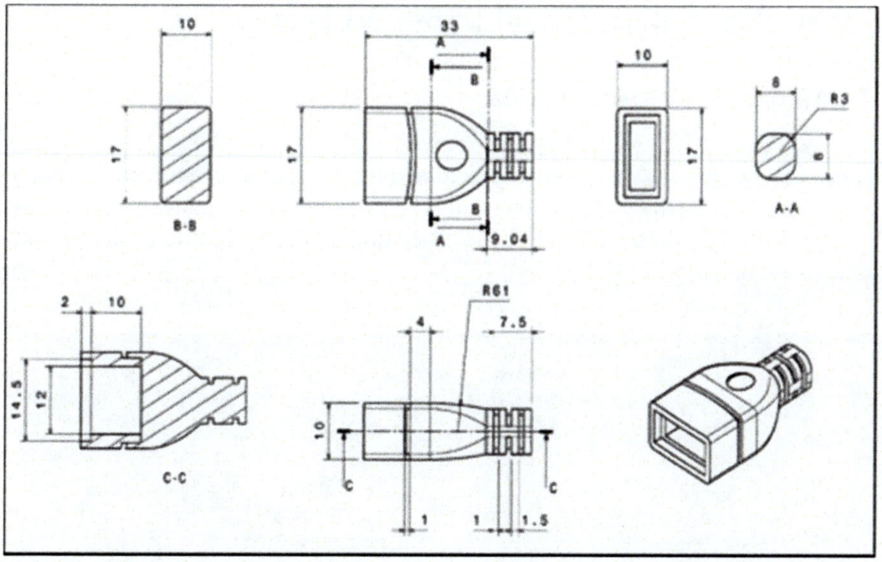

Fig. 3 Design assembly of USB female

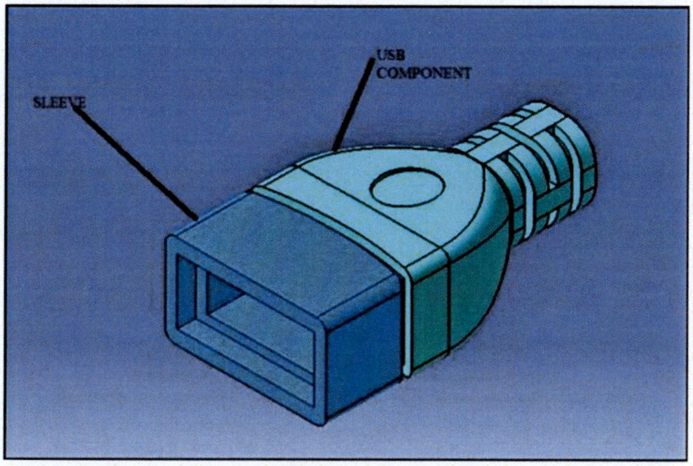

Fig. 4 USB female shield 3D model created in FUSION 360

3 Finite Element Analysis of USB-A 3.1 Female

3.1 Simulation Model of FEA

The FEA model for a USB-A female shield and mesh generation has done by using SOLID90 and SOLID87 units, as shown in Fig. 6. Two extreme boundary conditions Fig. 5 have been considered for the USB. The temperature and thermal variables of materials are listed in Table 1.

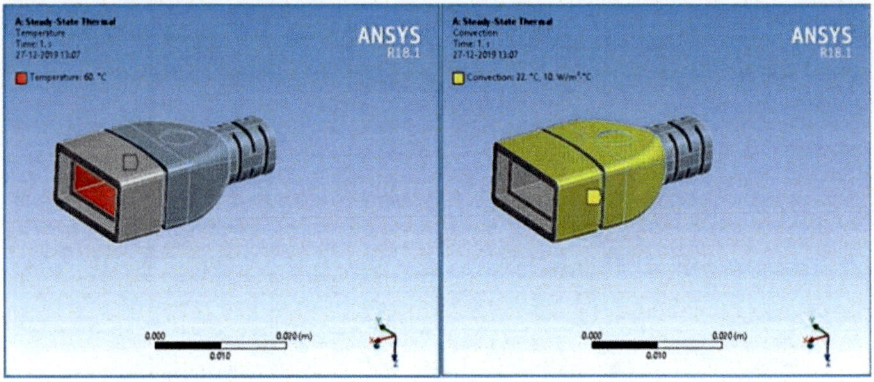

Fig. 5 Boundary conditions

Table 1 Properties of plastic material used for 3D printing and temperature boundary condition [10]

Parameters	Location	Temperature generated
Heat conductivity coefficient (W/m^2 °C)	Contact in micro USB: phosphor bronze	71
	LPC	0.44
	Resin of test PCB	0.2
	Circuit of test PCB	118
	Contacts in plug: brass	80
	Metal shell of plug: stainless steel	80
Melting point (°C)	LPC	350
Temperature BC	At contact with female socket	

3.2 Temperature Rise Due to Contact Resistance

Initially, the heat caused by the deteriorated contacts are evaluated by the FEA and experimental simulation. The heat power of GND contact was set at $q1$, as shown in Eq. (1)

$$q1 = I^2 R\,\text{GND} \tag{1}$$

3.3 Steady-State Thermal Analysis of USB-A Female Shield

As per the usual procedure to be adopted for the analysis of failure mechanism, two different heat generators are considered at USB and the heat generated in friction due to contact resistance in between the USB and a female socket. To study the thermal effects of the heat sources, the Finite Element Method (FEM) is adopted for building a simulation model of a USB-A female shield by using ANSYS 18.1 software.

The applied loading condition to this model is the heat generated inside the connector. It can be measured by Joule's heat produced due to surface contact resistance. When the USB male is connected with USB female, a contact fitting is available, which results into increased contact resistance [3]. According to Joule's law, heat produced by I^2R will develop at a time, which is raising surface temperature [3].

A meshed FEM model is as shown in Fig. 6. For thermal analysis, thermodynamic

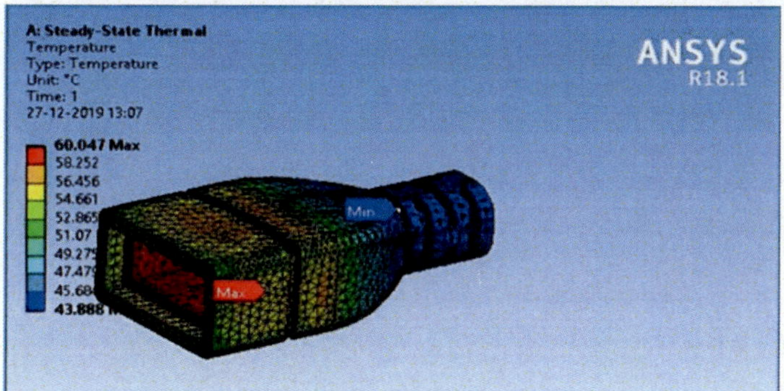

Fig. 6 Temperature condition

boundary conditions like heat generation and temperature are applied. The results of the temperature and total heat flux analysis are as shown in Figs. 6 and 7, respectively [3].

Considering the source of heat generation as heat fluxes (HF) can be defined as the heat flow rate per unit area (W/m^2). Following Eq. (2) can well explain the fact [3].

$$\mathrm{HF} = \frac{\mathrm{Power}}{\mathrm{Area}} = \frac{I^2 R}{S} \tag{2}$$

For the Finite Element Analysis, the material properties are defined. The specifications of material properties that are generating heat and USB shield material properties are given in Table 2.

In the above equation, the normal connector and contact area is $S = 0.25 \times 10^{-6}$ m^2 Joule heat. The Heat Flux obtained from experiments are given in Table 3 [3]. As per the USB standards given, a maximum 100 mA current can be supplied by a USB 1.0 port at 5 V, a maximum 500 mA current can be supplied by a USB 2.0 port at 5 V, and a maximum 900 mA current can be supplied by a USB 3.0 port at 5 V [4]. If a current carrying conductor is in contact with other current carrying conductor

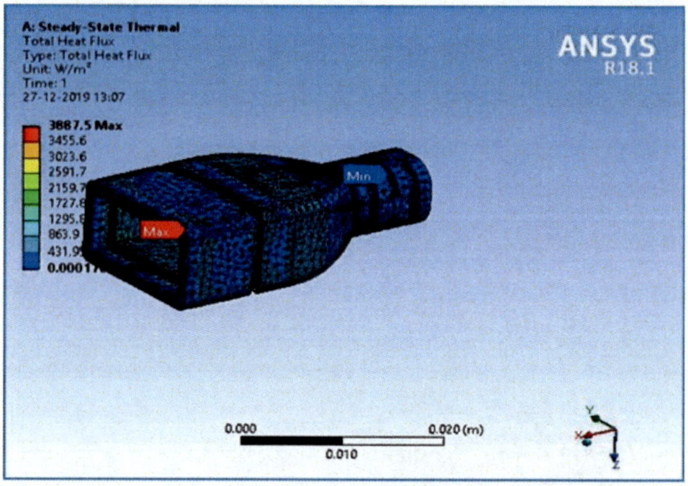

Fig. 7 Total heat flux

Table 2 The material properties of the connector

Material	Thermal conductivity W/(m°)	Specific heat J/(kg °C)	Density (kg/m^3)	Max. operating temperature (°C)
45 steel	45	46	7850	800
ABS	0.2	1.47	1050	60

Table 3 The heat flux obtained from experiments

Iteration 1	Heat flux (W/m^2)
1	3887.5
2	431.99

having some amount of contact force, then both of them have an influence on their [5].

Transmission of heat can be possible from three different ways like conduction, radiation, and convection. The conduction of heat is carried out by the differential Eq. (3) [3],

$$2\lambda A T q \nabla = q, \tag{3}$$

In Eq. (3), λ is coefficient of thermal conduction; A is area of cross section; T is temperature; q is heat production rate. As FEM is a numerical method, it solves equations but FEM software do it in the form of matrix equilibrium, Eq. (4) [3].

$$\{q\} = [K]\{T\}, \tag{4}$$

In Eq. (4), $[K]$ is heat conduction matrix; $\{T\}$ is temperature vector; $\{q\}$ is heat flow vector. By using these equilibrium equations with FEM, the distribution of temperature inside the connector is obtained [3].

The heat transmission through convection is calculated by using Eq. (5)

$$q = \alpha A(T_{\mathrm{F}} - T_{\mathrm{W}}), \tag{5}$$

In Eq. (5), α is coefficient of heat convection; A is area of cross section; T_{F} and T_{W} are temperatures of the fluid (air) and wall, respectively; q is heat flow [1].

4 Prototyping of USB-A Using 3D Printing Machine

Prototyping is the fourth next step of design thinking as well as design sprints. Rapid prototyping, or 3D prototyping, cannot be limited only up to 3D printing. It is actually known to a process of iteration for final and perfect design through a design on 3D modeling and then quickly physical prototypes manufacturing. It allows making small changes at every step again and again to improve it till satisfied prototype turns out to be perfect (Figs. 8 and 9).

Some traditional processes like injection molding, wood prototyping, and prototyping techniques require weeks or even months to manufacture each iteration. And that is not to mention the extra expense compared to other methods like 3D printing or laser cutting.

Fig. 8 3D printed USB female shield

Fig. 9 3D printing in process

4.1 Material Used for 3D Printer Filament

Acrylonitrile Butadiene Styrene (ABS), a thermoplastic material used for 3D printer Filament.

Distinct Characteristics—ABS is one of the mostly used and easily accessible, cheap materials for 3D printing.

Table 4 Comparison of calculated results and FEM results (°C)

Iterations	Calculated results	FEM results	Difference	Deviation
1	54.58	60.04	5.46	5.40
2	40.70	40.88	0.28	0.3

5 Results and Discussion

The results given by FEM are always greater than calculated results. Comparison of the calculated values and FEM values for iteration 1 and iteration 2 is given in Table 4. The deviation in the comparison is 5.40 and 0.3, respectively, for two experiments.

A 3D printed model has been developed for further mold design which is good at dimensions and can be used.

6 Conclusion

1. From the FEM analysis and calculated results, it has been seen that they are having a better consistency.
2. The deviation between calculated and FEM results is 5.40 which is less than 10%. So the FEM results are acceptable.
3. Prototype made with 3D printing is found well at all its dimensions, strengths, etc.

References

1. Volm D, Walesch R, Bachmann A (2004) Simulation of the heat transfer mechanism in relays to improve contact reliability at freezing temperatures. In: Proceedings of the 50th IEEE holm conference on electrical contacts and the 22nd international conference on electrical contacts
2. K Gaska X Xu S Gubanski R Kadar 2017 Electrical, mechanical, and thermal properties of LDPE graphene nanoplatelets composites produced by means of melt extrusion process Polymers 9 1 11 https://doi.org/10.3390/polym9010011
3. X Wang LJ Xu 2007 Finite element model analysis of thermal failure in connector J Zhejiang Univ Sci A 8 3 397 402
4. He F (2015) May. USB Port and power delivery: an overview of USB port interoperability. In: 2015 IEEE symposium on product compliance engineering (ISPCE, IEEE. 1–5 (2015). https://doi.org/10.1109/ISPCE.2015.7138710
5. Delshadpour S, Janssen JHJ, Kulkarni A (2018) Thermal and life analysis for USB PD chip with integrated load switches. In: 2018 24rd international workshop on thermal investigations of ICs and systems (THERMINIC). IEEE, 1–6 (Sept 2018)

Comparison of Line Contact and Point Contact Behavior in Rolling Element Bearing Using Statistical Analysis of Vibration Signal

Sumit Shoor and Manpreet Singh

1 Introduction

Bearing is vital component of any rotating machinery so it is important that to prevent the catastrophic failures bearing should perform safely in the machine. Non-destructive techniques are classified as thermal imaging, ultrasonic testing, wear analysis, and acoustic/vibration analysis to monitor bearing characteristics and later these could be used to design the maintenance strategies for its smooth functioning. Out of these techniques acoustic/vibration analysis are developing at higher pace which includes data acquisition, feature extraction, and condition classification because of their non-invasive and high reactivity to the incipient faults [1, 2]. In rotating components defect or any other characteristics are having unique feature in general and it can be related with the unique vibration characteristic [3]. Many authors have used vibration analysis techniques for detecting defect or other types of characteristics of rotating components of machinery [20]. Singh et al. uses motor current signature analysis, fast Fourier transform, maximum relative wavelet energy for successfully doing bearing fault detection [4]. Singh and shoor-et al. use statistical analysis of vibration signal for fault classification of bearing. Among all the parameters, they found Shannon entropy as best parameter for fault classification [5]. Many authors use wavelet transform, ANN, statistical parameters for identification and classification of defect [6–8]. Fault diagnosis is classified as extraction of feature and classification of defect from extracted features [9]. In another study, time domain features crest factor, kurtosis, skewness, and shape factor are used for classification of fault and uses APF-K algorithm. Mathematical model was also presented to analysis the nonlinear dynamic behavior having defect on rolling element [10].

S. Shoor · M. Singh (✉)
School of Mechanical Engineering, Lovely Professional University, Phagwara, Punjab 144411, India
e-mail: Manpreet.20360@lpu.co.in

A. Parey et al. (eds.), *Recent Trends in Engineering Design*, Lecture Notes in Mechanical Engineering, https://doi.org/10.1007/978-981-16-1079-0_4

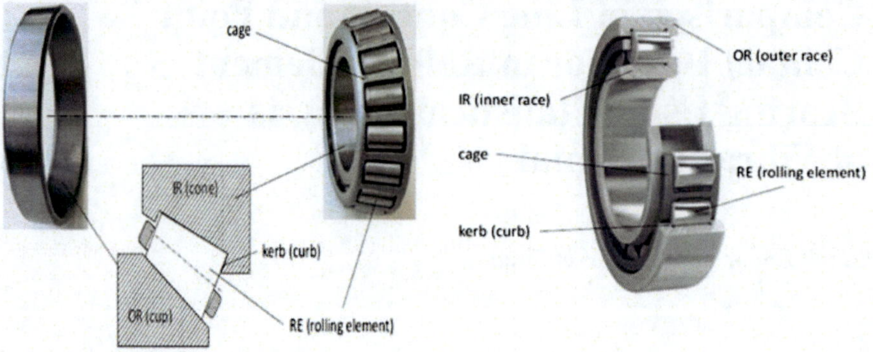

Fig. 1 Components of taper roller bearing and ball bearing

Study of vibration signature has been paid less attention while studying tribological properties of the coatings. However, it is important to study them because point contact characteristics might change to line contact or surface contact due to wear out of mating surface. Most prominent methods for coating used are thermal spray coating, Physical Vapor Deposition (PVD), Chemical Vapor Deposition (CVD), Physical Chemical Vapor Deposition (PCVD) [11], and plasma thermal spray [12] are highly used by the researchers for layer deposition (Fig. 1).

In this paper, attempt has been made to study the effect of line contact in taper roller bearing and point contact in ball bearing and further coating of AL2O3 deposited through thermal spray process. To elaborate the difference between line contact and point contact statistical analysis is performed on raw vibration signals and its spectral contents and results are presented in context with the reasoning.

2 Theoretical Background

These statistical techniques are widely used in industry for detection of fault and any variation in this parameter is an indication of fault and shown some trend.

2.1 Root Mean Square (RMS)

RMS Value is related to power of signal [13]. If the vibration level is increased then appearance of defect can easily be detected. RMS value can be used for detection of fault if it is present. Result of RMS in case of defective bearing and normal bearing can be compared for detection of fault. Theoretically, RMS can be calculated by formula [14].

$$\text{RMS} = \sum_{K=1}^{n} X^2K \Big/ n \tag{1}$$

X_K denotes signal amplitude and n denotes total data points.

2.2 Standard Deviation

Standard deviation is represented by letter σ which is used to measure the variation of data from mean value [15]

$$\sigma = \sum_{K=1}^{n} (X_K - \overline{x})^2 \Big/ n \tag{2}$$

where $\overline{x}$ denotes the mean amplitude value of the signal. When there is defect then the peak of amplitude will increase and it will cause increase in value of standard deviation and cause more deviation form mean. If mean is zero then standard deviation is equal to RMS value of signal.

2.3 Skewness

Skewness indicates symmetry of data. If skewness of data calculated is negative then curve is shift toward left side and if positive then curve shifts toward right side and zero represents the symmetry of curve [16] (Fig. 2)

$$\text{Skewness} = 1/n \sum_{K=1}^{n} (X_K - \overline{x})^3 \Big/ \sigma^3 \tag{3}$$

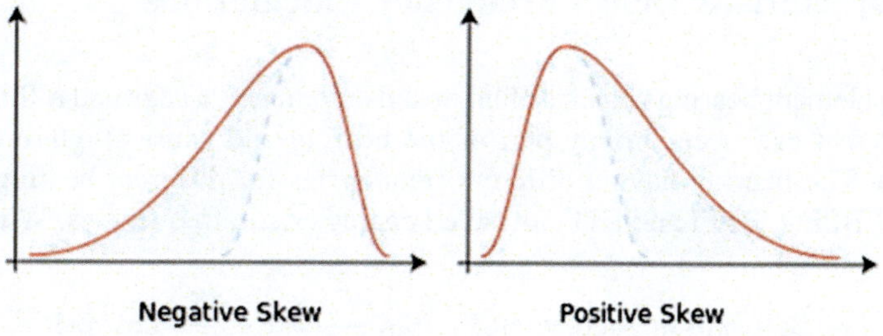

Fig. 2 Probability density with non-zero skewness

2.4 *Kurtosis*

Kurtosis is a statistical parameter which gives impulsiveness. Higher value of kurtosis gives higher value at mean and longer tail. Kurtosis value increases with increase in defect due to impulse created when rolling element passes through the defect. Hence, it can be used for detecting faults in rolling element bearing [16, 17].

$$\text{Kurtosis} = {}^{1/n \sum_{k=1}^{n} [x - \overline{x}]^4} \Big/ {}_{[1/n \sum_{k=1}^{n} [x - \overline{x}]^2]^2} \tag{4}$$

2.5 *Fast Fourier Transform (FFT)*

For converting time-domain signal to frequency domain, FFT is used [18]. The FFT for a time signal, $x(t)$ is represented by Eq. (5): here, X(f) represents Fourier transform, f indicates global frequency, and t represents the time. The signal $x(t)$ can be written as (6)

$$x(f) = \int_{-\infty}^{\infty} x(t)e^{-i2\pi ft} \tag{5}$$

$$x(f) = \frac{1}{2\pi} \int_{-\infty}^{\infty} x(F)e^{-i2\pi ft} \tag{6}$$

3 Characteristic Defect Frequency Calculations

A rolling element bearing consists of inner and outer races, a cage and rolling element (ball). Defect can occur in any part of the bearing and causes high amplitude of vibration. The bearing having different frequencies for different bearing elements like BSF, BPFO, BPFI and FTF are called characteristic frequencies. When,

$$\text{Ball spin frequency (BSF)} = f_s \frac{P_d}{2B_d}\left(1 - \frac{B_D^2}{P_D^2}\cos^2 \emptyset\right) \tag{7}$$

$$\text{Outer race frequency (BPFO)} = f_s \frac{N_B}{2}\left(1 - \frac{B_D}{P_D}\cos \emptyset\right) \tag{8}$$

$$\text{Inner race frequency (BPFI)} = f_s \frac{N_B}{2}\left(1 + \frac{B_D}{P_D}\cos\emptyset\right) \tag{9}$$

$$\text{Fundamental train frequency(FTF)} = \frac{F_S}{2}\left(1 - \frac{B_D}{P_D}\cos\emptyset\right) \tag{10}$$

Pd = Bearing pitch diameter, mm; Bd= Ball diameter, mm; Nb = number of balls.

Ø = contact angle; Fs=shaft speed or inner race speed, rpm.

4 Experimental Setup and Data Acquisition System

In this paper, bearing test rig is used for conducting the experiment. For conducting experiment AC current motor is used having a capacity 0.75 kW (Make: Crompton and speed 1440 rpm) drives the main shaft. Taper Roller Bearing and Ball Bearing are used for conducting experiment (Make: NBC, Model: 30205) and were used to support two shaft as shown in Fig. 3. Stepper pulley is used for providing the variation in speed. Speed which can be varied is 1.114, 1.955, 2.981 m/s, and also a provision is given to provide load on the Bearings. For measuring speed of shaft, optical tachometer is used and for acquiring vertical acceleration, PCB make accelerometer is used which is having a sensitivity of 1000 mV/g and placed on top of bearing casing 2. Data acquisition system of national instrument is used to capture vibration data. For further processing signal in Matlab.

Environment, a provision is given to store the data in hard disc.

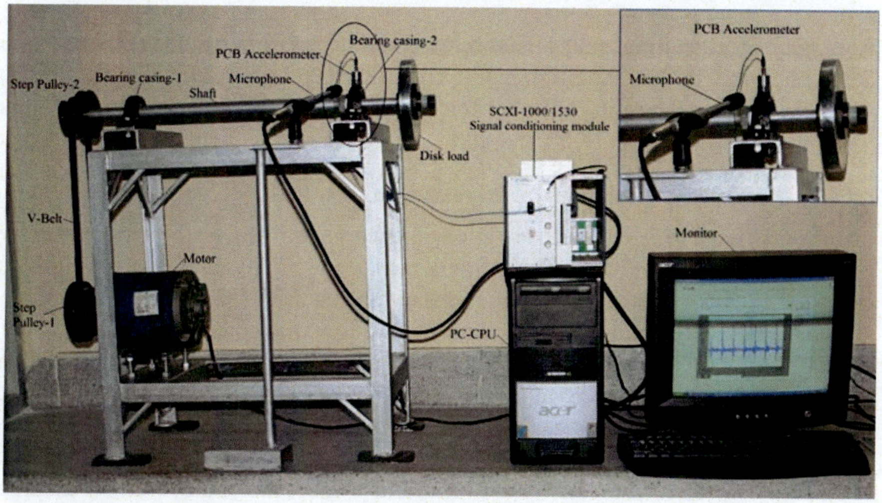

Fig. 3 Photograph of test rig for bearing

In this paper, dynamic behavior of two different bearings having the same coating of AL2 O3, one having a line contact (C1) and other having a point contact (C2) has been studied. The first bearing 30205 NBC model tapper roller bearing and another ball bearing (NBC 2Nu05) were used. The linear speed of rolling element for both bearings have been maintained the same by varying the RPM. Firstly, statistical analysis has been carried out on vibration signature originating from these two bearings. The results depict that the bearing having a line contact is yielding higher value of RMS, STD Deviation. The main reason lies that the line contact is having higher value of amplitude as well as randomness in comparison to bearing having a point contact. However, the kurtosis value is coming out to be higher in case of bearing having point contact that means the bearing having a point contact is having a higher value of peakedness. The same statistical analysis is carried on spectral contact of both signals. The same trend has been observed for RMS and STD Deviation as in the case of raw signal but better variation of parameter is observed for statistical analysis of spectral. The randomness for spectral component also has been calculated and bearing having a line contact is showing higher value in comparison to bearing having point contact. This might be because bearing having line contact is having better energy level randomness. The statistical parameter also put them in line for their effectiveness in elaborating the difference in line contact and point contact characteristics for their percentage rise in value for line/Point contact. The effectiveness value is coming out to be maximum for randomness measured for spectral content. It can be stated that while that statistical parameter can be used for differentiating point and line contact in bearing.

5 Result and Discussion

Rolling element bearings play a vital role in rotating machinery and received a great attention in the field of condition monitoring. In this present study, the ball bearing and taper roller bearing are selected for analysis. The vibration signal is recorded for linear speed of 1.14, 1.955 and 2.981 m/s. The signal captured were analyzed through statistical parameters RMS, skewness, kurtosis, etc. During experimentation, the bearings are run at same linear velocity by maintaining the same rpm. Statistical parameters like RMS, standard deviation, skewness, kurtosis, and Shannon entropy are mostly used in condition monitoring which are used for analyzing the signal to compare the line contact with point contact with increase in linear speed. Value of statistical parameter with increase in linear speed for point contact and line contact is presented along with difference in percentage in Table 1. RMS, STD, and Shannon entropy have witnessed more values for line contact bearing than that of point contact bearing at same speed, whereas kurtosis and logarithmic energy has witnessed more values for point contact bearing. At low speed, kurtosis showed maximum variation to elaborate different characteristics of line and point contact; however, for medium and high-speed STD and Shannon entropy, respectively, are proven to be best. Here, we have not considered the skewness because it has not shown any fixed trend

Table 1 Statistical parameter showing percentage variation of ball bearing and taper roller bearing

Statistical parameters for raw signal	C1 at 1.114 (m/s)	C2 at 1.114 (m/s)	Difference in percentage	C1 at 1.955 (m/s)	C2 at 1.955 (m/s)	Difference in percentage	C1 at 2.981 (m/s)	C2 at 2.981 (m/s)	Difference in percentage
RMS	0.2084	0.098	52.975	0.638	0.1009	84.172	0.9438	0.0929	90.15893
STD	0.2084	0.0994	52.303	0.638	0.1006	84.219	0.9437	0.0925	90.19816
Skewness	−0.0 296	0.0525	43.619	−0.09	−0.064	28.888	−0.1693	−0.004	97.637
Kurtosis	4.3506	12.6	65.471	3.057	7.639	59.982	2.27	5.986	62.078
Shannon entropy	4.928×10^3	1.741×10^3	64.671	2.836×10^3	1.847×10^3	34.873	2.1123×10^4	1.6529×10^3	92.174
Logarithmic energy	-2.26×10^5	-3.063×10^5	26.216	-1.112×10^5	-3.003×10^5	62.97	-6.318×10^4	-3.079×10^5	79.48

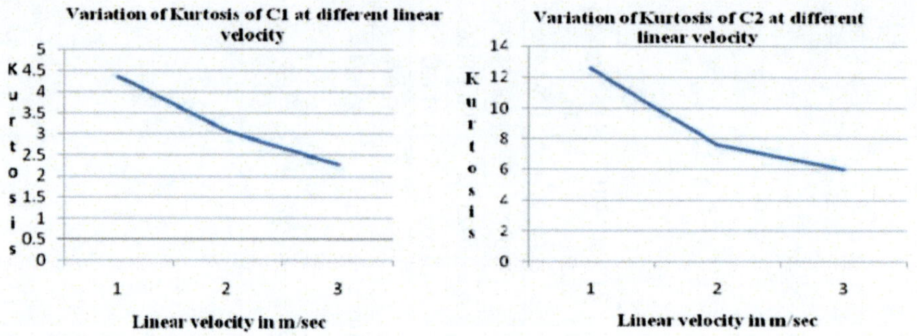

Fig. 4 Plot of Kurtosis at different linear velocities for the ball bearing and taper roller bearing

for different characteristics. It can be observed from Fig. 4 that only kurtosis is showing trend while studying the speed variation for the same case as value of kurtosis decreases with increase in speed for both the cases of C1 and C2. The sensitivity index [19] has been calculated for both the case of C1 and C2 and it is coming out to be 47.82% and 52.49%, respectively. The kurtosis has proved to be slightly more sensitive for case C2.

5.1 Frequency Spectrum Analysis

Spectral parameter generated with FFT was also analyzed. The statistical parameters are calculated in the same way of spectral components for each case and the calculated values are shown in Table 2. Percentage variation was also calculated to check the effectiveness of different statistical parameters for bearing having point contact and line contact with increase in RPM. It has been found that the Shannon entropy of spectral component has shown maximum variation for depicting the characteristics of C1 and C2 as in all the cases percentage variation is coming out to be maximum. However, no trend has been observed with respect to speed variation with any of the statistical parameters.

6 Conclusion

In this study, statistical parameters were proposed to compare the kinematic characteristics of two bearings: one having line contact (taper roller bearing and another having point contact (ball bearing) [NBC: model 30205 for taper roller bearing and ball bearing [NBC 2NU05]. Vibration signal were acquired and statistical analysis was carried out on both raw signal and its spectral components. To depict the line and point contact kinematic characteristics mixed response has been observed for

Table 2 Statistical parameter along with percentage in variation of spectral and content of ball bearing taper roller bearing

Statistical parameters for spectral contents	C1 at 1.114 (m/s)	C2 at 1.114 (m/s)	Difference in percentage	C1 at 1.955 (m/s)	C2 at 1.955 (m/s)	Difference in percentage	C1 at 2.981 (m/s)	C2 at 2.981 (m/s)	Difference in percentage
RMS	46.59	22.312	52.109	142.5	22.6	84.164	211	20.76	90.161
STD	32.9	16.28	50.516	94.24	17.7	81.218	145.82	16.08	88.972
Skewness	2.597	3.311	21.564	1.417	4.33	67.236	5.304	4.307	18.797
Kurtosis	18.39	25.118	26.785	6.294	37.6	83.261	160.71	36.466	77.309
Shannon entropy	-9.622×10^{8}	-1.899×10^{8}	80.263	-1.1012×10^{10}	-2.048×10^{8}	98.140	-2.6878×10^{10}	-1.687×10^{8}	99.372
Logarthmic energy	2.769×10^{5}	2.299×10^{5}	16.973	4.1591×10^{5}	2.153×10^{5}	48.233	4.585×10^{5}	2.113×10^{5}	53.914

statistical analysis of raw signal as at low speed kkurtosis showed maximum variation of 65.471%; however, for medium- and high-speed STD and Shannon entropy have shown maximum variation of 84.219% and 92.174%, respectively. In the statistical analysis of spectral components SShannon entropy has shown maximum variation and for the speed 2.981 m/s it was 99.3723%. The Shannon entropy of spectral components is suggested for depicting the kinematic characteristics of line and point contact bearings. In elaborating the speed variation, only kurtosis has shown decreasing trend in statistical analysis of raw signal with increase of speed and hence found suitable to deal with the speed characteristics.

References

1. Tandon N, Choudhury A (1999) A review of vibration and acoustic measurement methods for the detection of defects in rolling element bearings. Tribol Int 32:469–480
2. Elforjani M, Mba D (2010) Accelerated natural fault diagnosis in slow speed bearings with acoustic emission. Eng Fract Mech 77:112–127
3. Mogal SP, Lalwani DI (2015) Experimental investigation of unbalance and misalignment in rotor bearing system using order analysis. J Meas Eng 3(4):114–122
4. Singh S (2015) A Rev Vib Anal Tech Rotating Mach 4(3):757–761
5. Singh M, Shoor S, Singh H (2018) Shannon entropy a better indices for local defect detection and to study the effect of variable loading condition for taper roller bearing. Int J Mech Eng Technol (IJMET) 9(7):198–208
6. Li J, Wang H, Wang X, Zhang Y (2020). Rolling bearing fault diagnosis based on improved adaptive parameterless empirical wavelet transform and sparse denoising. Measurement 152. Article 107392
7. Ali JB, Fnaiech N, Saidi L, Chebel-Morello B, Fnaiech F (2015) Application of empirical mode decomposition and artificial neural network for automatic bearing fault diagnosis based on vibration signals. Appl Acoust 89:16–27
8. Dyer D, Stewart RM (2019). Detection of rolling element bearing damage by statistical vibration analysis 100:229–235.
9. Lu N, Zhang G, Xiao Z, Malik OP (2019) Feature extraction based on adaptive multiwavelets and LTSA for rotating machinery fault diagnosis, vol 2019. https://doi.org/10.1155/2019/1201084
10. Pandya DH, Upadhyay SH, Harsha SP (2013) Expert systems with applications fault diagnosis of rolling element bearing with intrinsic mode function of acoustic emission data using APF-KNN. Expert Syst Appl 40(10):4137–4145
11. Pahade VS, Chavan PS, Baisane VP (2016) A review paper on Vapour deposition coating. Int J Eng Appl Sci (IJEAS) 6:75–78
12. Lashmi PG, Ananthapadmanabhan PV, Unnikrishnan G, Aruna ST (2020) Present status and future prospects of *plasma sprayed* multilayered *thermal* barrier coating systems. J Eur Ceram Soc 40(8):2731–2745
13. Algule SR, Hujare DP (2015) Experimental study of unbalance in shaft rotor system using vibration signature analysis. 3(4):124–130
14. Igba J, Alemzadeh K, Durugbo C, Thor C (2016) Analysing RMS and peak values of vibration signals for condition monitoring of wind turbine gearboxes. Renew Energy 91:90–106
15. Jena DP, Singh M, Kumar R (2012) Radial ball bearing inner race defect width measurement using analytical wavelet transform of acoustic and vibration signal. Meas Sci Rev 12(4):141–148

16. Cambow R, Singh M, Bagha AK, Singh H (2018) To compare the effect of different level of self-lubrication for bearings using statistical analysis of vibration signal. Mater Today Proc 5(14) Part 2:28364–28373
17. Aye SA (2011) Statistical approach for tapered bearing fault detection using different methods. In: Proceedings of the world congress on engineering, vol III WCE 2011, London, pp 6–9, 6–8 July 2011
18. Singh G, Kumar R, Singh M, Singh J (2017) Detection of crack initiation in ball bearing using FFT analysis. Int J Mech Eng Technol 8(7):1376–1382
19. Pannell DJ (1997) Sensitivity analysis of normative economic models: Theoretical framework and practical strategies. Agric Econ 16(2):139–152
20. Kumar R, Jena DP, Singh M (2010) Identification of inner race defect in radial ball bearing using acoustic emission and wavelet analysis. In: ISMA2010 International conference on noise and vibration engineering, Leuven, Belgium, pp 2883–2891, 20–22 Sept 2010

Vibration Analysis for Failure Detection of Bearing and Gear Assembly

Tejbir Kaur and Gagandeep

1 Introduction

Bearings and gears are an essential part of any machine. Most of the bearings are used to support rotating shafts in machines and give minimum friction to the rotating shaft. The gear is a tooth element; these teeth are attached to a wheel that rotates and transmits motion. Vibration analysis is an extensively used technique to predict defects in rolling components such as bearings and gears[17, 19]. Bearings and gears produce vibration and sound due to the presence of a fault in them due to wear, improper lubrication, fatigue, excessive load, backlash, manufacturing fault, mechanical looseness, and failure of the transmission system.

Tandon and Choudhury [1] detected localized and distributed categories of defects in rolling element bearings and performed vibration analysis in both time and frequency domains using an accelerometer. Hoshi [2] developed a simple, low-cost setup for online monitoring of ball bearing on the machine tool spindle. The vibration analysis technique is used for analyzing the initiation and the progress of geometrical damage of the ball bearing surface and it is also used to predict the life of ball bearing assembly. Hariharan et al. [3] used vibration analysis for misaligned shaft-ball bearing systems and predicted the vibration spectrum and compared the experimental results with ANSYS-based numerical results. Patidar and Soni [4] studied different methodologies to analyze the bearing vibration and used these techniques to predict rolling element bearing faults. Vernekar et al. [5] developed a faulty engine setup for experimental studies to capture vibrational values from a healthy and faulty gear setup. The vibrational signals are acquired from the I.C. engine using an accelerometer, for healthy and faulty gear conditions. Two vibration signal analysis techniques, continuous wavelet transform, and vibration signal analysis was used for fault determination. Gligorijevic et al. [6] presented a new

T. Kaur (✉) · Gagandeep
Punjab Engineering College, Department of Mechanical Engineering, Chandigarh, India

A. Parey et al. (eds.), *Recent Trends in Engineering Design*, Lecture Notes in Mechanical Engineering, https://doi.org/10.1007/978-981-16-1079-0_5

automated vibration signal processing technique and wavelet transform in the packaging materials industry for early fault prediction in the rolling element bearing. The vibrational signals are acquired from various cases such as for healthy conditions, with ball faults, with inner and outer race faults. Lessmeier et al. [7] made the experimental setup for condition monitoring of bearings in combination with an extensive description of the bearing damage, the dataset generation by experiments, and results of data-driven classifications were used as a diagnostic method. Gupta and Pradhan [8] examined the latest research techniques on defects in bearings, vibration source in bearings, and vibration analysis techniques such as frequency-domain analysis, time-frequency domain analysis, and time-domain analysis. Generally, in any machine or equipment, source of vibration is a defected element of machine-like bearings or gears, and vibration signals acquired from these were used to determine fault [9, 10]. The presence of local defect was estimated by calculating the statistical parameters from vibration and acoustic signals. The healthy setup statistical parameters were compared with the faulty setup to predict the fault. Shannon entropy (S.E.) showed a better variation among all the faulty bearing parameters than the healthy bearing for both the signals[9].

1.1 Vibration Analysis

Vibration is a system response that causes the system to oscillate due to an external or internal force. Vibration analysis is the process of measuring the vibration level of a machine [11, 13, 16]. The acquired vibration data can be used to analyze the machine's health and its components, such as bearings and gears. The vibration analysis technique is used to reduce maintenance costs and equipment downtime by detecting equipment failures. Generally, vibration is measured using an accelerometer. An accelerometer is attached to the component and generates voltage signals (corresponding to the machine or component vibration)[12, 14, 15]. These signals are collected and stored by a vibration data acquisition system (DAQ) connected to the accelerometer. Further, the acquired vibration signals are analyzed in the time and frequency domain.

1.1.1 Time-Domain Analysis

The time-domain analysis is the variation of the output vibrational signals of the system concerning time. A time-domain representation shows how vibration signals change with time[18]. Most of the machine/equipment vibration issues are detected using this analysis and checks the system's stability and accuracy. The time-domain study examines some statistical parameters from vibrational data, and using these values, bearing and gear health conditions can be predicted. An easy way for this analysis is to measure RMS value, peak value, Kurtosis value, mean value (average value), standard deviation, and variance of vibration data. They are defined as follows.

$$\text{Mean}, \mu = \sum X_i/N \tag{1}$$

where $\sum X_i$ is sum of all the individual values and N is total number of items.

$$\text{Peak value} = \max(X_i) \tag{2}$$

where X_i is number of values i to N

$$\text{Variance} = \sum (X_i - \mu)^2/N \tag{3}$$

$$\text{Standard Deviation}, \sigma = \sqrt{\left[\sum (X_i - \mu)^2/N\right]} \tag{4}$$

$$\text{Root mean square value, RMS} = \sqrt{\sum (X_i)^2/N} \tag{5}$$

$$\text{Upper bound value, UB} = \max X_i + 0.5(\max X_i - \min X_i)/(N-1) \tag{6}$$

$$\text{Kurtosis value} = \sum (X_i - \mu)^4/N \times \text{RMS} \tag{7}$$

$$\text{Crest factor} = \max(X_i)/\text{RMS} \tag{8}$$

$$\text{Skewness value} = \sum (X_i - \mu)^3/N\sigma^3 \tag{9}$$

1.1.2 Frequency Domain Analysis

Frequency domain analysis is another way to investigate the vibration signals, which shows the amount of signal over a frequency range within each specific frequency band. Generally, the signals stored by DAQ are in the time domain; these signals are converted into the frequency domain using FFT (Fast Fourier transform). In frequency-domain analysis, frequency is plotting against the amplitude of acceleration. Figure 1 shows the time and frequency-domain signals.

2 Experimental Studies

The gear-bearing assembly test setup was developed for experimental studies. The detail about the experimental setup and procedure is discussed as following.

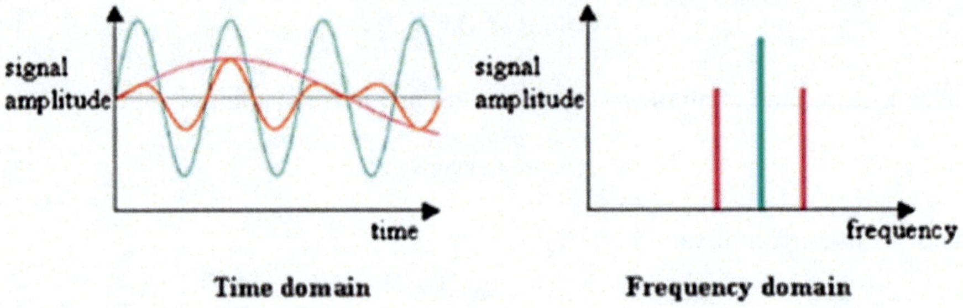

Fig. 1 Signal representation in time and frequency domain

2.1 Experimental Setup Description

Figure 2 shows the line sketch of the experimental setup. The setup was developed for fault detection in gears and bearings using the vibration analysis technique. A 1440 rpm (0.25 HP) constant speed motor is connected to the main shaft having a length of 609.6 mm and 20 mm diameter using jaw coupling. Both ends of the shaft were supported by bearings UCP204 having a diameter of 20 mm. One more shaft with length 304.8 mm and diameter 20 mm is used, which is connected to the main shaft using spur gears having a diameter of 20 mm with 40 teeth to transmit power and motion. A tri-axial accelerometer has 300 mv/g sensitivity with a data acquisition system (DAQ) that stores the vibrational signals at a rate of 800 samples/s.

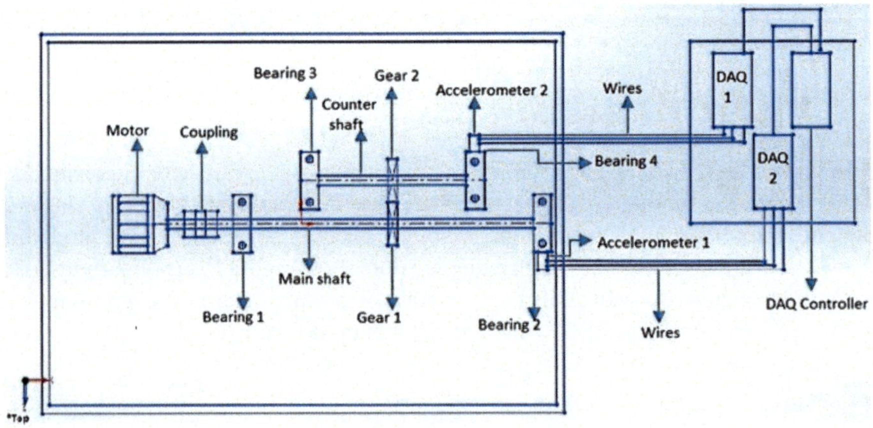

Fig. 2 Line sketch of the experimental setup

2.2 *Experimental Procedure*

Figure 3 shows the algorithm for vibrational signal monitoring and analysis in which the initial step was to measure the vibrations of the machine/system in X-, Y-, and Z-directions using a tri-axial accelerometer mounted at 90° on the bearing housing, for various cases such as (Fig. 4)

- Healthy setup
- Faulty bearing setup
- Faulty gear setup

The detected vibrational signals were stored by data acquisition system (DAQ) in the micro S.D. card in the form of time-voltage signal values. These vibrational signal values converted into time-acceleration values using the relation given below

$$\text{acceleration}\left(\text{m/s}^2\right) = \frac{\left[\text{output reading(mv)} - \text{zero signal output(mv)}\right]}{\text{senstivity(mv/g)}} \tag{10}$$

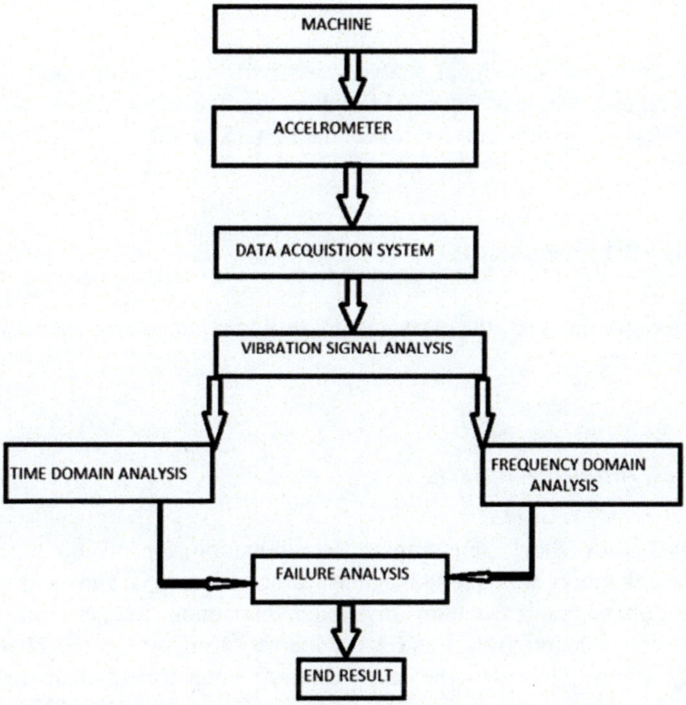

Fig. 3 The algorithm for vibrational signal monitoring and analysis

Fig. 4 **a** Experimental setup. **b** tri-axial accelerometer. **c** Faulty bearing. **d** Faulty gear

Further, the vibrational signal analysis was performed in time and frequency domain to compute the peak value, RMS value, average value, standard deviation, and other statistical parameters to estimate the system's fault.

3 Result and Discussion

In this section, we analyzed the experimental results in time and frequency domain for all three cases.

3.1 *Time-Domain Analysis*

The below figures show time-domain representation for all the three cases in which acceleration was plotted against time. Figure 5a shows the healthy setup time-domain representation in which maximum acceleration value is 1.716667(m/s^2), 1.006667(m/s^2) and 3.101628(m/s^2) in X-, Y-, and Z-directions, respectively. Figure 5b shows the faulty bearing setup time-domain representation in which maximum acceleration value is 2.603333(m/s^2), 1.698512(m/s^2), and 4.166867(m/s^2) in X-, Y-, and Z-directions, respectively, and Fig. 5c shows the faulty gear setup time-domain representation in which maximum acceleration value

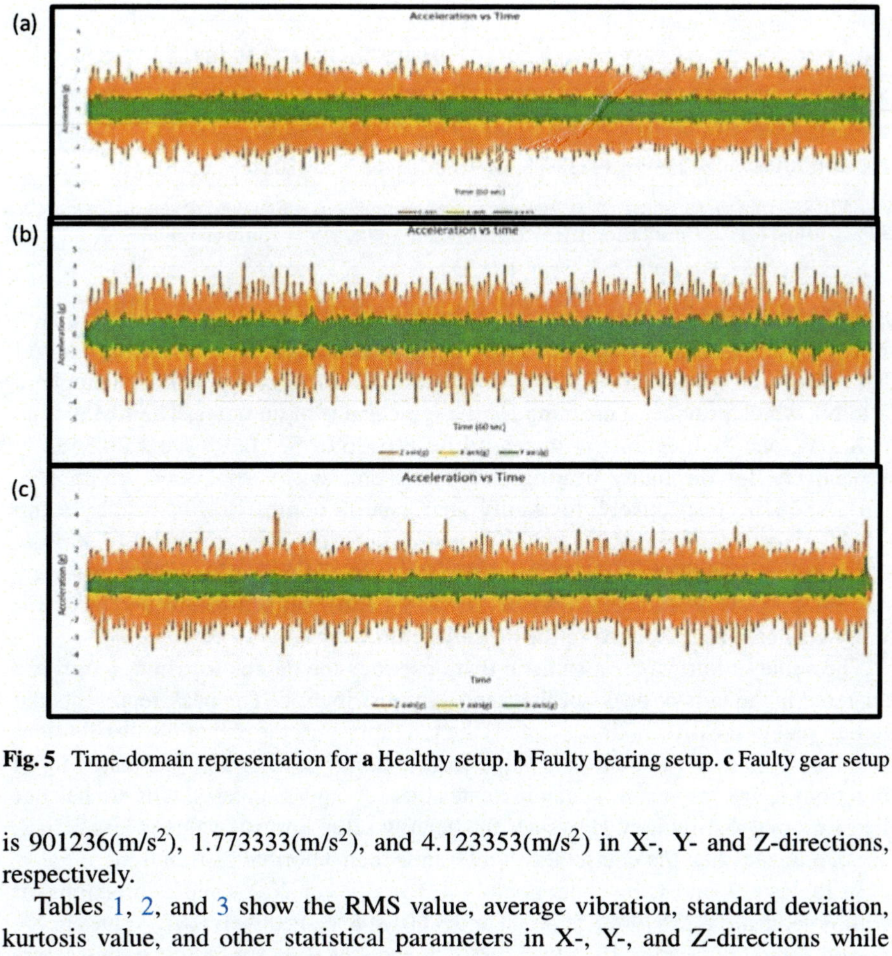

Fig. 5 Time-domain representation for **a** Healthy setup. **b** Faulty bearing setup. **c** Faulty gear setup

is 901236(m/s^2), 1.773333(m/s^2), and 4.123353(m/s^2) in X-, Y- and Z-directions, respectively.

Tables 1, 2, and 3 show the RMS value, average vibration, standard deviation, kurtosis value, and other statistical parameters in X-, Y-, and Z-directions while

Table 1 Healthy setup results

Axis	μ	σ	PV	C.F	RMS	KV	UB	Skewness
X	0.22181	0.1794	1.7166	4.7862	0.225375	2.739268	2.216796	0.814197
Y	0.17132	0.1302	1.0066	4.6722	0.215435	2.880868	1.500658	0.821229
Z	0.65708	0.4889	3.1012	4.1154	0.753648	2.911623	3.601895	0.799080

Table 2 Faulty bearing setup results

Axis	μ	σ	PV	C.F	RMS	KV	UB	Skewness
X	0.31121	0.26215	2.60333	7.9395	0.40683	4.30727	3.103596	1.37458
Y	0.25502	0.20615	1.69852	5.1535	0.32797	3.58001	2.198625	1.32489
Z	0.75624	0.60317	4.16687	7.2407	0.96724	3.47101	4.668564	1.42489

Table 3 Faulty gear setup results

Axis	μ	σ	PV	C.F	RMS	KV	UB	Skewness
X	0.32324	0.12208	1.90123	5.6582	0.284958	4.954268	2.401356	1.27589
Y	0.20461	0.17658	1.77333	6.5613	0.270271	4.556387	2.27335	1.22854
Z	0.70135	0.51236	4.12335	5.7569	0.783648	3.358084	4.62362	1.45861

Note RMS = root mean square, μ = Average vibration(m/s^2), σ = Standard deviation (m/s^2), PV = Peak value, C.F. = Crest factor, UB = Upper bound value, KV = Kurtosis value

using healthy, faulty bearing and faulty gear setup. These results were calculated from the time-domain acceleration responses measured using the accelerometer and data acquisition system. The RMS value is used to measure the magnitude of numbers (1 to N), which provides a meaning for the typical size of numbers. The RMS value in X-, Y-, and Z-directions is increased by 80.536%, 52.2069%, and 28.3482%, respectively, for the faulty bearing case and increased by 26.4373%, 25.4536%, and 3.98064%, respectively, for faulty gear case in comparison of healthy setup values. Average vibration value in X-, Y-, and Z-directions is increased by 40.3064%, 48.8568% and, 15.0908%, respectively, for the faulty bearing case and increased by 45.7299%, 19.4323%, and 6.73608% in X-, Y-, and Z-directions, respectively, for faulty gear case as compared to the healthy setup.

The value of kurtosis is a statistic that describes the degree to which scores are clustered in the tails or peaks of the frequency distribution. The peak represents the highest part of the distribution, whereas the tail represents the end of the distribution; in Tables 1, 2, and 3 the kurtosis value for the faulty bearing and gear cases in all directions is greater than 3, which indicates that for a given dataset, tails are heavier than a normal distribution. However, the healthy setup kurtosis value is less than 3, which indicates that the dataset has lighter tails than a normal distribution.

In Tables 1, 2, and 3, the upper bound (U.B.) values in -X, Y-, and Z-directions for healthy setup are 2.216796, 1.500658, and 3.601895, respectively. U.B. values in X-, Y-, and Z-directions are 3.103596, 2.198625, and 4.668564 for faulty bearing setup, respectively, and 2.401356, 2.273356, and 4.623635 for faulty gear setup, which is greater than healthy set. U.B. is a value that is greater than every element of a set of data. U.B. value is used to set threshold values. For a system, the acceleration value is greater than the U.B. value of the system, then employed machinery/system will have defects in its components.

In Tables 1, 2, and 3, the crest factor and the skewness values in X-, Y-, and Z-directions were compared with healthy bearings to conclude about the defective bearing and gear setup. The skewness values in X-, Y-, and Z-direction were increased by 68.827%, 61.330%, and 79.746% for faulty bearing setup in comparison to healthy setup and increased by 50.617%, 54.878%, and 81.544% for faulty gear setup in comparison of healthy setup. Bearing and gear with damages showed higher crest and skewness value as compared to the healthy setup.

3.2 Frequency-Domain Analysis

In this analysis, time-domain signals were converted into discrete frequency components using fast Fourier transform (FFT). FFT is a fast computation algorithm for the calculation of discrete Fourier transform (DFT). In standard DFT, as the value of N increases, it will require a very high number of computations. DFT implementation is very slow, so FFT is a technique to solve DFT in a faster way. To develop an FFT algorithm, a MATLAB Simulink model was created. In this model, the input was experimental time-domain vibrational signals, and the output obtained was the frequency spectrum. In a frequency spectrum plot, the x- and y-axis show frequency and amplitude, respectively. This technique can detect certain frequency components very easily, which is an advantage over the time domain. Because all these components are connected, the impact of the faulty component on other components is significant. Although each component has its own frequency of vibration, the magnitude of that frequency is still important to understand the severity of vibration. The total vibration captured by the accelerometer will increase the impact of all the faults in other components of the system. Therefore, the frequency analysis of individual components may not provide the true scenario for the condition of the machine, hence multiple faults should be studied. In this work, the same method is tried and finds significant results.

In the faulty setup frequency spectrum, when the large magnitude deviations have been observed, then the corresponding fault conditions were presented. The severity of the entire bearing and gear defect was identified using the frequency spectrum. The frequency magnitudes for faulty bearing and gear setup were found larger as compared to the healthy setup. Figure 6a is for healthy setup, and Fig. 6b, c is for

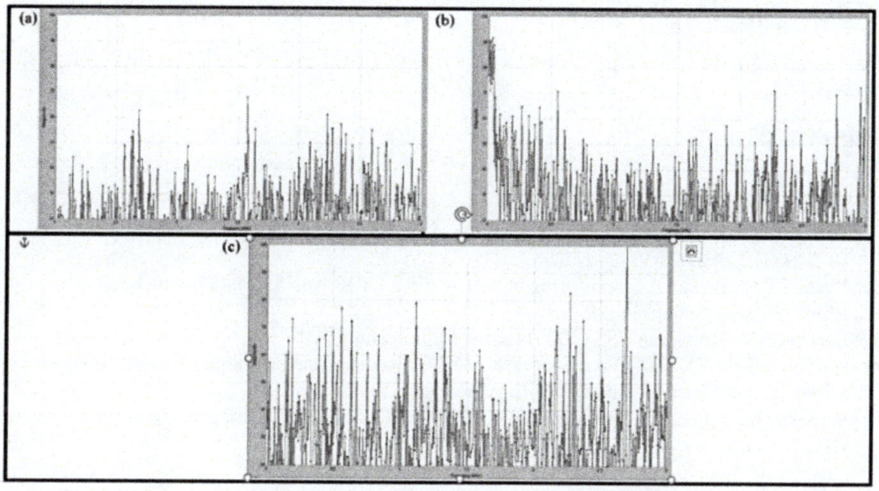

Fig. 6 Frequency-domain representation for **a** healthy setup. **b** Faulty bearing setup. **c** Faulty gear setup

faulty bearing and faulty gear. The comparison shows that the gear and bearing vibration frequency amplitude changes appreciably due to the presence of faults; there are a greater number of peaks observed in Fig. 6b, c between 0 and 3 kHz as compared to a healthy setup.

4 Conclusion

The time- and frequency-domain analysis of vibrational signals were carried out for the fault detection of bearings and gears. The vibrational analysis was performed for a healthy setup, faulty bearing setup, and faulty gear setup. Afterward, the vibration acceleration levels have been studied for all the aforementioned three cases. The time-domain analysis technique uses scalar indicators like average vibration, upper bound value, peak value, RMS value, and kurtosis for detection of the fault. FFT was used for frequency-domain analysis and to transform time-domain signals into frequency-domain signals using MATLAB. The RMS value in X-, Y-, and Z-directions is increased by 80.536%, 52.2069%, and 28.3482%, respectively, for the faulty bearing case and increased by 26.4373%, 25.4536%, and 3.98064%, respectively, for faulty gear case in comparison of healthy setup values. In the frequency-domain analysis, faulty bearing and gear frequency spectrum was compared with a healthy setup spectrum, and large magnitude deviations were observed. All the result has been compared, and we found that the vibrational acceleration and other statistical parameters are much larger for both faulty cases and magnitude of frequencies also larger than healthy setup at given frequencies. So, the experimental results indicate that the proposed method is effective in bearing and gear fault diagnosis. This research study can be extended to online monitoring of bearing and gear using IoT (Internet of Things).

References

1. Tandon N, Choudhury A (1999) A review of vibration and acoustic measurement methods for the detection of defects in rolling element bearings. Tribol Int 32(8):469–480. https://doi.org/10.1016/S0301-679X(99)00077-8
2. Hoshi T (2006) Damage monitoring of ball bearing. CIRP Ann 55(1):427–430. https://doi.org/10.1016/S0007-8506(07)60451-X
3. Hariharan V, Srinivasan PSS (2009) Indian J Sci Technol
4. Patidar S, Soni PK (2013) An overview on vibration analysis techniques for the diagnosis of rolling element bearing faults. Int J Eng Trends Technol 6
5. Vernekar K, Kumar H, Gangadharan KV (2014) Gear fault detection using vibration analysis and continuous wavelet transform. Proc Mater Sci 5:1846–1852. https://doi.org/10.1016/j.mspro.2014.07.492
6. Gligorijevic J, Gajic D, Brkovic A, Savic-Gajic I, Georgieva O, Di Gennaro S (2015) Online condition monitoring of bearings for improved reliability in packaging materials industry.

In: Proceedings of 2nd international electronic conference on sensors and applications, Sciforum.net, p. E006. https://doi.org/10.3390/ecsa-2-e006
7. Lessmeier C, Kimotho JK, Zimmer D, Sextro W (2016) Condition monitoring of bearing damage in electromechanical drive systems by using motor current signals of electric motors: a benchmark data set for data-driven classification, p 17
8. Gupta P, Pradhan MK (2017) Fault detection analysis in rolling element bearing: a review. Mater Today: Proc 4(2):2085–2094. https://doi.org/10.1016/j.matpr.2017.02.054
9. Singh M, Shoor S, Singh H (2018) Shannon entropy a better indices for local defect detection and to study the effect of variable loading conditions for taper roller bearing. Int J Mech Eng Technol (IJMET) 9(7):198–208 (July 2018, p. 12)
10. Cambow R, Singh M, Baghab AK, Singh H (2018) To compare the effect of different level of self-lubrication for bearings using statistical analysis of vibration signal. Mater Today Proc 5:28364–28373
11. Yang Z, Merrild UC, Runge MT, Pedersen G, Børsting H (2009) A study of rolling-element bearing fault diagnosis using motor's vibration and current signatures. IFAC Proc Vols 42(8):354–359. https://doi.org/10.3182/20090630-4-ES-2003.00059
12. Hariharan V, Srinivasan PSS (2010) Condition monitoring studies on ball bearings considering solid contaminants in the lubricant. Proc Inst Mech Eng Part C. J Mech Eng Sci 224(8):1727–1748. https://doi.org/10.1243/09544062jmes1885
13. Saruhan H, Sandemir S, Çiçek A, Uygur I (2014) Vibration analysis of rolling element bearings defects. J Appl Res Technol 12(3):384–395. https://doi.org/10.1016/S1665-6423(14)71620-7
14. El-Thalji I, Jantunen E (2015) A summary of fault modelling and predictive health monitoring of rolling element bearings. Mech Syst Signal Process 60–61:252–272. https://doi.org/10.1016/j.ymssp.2015.02.008
15. Shelke SV, Thakur AG, Pathare YS (2016) Condition monitoring of ball bearing using vibration analysis and feature extraction, vol 03(02), p 5
16. M. Cerrada et al (2018) A review on data-driven fault severity assessment in rolling bearings. Mech Syst Sig Process 99:169–196. https://doi.org/10.1016/j.ymssp.2017.06.012
17. Kumar S, Goyal D, Dang RK, Dhami SS, Pabla BS (2018) Condition based maintenance of bearings and gears for fault detection—a review. Mater Today: Proc 5(2):6128–6137. https://doi.org/10.1016/j.matpr.2017.12.219
18. Kumar KMA, Manjunath TC (2017) Vibration signal analysis using time and time frequency domain: Review. In: IEEE international conference on power, control, signals and instrumentation engineering (ICPCSI), Chennai, pp 1808–1811. https://doi.org/10.1109/icpcsi.2017.8392027
19. Yang B, Lei Y, Jia F, Xing S (2019) An intelligent fault diagnosis approach based on transfer learning from laboratory bearings to locomotive bearings. Mech Syst Sig Process 122:692–706. https://doi.org/10.1016/j.ymssp.2018.12.051

Shrinkage Optimization and Development of Predictive Model for Injection-Molded Part Using Numerical Simulation Combined with Artificial Neural Network

Dattatray Chopade, Rajesh Metkar, Nitish Sinha, and Avinash Panchal

1 Introduction

There is a very huge demand for injection-molded products over the past few decades. Initially, the focus of the injection molding process was on producing products with high volume and low to moderate quality requirements. Nowadays, the focus is shifted to produce precision molded items having high quality, such as automotive dashboards, medical equipment, and aerospace applications [1]. In the plastic injection molding industry, the determination of the optimal process parameters largely affects the production output, quality, and cost of production.

In the present paper, the emphasis is given on studying and understanding the shrinkage behavior and its optimization in the injection molding plastic gear. The amount of shrinkage at a particular location is largely dependent on processing parameters like mold temperature, melt temperature, packing time, packing pressure, etc. [2]. Molded plastic gears are used for this study because they are good alternatives to metal gears in lightly loaded drives. In applications such as windshield wiper drives, food processors, and even watches, plastic gears transmit power quietly and often without lubrication [1]. Due to the variations in properties and different response to different environmental conditions, the use of plastic gears was limited. Previously, they were used in drive trains with low power delivery. Recently, better control over process parameters and advanced design practices boosted the capacity of plastic gears multiple times [3].

Previously, many successful attempts were carried out to optimize the process parameters associated with injection molding process [2–14]. A lot of research work has been completed on shrinkage minimization in various plastic injection

D. Chopade (✉) · N. Sinha · A. Panchal
Department of Mechanical Engineering, G H Raisoni Institute of Business Management, Jalgaon, India

R. Metkar
Department of Mechanical Engineering, Government College of Engineering, Amravati, India

A. Parey et al. (eds.), *Recent Trends in Engineering Design*, Lecture Notes in Mechanical Engineering, https://doi.org/10.1007/978-981-16-1079-0_6

molding processes using different optimization methods like gray rational analysis, neural network, response surface methodology, firefly algorithm, genetic algorithm, particle swarm algorithm, etc. Sometimes a combination of two or more optimization methods is used [5, 7–10, 12–14]. Various researchers have carried out work on gears used for different applications. Injection-molded plastic spur gear specimens testing using different teeth module with a less cost [2], use of different location of gates, gating systems, and process parameters to analyze the spur gear performance, using grey-based Taguchi optimization method [3], parametric optimization for warpage in injection-molded plastic bevel gear [9] are some of the attempts made for better part quality of gears.

Bevel gears have applications in various drives. During the manufacturing of these small gears, process parameters need to be controlled to achieve desired dimensional accuracy and finish. In this paper, an attempt is made to optimize the shrinkage percentage in critical parts of the plastic bevel gear made up of nylon66 to reduce the defects arising due to variable shrinkages at different locations.

2 Experimental Details

A bevel plastic gear (shown in Fig. 1) (diameter: 16.4 mm, number of teeth: 20, gear thickness: 3.52 mm) is used as a model for this study. The plastic material used for the gear model is nylon-66 A125 (Manufacturer-Unitika). As the gears are smaller in size, a multi-cavity mold design will be suitable. Hence, 8-cavity mold layout is selected in this study. Mesh type used for the model is 3D mesh. A commercial software package Moldflow Plastic Insight (MPI) is used to perform the simulations. It is a specialized software to simulate different molding methods for solving temperature, flow, and pressure fields. Figure 1 shows the meshed model used for simulation with cooling channels.

The abbreviations 'MeT', 'MoT', 'PP', 'IT' and 'CT' are used for representing the process parameters that are selected for shrinkage analyses, viz. melt temperature,

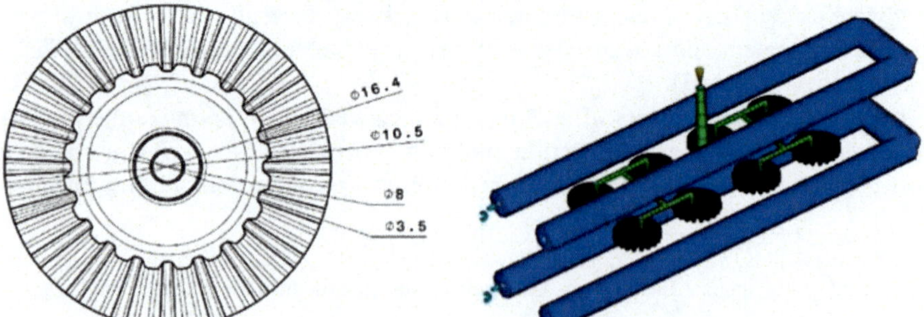

Fig. 1 Gear model and finite element model with feed and cooling system used for this study

Table 1 Three levels and corresponding values of parameters

Level	Process Parameters				
	MeT (°C)	MoT (°C)	PP (MPa)	IT (s)	CT (s)
1	280	75	30	0.5	15
2	290	85	40	1	20
3	300	95	50	1.5	25

mold temperature, packing pressure, injection time, and cooling time, respectively. Three levels of process parameters are selected and their values are taken from material database provided in MPI (see Table 1). The design of experimental runs is done using the Taguchi method. Taguchi L27 (3^5) orthogonal array is chosen which consists of 27 experimental runs.

All 27 combinations of process parameters are used one by one as input in the finite element analysis of this system. 'Fill + Pack + Cool' sequence of analysis is used in this case. The packing time is kept constant (5 s) for all the cases. Average shrinkage values obtained after simulation are tabulated in Table 2.

Other machine parameters and process parameters are kept at the default settings. The results of the volumetric shrinkage for one of the analyses with shrinkage at different locations across its cut section are shown in Fig. 2. The color scale shows the maximum shrinkage value of the part and its location. The value of average shrinkage in part can be obtained from the report generated by the software for each simulation.

The prime motive behind this study is to minimize the average volumetric shrinkage to the greatest extent possible. As a large value of volumetric shrinkage at a particular location gives rise to sink marks, it is desirable to keep it to a minimum value. Another objective is to develop a prediction model for the calculation of shrinkage to decrease the cost and time involved in monotonous finite element analyses. Hence, using the Matlab neural network toolbox, a feed-forward backpropagation type neural network model is developed.

3 Taguchi Method

Table 2 shows the obtained shrinkage values and their S/N ratios. S/N ratios show the effect of variation in a parameter value on the output. As the aim of this study is to keep variable shrinkage to the minimum value, 'smaller the better' quality characteristic is selected while computing the S/N ratios. For smaller the better S/N ratio,

$$\text{S/N}\ Ratio = -10\log(MSD),\ where\ MSD = \frac{1}{n}\sum_{i=1}^{n} y_i^2 \tag{1}$$

Table 2 Set of process parameters with their shrinkage values obtained after simulation

Exp. No	Process parameters					Shrinkage value after simulation (%)	S/N ratio
	MeT	MoT	PP	IT	CT		
1	28	75	30	0.5	15	8.1278	−18.1995
2	28	75	30	0.5	20	8.1278	−18.1995
3	28	75	30	0.5	25	8.1278	−18.1995
4	28	85	40	1.0	15	8.0367	−18.1016
5	28	85	40	1.0	20	8.0514	−18.1174
6	28	85	40	1.0	25	8.0754	−18.1433
7	28	95	50	1.5	15	8.0347	−18.0994
8	28	95	50	1.5	20	8.0323	−18.0968
9	28	95	50	1.5	25	8.0347	−18.0994
10	29	75	40	1.5	15	7.9853	−18.0458
11	29	75	40	1.5	20	7.9853	−18.0458
12	29	75	40	1.5	25	7.9853	−18.0458
13	29	85	50	0.5	15	8.3864	−18.4715
14	29	85	50	0.5	20	8.3964	−18.4819
15	29	85	50	0.5	25	8.4159	−18.5020
16	29	95	30	1.0	15	8.6887	−18.7791
17	29	95	30	1.0	20	8.6597	−18.7501
18	29	95	30	1.0	25	8.6345	−18.7247
19	30	75	50	1.0	15	8.4619	−18.5494
20	30	75	50	1.0	20	8.3809	−18.4658
21	30	75	50	1.0	25	8.4619	−18.5494
22	30	85	30	1.5	15	8.6847	−18.7751
23	30	85	30	1.5	20	8.6847	−18.7751
24	30	85	30	1.5	25	8.6847	−18.7751
25	30	95	40	0.5	15	8.9398	−19.0266
26	30	95	40	0.5	20	8.9548	−19.0411
27	30	95	40	0.5	25	8.9548	−19.0411

The computed S/N ratios are tabulated in Table 3. Figure 3 shows the graph plotted using the S/N ratio data given in Table 3.

For an optimum combination of process parameters, the level with the highest value for each factor should be selected. Hence, MeT:Level-1, MoT:Level-1, PP:Level-3, IT:Level-3, and CT:Level-1 become the optimized combination of process parameters for this model. The difference values in Table 3 denote the most influencing process parameter resulting in shrinkage. MeT is found to be the most significant process parameter followed by MoT, IT, and PP. CT had the least effect on the shrinkage of the bevel gear.

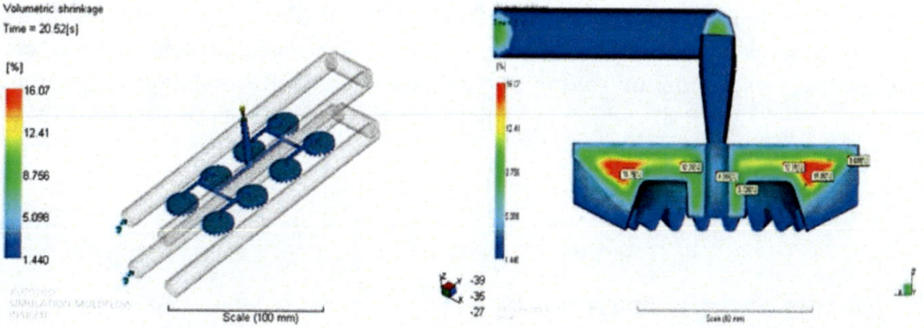

Fig. 2 Volumetric shrinkage analysis result of the model with its cut section showing values of shrinkage at various locations

Table 3 The response table of S/N ratios

Parameters	MeT	MoT	PP	IT	CT
Level 1	−18.1396	−18.2556	−18.5753	−18.5736	−18.4498
Level 2	−18.4274	−18.4603	−18.4009	−18.4645	−18.4415
Level 3	−18.7776	−18.6287	−18.3684	−18.3065	−18.4534
Difference	0.6380	0.3731	0.2069	0.2651	0.0119

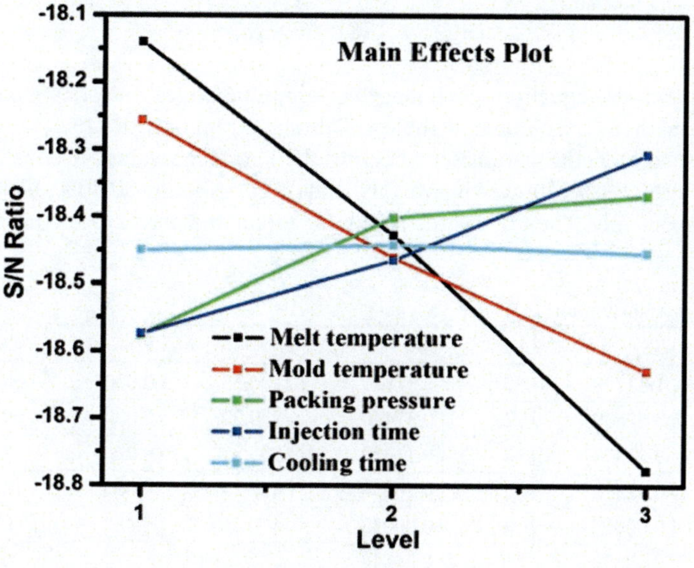

Fig. 3 Main effects plot showing effect of the processing parameters on shrinkage

The 27 simulation runs given in Table 2 do not include the above optimum combination of process parameters. Therefore, to validate the results, a confirmation test is conducted using optimum setting of parameters. The corresponding S/N ratio is found to be −17.7622 dB which is higher than S/N values obtained in Table 2. It confirms that the results obtained are better and the shrinkage value is optimized (see Table 5).

4 Results and Discussion

4.1 *Anova*

The data obtained from the finite element analyses are used to conduct the ANOVA analysis by Minitab V19 and the results are tabulated in Table 4. 95% confidence interval and 5% significance level are used for the ANOVA analysis. The parameter having 'P-value' less than 0.05 has a notable effect on the shrinkage. It is seen from the 'P-value' column of Table 4 that MeT, MoT, PP, and IT have a notable effect on the volumetric shrinkage. The mathematical relation to calculate percentage contribution (PC) of each parameter to shrinkage is given by

$$(\mathrm{PC})_x = (\mathrm{SS})_x/(\mathrm{SS})_{\mathrm{total}} \tag{2}$$

Using the above equation, percentage contributions for all factors are calculated and are stated in the last column of Table 4. The results highlight that melt temperature is the most influencing parameter affecting the shrinkage value. It contributes to shrinkage by 60.83%, followed by MeT, IT, and PP contributing 20.80%, 10.75%, and 7.37%, respectively. The CT has the negligible effect and percentage contribution.

Table 4 The ANOVA table

Factor	SS	Df	MS	F	P	PC (%)
MeT	1.83778	2	0.91889	2106.33	0.000	60.83
MoT	0.62840	2	0.31420	720.22	0.000	20.80
PP	0.22278	2	0.11139	255.33	0.000	7.37
IT	0.32472	2	0.16236	372.17	0.000	10.75
CT	0.00067	2	0.00033	0.76	0.482	0.02
Error	0.00698	16	0.00044			0.23
Total	3.02131	26				100.00

Table 5 Comparison of shrinkage before and after optimization

Process setting	MeT	MoT	PP	IT	CT	Simulated shrinkage (%)	ANN shrinkage (%)
MPI recommended	290	90	40	1	20	8.5292	8.5152
After optimization	280	75	50	1.5	15	7.7288	7.7354
Change rate (%)						−9.38	−9.15

4.2 *Neural Network*

A neural network has input and output layers between which one or more hidden layers are placed. The layers consist of processing units called neurons that are connected by variable weights. Neuron in the succeeding layer receives input from neurons in the preceding layer. MATLAB® neural network toolbox is used for this study to develop a feed-forward backpropagation type neural network model having 5–10–10–1 architecture. Here input layer has 5 neurons corresponding to 5 input parameters, both hidden layers have 10 neurons each, and output layer has one neuron representing shrinkage value. The network is created and trained using the average shrinkage data obtained by finite element analyses. Shrinkage values are calculated by trained neural network using optimal process setting obtained by the Taguchi method and MPI recommended process setting as input to the neural network.

Transition function shown below governs the output of the neurons contained in the previous layer and gives the input to each neuron in succeeding layer:

$$(\mathrm{total})_{\mathrm{i}} = \sum_{j=0}^{N} \mathrm{w}_{ij}\mathrm{x}_{j} \tag{3}$$

where $(\mathrm{total})_i$ denotes the total input of the ith neuron in the computing layer; N denotes the number of the neuron in the succeeding layer; w_{ij} represents the connection weight of the jth neuron in the succeeding layer and ith neuron in the computing layer and x_j represents the output of the jth neuron in the forward layer. Table 5 gives a comparison of the predicted values and finite element simulation values of the average shrinkage for the optimal setting and Moldflow recommended setting.

5 Conclusions

The optimal values of melt temperature, mold temperature, packing pressure, injection time, and cooling time for minimum shrinkage are found to be 280 °C, 75 °C, 50 MPa, 1.5 s, and 15 s, respectively. Melt temperature is identified as the largest influencing process parameter contributing 60.83% to the volumetric shrinkage.

Mold temperature, injection time, and packing pressure are other process parameters having substantial effect. Cooling time has the least effect on the shrinkage. 9.38% reduction in average shrinkage is obtained as compared to MPI recommended process setting. The neural network formulated is found to predict shrinkage values with good accuracy. The optimization methodology and development of the neural network-based predictive model used in this paper can help in reducing the cost and time associated with repetitious finite element analyses.

References

1. Machine Design. https://www.machinedesign.com/motors-drives/article/21833583/plastic-gears-more-durable-than-ever-plastics-get-in-gear. Last assessed on 2020/08/06
2. Mehat N, Kamaruddin S, Othman A (2013) Investigating the effects of injection moulding process parameters on multiple tensile characteristics of plastic spur gear via experimental approach. Adv Mater Res 748:544–548
3. Mehat N, Kamaruddin S, Othman A (2013) Modeling and analysis of injection moulding process parameters for plastic gear industry application. ISRN Ind Eng 1–10. ID 869736
4. Saad M (2020) Experimental-based optimization of injection molding process parameters for short product cycle time. Adv Polym Technol 2020:1–15
5. Ryu Y, Sohn J, Yun C, Cha S (2020) Shrinkage and warpage minimization of glass-fiber-reinforced polyamide 6 parts by microcellular foam injection molding. Polymers 12, 889:1–18
6. Wang Y, Zhu Z, Tang L, Jiang Q (2020) Research on the molding design and optimization of the molding process parameters of the automobile trunk trim panel. Adv Mater Sci Eng 1–19. Article ID 5629717
7. Ahmad A, Wahab M, Shah A, Kamarudin K, Hehsan H (2019) Optimization of processing parameters for plastic injection moulding process towards moulded part shrinkage. AIP Conf Proc 2129(020168):1–5
8. Öktem H (2012) Optimum process conditions on shrinkage of an injected-molded part of DVD-ROM cover using Taguchi robust method. Int J Adv Manufv Technol 61:519–528
9. Altan M (2010) Reducing shrinkage in injection moldings via the Taguchi, ANOVA and neural network methods. Mater Des 31:599–604
10. Chopade D, Metkar R, Hiwase S (2015) Experimental evaluation and optimization of warpage in injection molded Nylon 66 bevel gear using numerical simulation and neural network. In: Tekwani, Bhavsar, Modi (eds) Multi-disciplinary sustainable engineering: current and future trends—NUiCONE 2015. Taylor & Francis Group, London, pp 247–252
11. Bhirud T, Metkar R (2017) Experimentation and optimization of shrinkage in plastic injection molded GPPS part. In: Iyer B, Nalbalwar S, Pawade R (eds) ICCASP/ICMMD-2016. Advances in intelligent systems research, vol 137, pp 107–112
12. Chang T, Faison E (2001) Shrinkage behavior and optimization of injection molded parts studied by the Taguchi method. Polym Eng Sci 41(5):703–710
13. Abasalizadeh M, Hasanzadeh R, Mohamadian Z, Azdast T, Rostami M (2018) Experimental study to optimize shrinkage behavior of semi-crystalline and amorphous thermoplastics. Iran J Mater Sci Eng 15(4):41–51
14. Kumar S, Singh A (2019) Warpage and Shrinkage analysis and optimization of rapid tooling molded thin wall component using modified particle swarm algorithm. J Adv Manuf Syst 18(1):85–102

Deposition of Metallic Coating on Plastic Substrate Material Using Cold Spray

Harpinder Singh Sandhu, Gurpreet Singh Phull, Ujjwal Anand, Jaiinde Preet Singh, and Piyush Gulati

1 Introduction

After many decades of the industrial revolution, metals have been replaced with composite materials and polymers, but these composite materials neither possess comparable properties of metals nor are the replacement of metals, but metallic deposition coating on the surface of these composite and polymeric materials may enhance the surface properties of the parent material and may be the metal replacement for certain applications. From basic operations such as cosmetic layering to more complicated applications such as electrical products, aerospace engineering, medical equipment's, plastic engineering, and lot more, there is a strong market for the development of metallic coating on polymers with the growing demand.

Surface engineering is characterized as the design structure of a mixed substratum system along with its surface property, providing a value that cannot be accomplished by either the surface or the substratum alone [1]. The system can consist of a coating or a layer of multiple coatings with or without an associate, a modification of the surface or a combination of the two or more substance on the substrate. Surfaces can be engineered further to improve their appearances and to protect them from the environmental hazards, therefore, enhancing the mechanical or physical performance of the substrate. The coating is generally a layer or multiple layering applied on a surface either for protection or for an aesthetic purpose to enhance the surface property of the parent substrate material. Coatings in various types of metals are most commonly

H. S. Sandhu (✉) · G. S. Phull · U. Anand · J. Preet Singh · P. Gulati
School of Mechanical Engineering, Lovely Professional University, Jalandhar-Delhi G.T. Road (NH-1), Phagwara, Punjab 144411, India

J. Preet Singh
e-mail: jaiinder.14740@lpu.co.in

P. Gulati
e-mail: piyush.14775@lpu.co.in

A. Parey et al. (eds.), *Recent Trends in Engineering Design*, Lecture Notes in Mechanical Engineering, https://doi.org/10.1007/978-981-16-1079-0_7

done on specific mechanical parts in order to secure the parent material against flexible conditions or on instruments such as the machining machines to improve their strength and reliability. The thermal spray methods are widely used for industrial applications when evaluating the production of metallic surface coatings. In such processes, the coating feedstock material is heated to its melting point, usually in the form of powder particles to the machining sheet, and then packed as individual droplets. Usually fuel gasses or an electric arc supply the energy required. As soon as the feedstock powder is injected, it sticks on the substrate, forming a splat and then it solidifies. According to its adhesion effect, the coating usually experiences large rounds of porosity caused by insufficient bonding of the spit material [2, 9, 12]. It also exhibits a high range of oxidation on the substrate substance due to the upcoming bounds of oxygen gas available at the atmospheric air. Another coating method is Cold Spray coating technology. Here, the feedstock powder metal is not allowed to melt, but is forced to a very high velocity considering gases as a medium, moving at the very high speed with the help of a spray gun, using processed gasses or commercially produced Nitrogen or Helium as a carrier gas. They are further carried to a regulated chamber to the converging/diverging type, de-Laval nozzle. With sudden collision of particles on substrate material, the high velocity kinetic energy results in plastic deformation leading in the building of the deposition layer by layer [3]. Cold Spray coating technique typically shows enhancement in some physical properties such as excellent mechanical characteristics of the coated surface and high bond strength, combining metals of stiff engineering operations as aluminum, titanium, tin, and some other metal alloys.

2 Discussion

In cold spray coating, feedstock particles, usually in the form of a powdery substance, are accelerated by the aid of processed gases to a very high speed and are allowed to strike at the substratum with the help of specially designed de-Laval type nozzles, without heat effect [1] (Fig. 1).

Now, we know all about the different types of thermal spray coating that are widely accessed all over the industrial sectors. Here is a graph showing the comparison between all the methods in comparison to their operating temperature and particle velocity.

Figure 2 depicts the variation shown by different thermal spray techniques at different parameters. Here, we have described the variation to the gas temperature and the velocity of the particle. The graph shows that the plasma arc spray coating has the highest integrity among the other variables while the HPCS, i.e., High Pressure Cold Spray stands least in the table. This shows deposition of metals is permitted on low temperature, whereas a high temperature is required in coating a substrate with the plasma arc thermal spray coating. This also reduces the chances of phase change in the cold spray process without changing the original shape of the sample

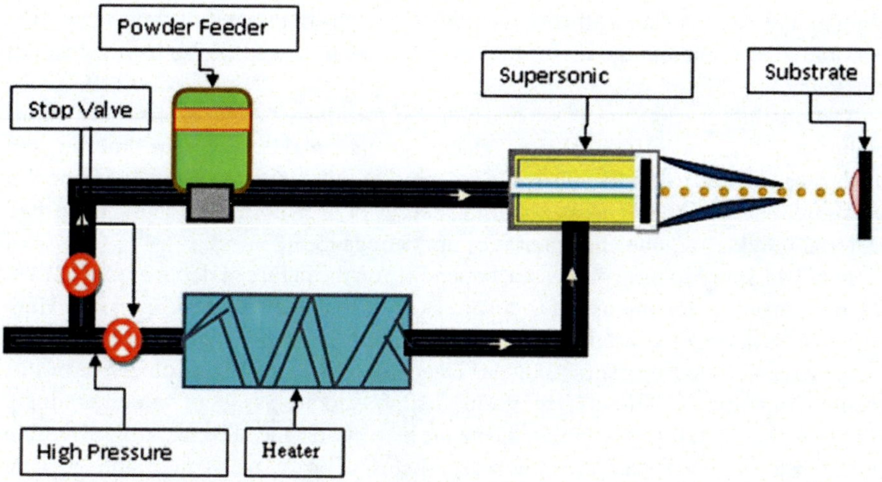

Fig. 1 A schematic diagram showing cold spray technology

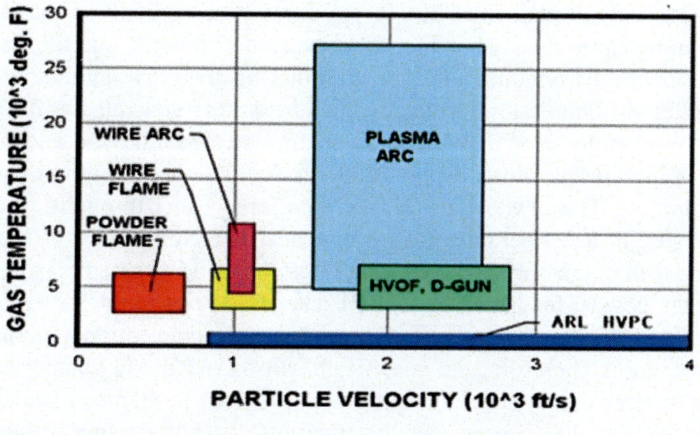

Fig. 2 Comparison between different processes of thermal spraying

substrate. This restricts the deposition over a wider span of the area compared with the other thermal spray methods, preparing a wrought-like microstructure [4].

Hence, it is clear from the above advantages that the cold spray coating is the most suited coating method when coating on a very low temperature and on the polymer/composite substrate.

Research for metallic coating on non-metallic surfaces such as glass fabrics, polymer plastics, or composite materials is being carried out with further advancement in modern manufacturing [5]. The essence of these materials descends the process of metallic deposition. However, it is necessary to understand the nature of the process in the cooling system, to preserve the fabric material from the peak temperature. With a stand-off distance of 250 mm from the exit of the nozzle, it is possible to coat without the use of the coolant system hence making low deposition efficiency.

The coating mechanism is represented by forming coating layers from semi-melted or melted droplets of the feedstock particles, maybe a wire or thin rod, and then place in a fire created by the spray weapon. Moving towards the phenomenon of the source energy either kinetic or thermal energy or even both, the coating deposited material provides unique characteristics of coatings [5, 8].

For Cold Spray coating processes, depending on the nature of deposition required, the CS process is distinguished to a series of processes as Kinetic Spraying, High Velocity Particle Consolidation, Cold Gas Dynamic Spraying, High Velocity Powder Deposition and the Supersonic Powder Deposition [4]. One major objective of this method is mentioned below. Here, we use a high velocity gas jet of velocity ranging between 300 and 1200 m/s generated with the help of a de-Laval converging/diverging type nozzle which forces the powder particles of sizes between 1 and 50 μm, bombarding them on the sample substrate, placed at a stand-off distance of nearly 1 inch from the end of throat, which further are spitted to constitute a layer or multiple layers of coatings [2, 6].

The kinetic energy generated due to the bombardment of these particles help them in coating layer by layer which plastically deforms the surface on sudden impact and makes the bonding adhesive with the substrate making metallic bonds, hence avoiding the depression formed by the temperature generated as it was in the case of conventionally used thermal spray coating methods like the release of gas, oxidation on high temperature, evaporation, residual stresses, melting of substrate, crystallization, etc. The powder particles are, therefore, pushed through a spray piston and accelerated at a temperature far lower than the melting point of the sample substratum by the high-speed gas jet, that in effect renders the coating deposition layer from the powder feedstock in its solid state and so no melting or solidification phase will occur via the powder particles as used in the conventional thermal spray coating procedure. Therefore, the specifications of Cold Spray coating technique is likely to be very stiff typically shaping near to 5 mm in diameter because of its small nozzle size of 10–15 mm^2 and at a stand-off distance ranging between 5 and 25 mm, production at a high intensity particle beam which further results in precise regulation of the cross-section of the surface of the substratum.

For Cold Spray coating process, none of the powder particles are melted due to the low processing temperature usually ranging below 800 °C, intakes a high velocity Oxygen Fuel, which may rise up to 3000 °C high [7]. The softer materials are not difficult to deform plastically and have light heat capacities, making it easy to maintain its existing shear instability, and has the advantage of using this technology.

For thermal spray coating, the feedstock is heated to its molten or semi-molten state using an external source. These heated particles are then, accelerated to a very high speed and then is spattered towards the specimen with the help of processed gases or high-speed jets. On collision with the specimen, an intermolecular particle force is produced with the top layer of the substrate and with the feedstock making a heavy thickness on the surface and hence forming a lamellar microstructure [8].

The outcome may alter some properties on deposition, depending upon the density of the layer formed, the adhesive cohesion between the deposited particles, etc.

The basic equipment used in cold spray process is the setup of a special high velocity gas jet powered with a converging/diverging type, de-Laval nozzle to accelerate the powder particles of size ranging 30–45 μm at a very high velocity of 300–1200 m/s which is bombarded onto the substrate at the exit distance of around 25 mm from the nozzle where the powder particle impacts and forms a layer of coating. The particles' Kinetic Energy ($KE = ½ mv^2$) shears the powder particles to maintain the strain energy from the bombardment to the substratum plastically, generates heat energy that boosts the specimen's temperature, and makes the material flexible that decreases the rate of strain hardening [9, 10]. At a certain temperature, the powder material is accelerated by the high-speed jet which enters and completes the coating formulation. Gases such as Helium and Nitrogen are heated to a temperature ranging between 330 and 800 °C and then is carried through the converging/diverging nozzle or de-Laval nozzle, further creating a supersonic gas jet. The mixture of the particle/powder carrier gas is transferred axially into the machine and is then carried before the nozzle gun towards a pre-chamber. Instead, the expanding steam, which is bombarded to the substratum as solid particles, accelerates rock. The substrate material is selected over a wide range of polymers and glass fibres which are capable of machining. Metallic coatings on the non-metallic surfaces such as plastics, fabrics, composites, or the polymer materials are feasible under a number of conditions over a variety of engineering applications. The property of such a substrate material essentially renders the process of metallic coating difficult. Any coatings were conducted successfully using thermal spray coating techniques by totally melting the feedstock content without harming the parent material (Fig. 3).

The research was performed on various materials such as polyamide-6, commercially used polycarbonate, polypropylene, polystyrene, and glass fibres composite material to classify the most suitable content. Thanks to its wide variety of quality,

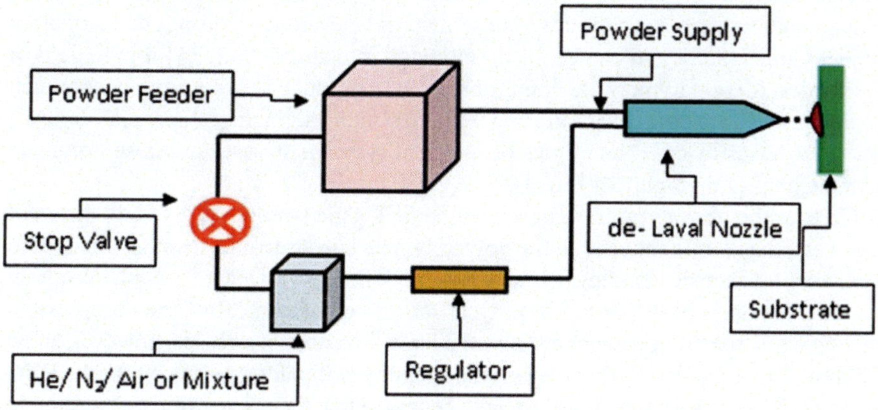

Fig. 3 Systematic diagram of cold spray gun machine

cost effective and improved machining performance trend, and to the best matched material that is the polycarbonate. The graph given below shows the comparison between different parameters of measurements while experimenting upon three major polymers such as PVC, Polypropylene PP, and Polycarbonate PC. These are the most widely used polymers which fit our essentials. We will in detail explore some of the basic factors which were practically performed on these three polymers and their fluctuation in the result procedure. The polymers are

2.1 Polyvinyl Chloride (PVC)

Polyvinyl chloride or PVC is the polymer generated by the polymerization of many vinyl chloride monomer chains together. It is extensively manufactured in tonnes across the world and is widely used in the manufacture of several daily life products such as pipes, cards, sinks, electrical wires, leather, etc. [11, 12].

2.2 Polypropylene PP

Also, polypropylene is a monomer created by multi-chain linkages. Often known as thermoplastic polymer, it is similar to low-density polyethylene, making it flexible and tough in strength [12].

2.3 Polycarbonate PC

Polycarbonate is the family of thermoplastic polymers in their chemical structure which contains carbonate groups. These materials are tough, strong, and some are optically transparent and can be easily moulded into shapes [13, 14]. In the field of electronics, these polymers are commonly used to produce various components such as diodes and capacitors, for building purposes such as glazing, the shelter houses, etc. Now, let us discuss the effects of different types of stresses on these polymers with the help of a graph [5] (Fig. 4).

Here, as the above graph suggests, while the tensile strength is acted on different polymers, the results shows that the polycarbonate substrate has the most strength in comparison to tensile loading while polypropylene has the least strength. While in the case of compressive strength, polycarbonate can bear most of the compressive loading while the polypropylene shows the least. The impact stresses, however, show different scenario. Polycarbonate has the maximum impact stresses while the PVC shows the least impact stress. So, we may conclude that the polycarbonate PC sample shows the least deflection during stress applied.

Figure 5 shows the relation between the operating temperature of different poly-

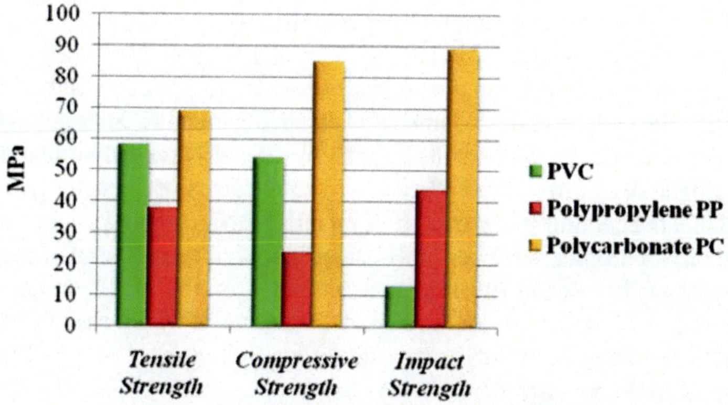

Fig. 4 Graph showing the different stress analysis done on various polymers

Fig. 5 Graph comparing the operating temperature analysis

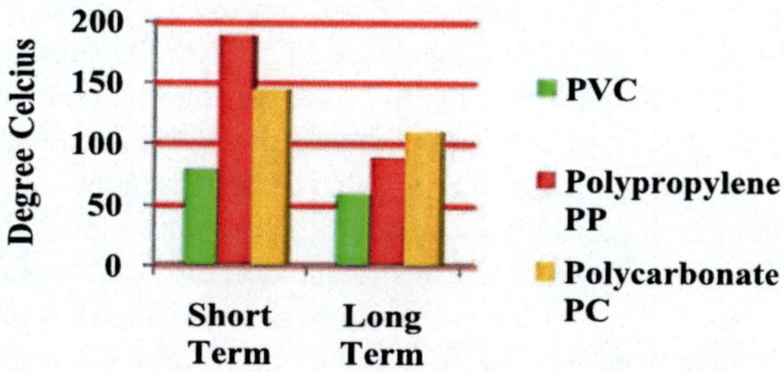

mers for both short and long ranges of operating temperature. Note that the operating temperature is the range of temperature in which the components significantly fail in performing their properties. The analysis is shown below [15].

As the graph represents the comparison, it is very much clear that the polypropylene has a fatal range of temperature as compared to the other two polymers when considered in short ranges while polycarbonate performs better in constant fields such as the long ranges. The amount of moisture content is also a major factor as it alters strength of the coating adhesion. As the graph shows, polypropylene has the maximum moisture content among the other two polymers. When comparing the cost among all the three polymers which is widely used for manufacturing purposes, we come to know that the polypropylene polymers are the most economical and cheaper as compared to the other two polymers while the polycarbonate shows the highest cost as the process of manufacturing polycarbonates is expensive. There are some other polymers such as Nylon, Polyester, Polytetrafluoroethylene (PTFE), and polyoxymethylene copolymers (POM-C) which have comparable properties, but lacks in some of the other useful parameter. So, we sorted these three polymers out of the bag [16]. So, we come to know that among the three chosen polymers, i.e., PVC, Polypropylene, and the Polycarbonate specimen, the most efficient material while comparing properties like stress analysis, operating temperature, the moisture absorption and cost comparison is polycarbonate PC material. Since it is easily available commercially, we would be using the polycarbonate PC material. A list of reports for applying the Cold Spray coating technique in advancing metallic coatings on polymer substrates is presented. Further results have shown the effects on the substrates used when spraying aluminium powder particle of average size approx to 25 μm. Further tests found that the coating was thin on non-metallic surfaces and was not in constant flow. Very small particles were certainly adhered to and the substrate listed a very heavy erosion of the plastic substratum. Consequently, the results suggested that mixing the feedstock powder with the polymer substratum would result in the deposition of metallic powder particles on hard non-metallic surfaces [14, 17]. The surface binding film also demonstrated strong layering properties (Fig. 6).

These studies also suggested the importance of the coating with Cold Spray method to form a layer of metallic coatings on non-metallic polymer surfaces. Certain engineering metals used in day-to-day manufacturing such as tin and aluminium are machined and processed into the powder particles, which were then sprayed onto the substrate polymer substance PC, while the substratum is covered at a pressure of about 30 bar, the velocity of the particle increases by about 2 times the pressure value of 5 bar taken near the mouth, and the resultant value increases by about half at the end of the throat. Results also demonstrate that due to its low individual weight, aluminium particles move much faster than nickel. In experiments with segment duration of 180 mm the supersonic de-Laval converging/diverging nozzle was used. By holding the coating deposition at a pressure of 5 bar on the PC, with a stand-off distance from the nozzle to be about 40 mm at a transverse speed of 8.3 mm/s on the substratum and by directing a rotational velocity of the particle feeder to the junction wheel at about 15 rpm. After a significant experiment, we are conscious that at the process conditions mentioned above, a thin layer of coating is deposited on the surface. In addition, a further sample was coated at pressure of 30 bar on the polycarbonate PC substratum at a stand-off distance of around 40 mm at a transverse

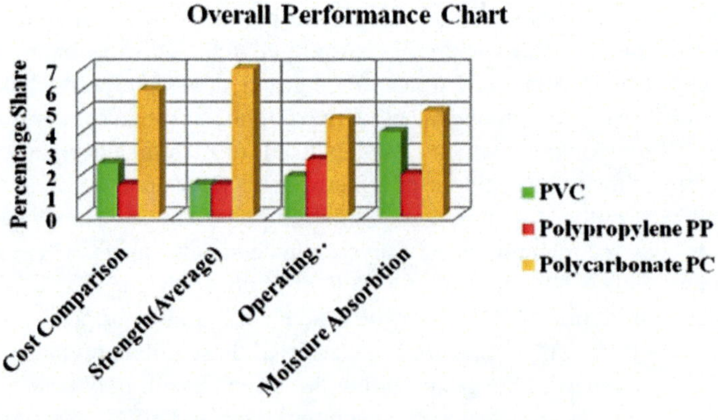

Fig. 6 Graphical representation of overall performance of the polymers

speed of 16.6 mm/s and the feeder speed of 18 rpm was taken. The outcome showed advancement in electrical conductivity, without any treatment on the sample. The resultant thickness of coating varied from 40 to 45 μm on the ABS/PC polymer specimen, while the sample substrate coated with the Tin powder particles resulted in the depletion of the substrate polymer. Although aluminium coated substrate produced inappreciable surface erosion, but there was no layer of coating deposition on the sample substrate. However, tin coated sample substrate was experienced effectively and the deposition on the sample substratum was achieved [18, 19].

3 Conclusion

Coating procedure Cold Spray is in turn a quick and cheap substance coating technique, using high-pressure gasses to accelerate solid particles at very high speeds. It adheres to and forms a sheet of insulation as the powder surface interacts with the ground. This technique is widely used on surfaces of similar and dissimilar characteristics in metal coating as well as on the non-metallic substances like polymers, glass fibres, etc. In order to proceed towards the metallic coatings on the non-metallic surfaces like the PC manufactured products, we have researched extending this surface coating cycle. Extensively, owing to its suitable properties, other electronic uses, such as aluminium and tin, had transformed it into the shape of powder particles and then sprayed onto the plastic substrate material, PC. Results have also shown that aluminium does not cause any significant harm to the surface of the soil due to its low physical weight. With those premises, however, the critical velocity cannot be attained. Therefore, it became possible to maintain the coating parameters and select an appropriate nozzle type to coat with Tin particles on selected substrate polymers. The thickness of the surface coating was measured (to be between 45 and 100 μm in the region) while the estimated average critical velocity was close to 310 m/s throughout the CFD analysis as some research paper previously studied. The tin powder's theoretical impact energy was calculated to be significantly lower than aluminium. In addition to the results achieved, no pre-treatment of the surface was required before spraying, and both sample coatings also showed good electrical conductivity. Cold Spray coating technology is advancement in the field of material science and it has some unique properties that overcome the parent family of Thermal Spray Coating. Instead, coating technology from Cold Spray is hoped to be the motive of modern research and expands the series of applications for the parent Thermal Spray Coating method as an eco-friendly mode of coating according to many safety standards authorities. Cold Spray coating technique from several prospective publications in the field of biomedical, aerospace, power plants, space industries, etc., has already labelled their existence to be ideal for surface coating deposition. Nonetheless, much further work is needed to determine the optimal parameters of Cold Spray coating technique such as gas composition, temperature regulation, nozzle configuration, and material selection with the measurement of appropriate critical velocity for various combinations of powder/substrate. Coating technique Cold Spray is essential for further research into the deposition of hard and brittle ceramic materials. The Cold Spray coating process has enormous development potential for more applications in industrial areas such as boiler factories to improve the longevity of boiler tubes by preventing corrosion failures at high temperatures under extreme chlorine and sulphate-based conditions. Depositions by the coating methods of Cold Spray can be useful for the prevention and acquisition of high temperature oxidation loss of boiler tubes, which are now operating at much less output than conventional power plants.

References

1. Pawlowski L (2008) The science of engineering of thermal spray coating. Wiley
2. Lupoi R, O'Neill W (2010) Surface & coating technology, vol 205, pp 2167–2173
3. Zhang D, Shipway PH, McCartney DG (2005) J Therm Spray Technol 14(1):109
4. Singh H, Sindhu TS (2012) Cold Spray Technol Future Coat Deposition Process 22:69–84
5. http://www.supplyservices.co.nz/engineering-plastics-properties-chart/
6. http://en.m.wikipedia.org/wiki/Electroplating
7. Fukumoto M, Hiroki T (2008) Deposition of copper fine particle by cold spray process
8. Sunil P, Saha GC (2017) Development of sustainable cold spray coatings and 3D additive manufacturing components for repair/manufacturing components
9. Cormier Y, Dupuis P, Jodoin B, Corbeil A (2013) Net Shape fins for compact heat exchanger produced by cold spray. J Therm Spray Technol 22:1210–1221
10. Alkhimov AP, Kosarev VF, Papyrin AN (2001) Method of cold gas-dynamic deposition. Dokl. Akad. Nauk SSSR, pp 54–65
11. Villafurerte J (2010) Current and future applications of cold spray technology, metalfinishing
12. Choudhuri A, Mohanty PS, Karthikeyan J (2008) Bio-ceramic Composite Coatings by Cold Spray Technology as industries paper 22241 manuscript 3762
13. Li C-J, Li W-Y (2003) Surf Coat Technol 167:278
14. Irissou E, Legoux J-G, Ryabinin AN, Jodoin B, Moreau C (2008) Review on cold spray process and technology: part-I-intellectual property. J Therm Spray Technol 17:495–516
15. Peat T, Galloway A, Toumpis A, McNutt P, Iqbal N (2017) The erosion performance of cold spray deposited metal matrix composite coatings with subsequent friction stir processing. Appl Surf Sci 396:1635–1648
16. Ajdelsztajn L, Zuniga A, Jodoin B, Lavernia EJ (2006) Cold gas dynamic spraying of a high temperature Al alloy. Surf Coat Technol 201:2109–2116
17. Champagne VK (2007) The cold spray materials deposition process: fundamentals and applications. Woodhead, Cambridge, UK
18. http://en.m.wikipedia.org/wiki/Polymer
19. Hanqing C, Phuong V, Yue S (2019) Investigation of cold spray on polymers by single particle impact experiments, vol 28, pp 135–143

A Review: Bio-compatible Thermal Spray Coating on Bio-implant

Harpinder Singh Sandhu, Gurpreet Singh Phull, Mandeep Singh Saini, Jai Inder Preet Singh, and Piyush Gulati

1 Introduction

In the arena of biomedical engineering, one of the most discussed issues is the repairing of hard tissues or bones. Bone damage and worn out joints have always been a challenging situation in orthopedics. To deal with such situations, bio-implants are used either as reinforcement or even sometimes as a replacement to the damaged tissues. Majorly, the bone damage is evident in the case of severe fractures or in the case of density reduction as a consequence of growing age, some disease, or trauma. Various biomaterials have been discovered till yet, that are used to serve the purpose of bio-implants. Established materials that are used commonly for bio-implant applications are reported as biopolymers such as bioactive plastics, bio-ceramics such as calcium phosphates, and metals such as titanium (commercially pure) and titanium alloys.

Due to the excellent balance between the mechanical and physical-chemical properties, titanium and its alloys have been recognized as the best materials for bio-implants. Although titanium is found to be an excellent material for loading applications, it has also been reported that the stiffness of titanium is higher than that of actual bones which evidently produces the stress shielding phenomenon at the implant site [1–3]. Stress shielding can be a severe problem and may discomfort the patients for the rest of their lives. So, to be prevented from stress shielding, several studies

H. S. Sandhu (✉) · G. S. Phull · M. S. Saini · J. I. Preet Singh · P. Gulati
School of Mechanical Engineering, Lovely Professional University, Phagwara, Punjab 144411, India

M. S. Saini
e-mail: mandeep.16853@lpu.co.in

J. I. Preet Singh
e-mail: jaiinder.14740@lpu.co.in

P. Gulati
e-mail: piyush.14775@lpu.co.in

A. Parey et al. (eds.), *Recent Trends in Engineering Design*, Lecture Notes in Mechanical Engineering, https://doi.org/10.1007/978-981-16-1079-0_8

have been made to develop titanium implants with a stiffness of the order of actual bones. And this is achievable by imposing porosity in them while manufacturing the titanium implants [4–6].

Titanium and its alloys are reported to have inertness toward the chemicals. Despite this being a desired trait in the ever-changing biochemical environment that prevails in the body, this property also affects the biocompatibility of the implants. It is evident from the cases that the muscles and ligaments do often find incompatibility barriers in attaching to the implants, and therefore may result in loosening events. Moreover, titanium implants may also cause inflammation and several immunogenic reactions at the implantation site. Hydroxyapatite, a calcium phosphate biomaterial, on the other hand, has been established as the most compatible bio-implant material in orthopedics and dentistry because of its osteoconductive and osteoinductive properties [7–12]. The bones and hard tissues in mammals are chemically similar to hydroxyapatite [13]. With adjacent bones, hydroxyapatite creates stable adhesive bonding and does not generate a pathological reaction. Instead of hydroxyapatite's excellent biocompatibility and osteointegration properties, its intrinsic mechanical properties such as brittleness and low fracture toughness limit its scope to small size implants that are subjected under low load conditions [14–16].

To achieve desired mechanical strength and simultaneously better osseointegration properties, researchers devised metal implants with hydroxyapatite coating over them [17, 18]. Various techniques are used to deposit hydroxyapatite coatings on bio-implant such as dip coating and sintering [19, 20], hot isostatic pressing [21], electrophoretic deposition [22], ion-beam sputtering coating [23, 24], biomimetic deposition [25, 26] but for commercial purposes, thermal spray processes such as Plasma Spraying [27] and High Velocity Oxy-Fuel spraying [28] are widely used. Nowadays, another thermal spray technique, cold spraying is gaining popularity in coating metal implants with bio-compatible coatings of hydroxyapatite.

2 Coating Methods and Technologies

By the use of thermal spray techniques, a thick layer of hydroxyapatite coating of range 50 μm to several millimeters can be deposited over metal implants with a high deposition rate and greater deposition efficiency. In thermal spraying, the hydroxyapatite powders are heated to a molten, semi-molten or even solid state and are accelerated to bombard over the surface of the metal implant. The bombarding particles, due to impact, get flatten and form tiny splats that imitate and adhere to the implant surface and on each other. As the sprayed particles strike upon the surface, they build up to form a laminar structure, splat by splat, thus building a coat [29].

In high-velocity oxy-fuel spraying, (HVOF) oxygen and fuel are served into a combustion chamber, where they are continuously ignited. This causes the development of high pressure (up to 1 MPa) which emanates through a nozzle as given in Fig. 1. The jet velocity at the exit of the barrel becomes supersonic. Hydroxyapatite powder is fed into the gas stream which accelerates up to 800 m/s. The stream of hot

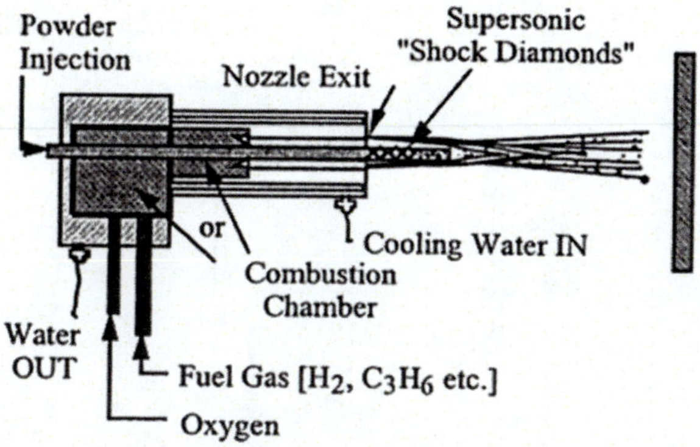

Fig. 1 Schematic of high velocity oxy-fuel spray apparatus [33]

gas and powders is focused on the implant surface. The powder partially melts in the stream and deposits low porosity and high bond strength coatings upon the implants.

In the plasma spraying process, a plasma jet is emanated from a plasma torch, in which hydroxyapatite powder is fed. The temperature in the jet usually reaches above the order of 10,000 K which melts and propels the powder toward the implant as given in Fig. 2. At the implant surface, the molten droplets get flattened and solidify rapidly to form a deposit. The deposits remain adherent to the implant surface as a coating. In cold spraying, the hydroxyapatite powder is heated much below the melting point and thus sprayed in the solid state over the implant surface. The powders are accelerated in supersonic gas jet and are deposited by the sake of velocity impact rather than deposition by solidification of molten or semi-molten droplets as shown in figure (Fig. 3).

Plasma spray is the most used technology for coating bio-implants with hydroxyapatite due to its many advantages such as lubrication, high deposition rates, low surface temperature, and variable coating porosity. The main problem in plasma-sprayed hydroxyapatite coatings is metastable crystalline products such as tetra calcium phosphate, tricalcium phosphate, oxyhydroxyapatite or oxyapatite, and amorphous calcium phosphate and bioactive phosphate such as CaO [30, 31]. High levels of amorphous calcium phosphate are undesirable because they are strong in vivo regeneration and may, therefore, contribute to the mechanical and resin instability of the coating [32].

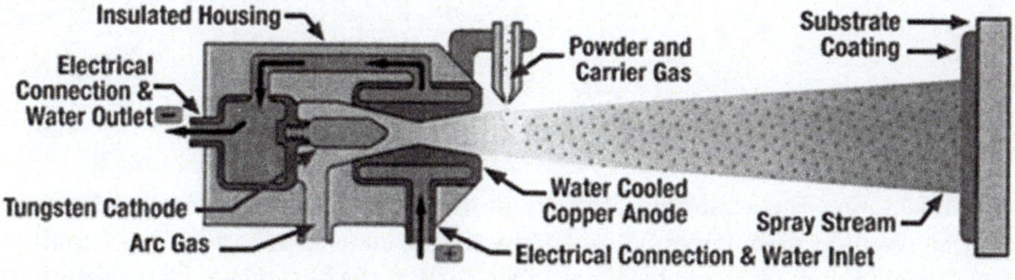

Fig. 2 Schematic of plasma spray apparatus [33]

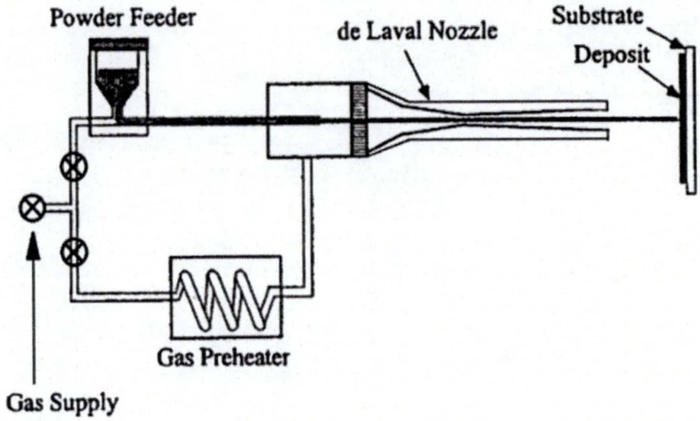

Fig. 3 Schematic of cold spray apparatus [34]

The deficiency of crystalline calcium phosphate can be overcome by hydroxyapatite coating that can be achieved by bringing down the temperature and rapid flame in HVOF. Although HVOF spray may deposit stable coatings but similar to plasma spraying, it also induces residual thermal stresses in the implant which shortens their life and their durability. Besides, non-combustible or partially combustible hydrocarbons can change the purity of the coating mixture.

Cold spray, on the other hand, is a low-temperature process that does not cause particle melting thus eliminating residual stresses and coating and/or implant oxidation. The lower heat supply to the implant reduces the cooling requirement and simultaneously retains the composition/phases of the materials. Cold spraying provides high hardness due to cold worked microstructure with lower defects. The cold spray has the limitations that it may require ductile binders to coat brittle hydroxyapatite powders which may alter the properties.

Charted in Table 1 are the parameters that are certainly required to achieve the processes. From Table 1, it can be seen that Plasma spray and HVOF spray require more gas temperatures, which is a critical parameter to achieve. Furthermore, the characteristics that can be achieved through the processes are charted in Table 2. For cold spray, the characteristics are more on the favorable side, in terms of thickness, production rate, hardness, and porosity. Bond strength that can be achieved is lower than the other two processes but within the acceptable region.

3 Conclusions

Bio-implantation is a vital field of study in the present scenario. Commonly used materials include biopolymers, bio-ceramics, and metals. In loadbearing applications, Titanium and its alloys have been identified as the best bio-implants material.

Table 1 Parameters of Coating Processes [33, 34]

Characteristics	Plasma spray	HVOF spray	Cold spray
Gas temperature (°C)	12,000–16,000	2300–3100	550–1000
Gas velocity (m/s)	360–490	800–1100	1000–1400
Gas pressure (bar)	–	3–12	25–30
Particle size (μm)	5–180	20–60	5–50
Particle velocity (m/s)	Up to 450	Up to 700	500–1200
Splat size (μm)	10–50	1–20	5–50
Nozzle distance (mm)	25–150	60–200	5–25

Table 2 Achievable characteristic [34]

Characteristics	Plasma spray	HVOF spray	Cold spray
Coating thickness (mm)	0.05–5	0.05–2.5	0.5–50
Spray rate (kg/h)	2–10	2.3–14	10–30
Coating hardness (HRC)	30–65	30–72	52–90
Coating porosity (%)	1–5	3–7	4–9
Bond strength (Mpa)	14–69	48–83	10–64

But they need to be covered with bio-ceramics such as calcium phosphate (such as hydroxyapatite) to improve their bioavailability. There are many ways to create a bio-compatible mesh, but strong bonding benefits everyone. Despite some shortcomings, such as residual stress induction and less stable coatings, plasma spraying is a widely used method to create thermal spray holes during installation. High velocity oxy-fuel spray provides a stable coating despite its own residual stresses induction and oxide impurities. The problem of residual stress is not available in cold spray and simultaneously it forms most stable coatings but it has the limitation that it may require binders to deposit the coatings.

References

1. Imwinkelried T (2007) Mechanical properties of open-pore titanium foam. J Biomed Mater Res Part A 81A(4):964–970

2. Oh I.-H, Nomura N, Masahashi N, Hanada S (2003) Mechanical properties of porous titanium compacts prepared by powder sintering. Scripta Mater 49(12):1197–1202
3. Ryan G, Pandit A, Apatsidis D (2006) Fabrication methods of porous metals for use in orthopaedic applications. Biomaterials 27(13):2651–2670
4. Dunand DC (2004) Processing of Titanium Foams. Adv Eng Mater 6(6):369–376
5. Spoerke ED, Murray NGD, Li H, Brinson LC, Dunand DC, Stupp SI (2008) Titanium with aligned, elongated pores for orthopedic tissue engineering applications. J Biomed Mater Res Part A 84(2): 402–412
6. Dewidar MM, Lim JK (2008) Properties of solid core and porous surface Ti–6Al–4V implants manufactured by powder metallurgy. J Alloy Compd 454(1–2):442–446
7. Lazic S, Zec S, Miljevic N, Milonjic S (2001) The effect of temperature on the properties of hydroxyapatite precipitated from calcium hydroxide and phosphoric acid. Thermochimica Acta 374(1):13–22
8. Afshar A, Ghorbani M, Ehsani N, Saeri M, Sorrell C (2003) Some important factors in the wet precipitation process of hydroxyapatite. Mater Des 24(3):197–202
9. Petit. R (1999) The use of hydroxyapatite in orthopaedic surgery: A ten-year review. Eur J Orthop Surg Traumatol 9(2):71–74
10. Oonishi H, Oomamiuda K (1998) Degradation/resorption in bioactive ceramics in orthopaedics. Biomaterial properties. Chapman & Hall, London, p 407
11. Kent JN, Quinn JH, Zide MF, Guerra LR, Boyne PJ (1983) Alveolar ridge augmentation using nonresorbable hydroxylapatite with or without autogenous cancellous bone. J Oral Maxillofac Surg 41(10):629–642
12. Kent JN (1986) Reconstruction of the alveolar ridge with hydroxyapatite. Dent Clin North Am 30:231–257
13. Suchanek W, Yoshimura M (1998) Processing and properties of hydroxyapatite-based biomaterials for use as hard tissue replacement implants. J Mater Res 94–117
14. Grootde K, Wolke JGC, Jansen JA (1998) Calcium phosphate coatings or medical implants. Proc Inst Mech Eng [H] 212(2):137–147
15. Geesink RGT, Grootde K, Klein CPAT (1987) Chemical implant fixation using hydroxylapatite coatings. The development of a human total hip prosthesis for chemical fixation to bone using hydroxyl-apatite coatings on titanium substrates. Clin Orthop Rel Res 225:147–170
16. Jeong KI (2012) Experimental study of osseointegration and stability of intentionally exposed hydroxyapatite coating implants. J Korean Maxillofac Reconstr Surg
17. Yap AU, Pek YS, Kumar RA, Cheang P, Khor KA (2002) Biomaterials 23:955
18. Taichi I, Yoshihiro S, Hidetaka K, Yasuhiko A, Satoru Y (2001) Nippon Koku Inpuranto Gakkaishi 14(557):561
19. Weng W, Baptista JL (1998) Alkoxide route for preparing hydroxyapatite and its coating. Biomaterials 19(1–3):125–131
20. Shi D, Jiang G, Bauer J (2002) The effect of structural characteristics on the in vitro bioactivity of hydroxyapatite. J Biomed Res Appl Biomater 63(1):71–78
21. Fu Y, Batchelor AW, Khor KA (1999) Fretting wear behavior of thermal sprayed hydroxyapatite coating lubricated with bovine albumin. Wear 230(1):98–102
22. Yildirim OS, Aksakal B, Celik H, Vangolu Y, Okur A (2005) An investigation of the effects of hydroxyapatite coatings on the fixation strength of cortical screws. Med Eng Phys 27(3):221–228
23. Ektessabi AM, Hamdi M (2002) Characterization of calcium phosphate bioceramic films using ion beam analysis techniques. Surf Coat Technol 153(1):10–15
24. Yang Y, Kim KH, Ong JL (2005) A review on calcium phosphate coatings produced using a sputtering process-an alternative to plasma spraying. Biomaterials 26(3):327–337
25. Kim HM (2003) Ceramic bioactivity and related biomimetic strategy. Curr Opin Sol State Mat Sci 7(4–5):289–299
26. Leeuwenburgh S, Layrolle P, Barre' re F, Bruijnde J, Schoonman J, Blitterswijkvan CA, Grootde K (2001) Osteoclastic resorption of biomimetic calcium phosphate coatings in vitro. J Biomed Mater Res 56(2):208–215

27. Fernandez J, Gaona M, Guilemany JM (2004) Tribological study of plasma hydroxyapatite coatings. Bioceramics 16 Key Eng Mat 254–256, 383–386
28. Ferna´ndez J, Guilemany JM, Gaona M (2005) High crystallinity degree hydroxyapatite coatings using HVOF spraying. In: Lugscheider E (ed) Conference proceedings ITSC, Basel, Switzerland, DVS/IIW/ASMTSS, pp 1219–1224
29. Champagne VK, Helfritch D, Leyman P, Grendahl S, Klotz B J Thermal Spray Technol 14(3):330
30. Sun L, Berndt CC, Gross KA, Kucuk A (2001) Material fundamentals and clinical performance of plasma-sprayed hydroxyapatite coatings: a review. J Biomed Mater Res: Appl Biomater 58(5):570–592
31. Sun L, Berndt CC, Khor KA, Cheang HN, Gross KA (2002) Surface characteristics and dissolution behavior of plasma sprayed hydroxyapatite coating. J Biomed Mater Res 62(2):228–236
32. Chean P, Khor KA (1996) Addressing processing problems associated with plasma spraying of hydroxyapatite coatings. Biomaterials 17(5):537–544
33. Hermanek FJ (2001) Thermal spray terminology and company origins. First Printing, ASM International, Materials Park, OH
34. Singh H et al (2012) Frattura ed Integrità Strutturale, 22 pp 69–84

Astrophysical Black Holes—A Review

Rashi Kaushik and Amit Kumar Thakur

1 Introduction

A black hole can be determined as a zone in space-time wherein the force of gravity is so powerful that even light could not escape through it. The general relativity theory explains that a sufficient packed mass is capable of distorting the space-time leading towards the formation of black holes. But under quantum-mechanical effects the emission of particles as well as the creation of black holes is possible if the black holes were similar to hot bodies. This thermal emission results in a slow decrement in the mass of black hole leading to its eventual disappearance, also any primeval black hole is having a mass less than 10^{15} g must have been evaporated by now [1]. The quantum mechanics plays an important role like matter fields but here the classical law seems to be violated in the context that black hole's boundary or the event horizon can never decrease since there pertains a generalized second law. Any black hole of angular momentum, charge and a given mass can possess a large number of varying unobservable intramural configurations which expresses the attainable preliminary contours of the matter which got collapsed to produce the hole [2]. According to the quantum-mechanical derivation, it has been found that the black holes emit radiation at the heat rate $hk/2\pi kc$ where k is the surface gravity which aids one to prove that the entropy is finite and equals to $c^3S/4Gh$ being S to the surface area since the black holes are unstable in thermal equilibrium excluding when there is an availability of additional energy less than ¼ of the black hole's mass. Also, the black holes possess negative specific heat [3]. A gravitationally fainting star having a mass M will shrink to a radius having an order $2GM/c^2$, at which the gravitational field is believed to be so strong that neither radiation nor anything else could escape to infinity. However, while considering the quantum effects it turns out that a black hole is not fully black as the radiation underpasses through the null hypersurface and absconds to infinity

R. Kaushik (✉) · A. K. Thakur
Department of Aerospace Engineering, Lovely Professional University, Phagwara, Punjab, India
e-mail: amit.25010@lpu.co.in

A. Parey et al. (eds.), *Recent Trends in Engineering Design*, Lecture Notes in Mechanical Engineering, https://doi.org/10.1007/978-981-16-1079-0_9

at a steady rate. The formation of black holes in the early years of the cosmogonic history imprints an exquisite event in the cosmological domain [3]. One of the most sought-after questions in the present era of cosmological study is to understand when the black holes first showed up at the cosmogonic scene and started to affect the evolution of the Universe. The early black holes were nearly lesser massive, forming from the fainting of Pop-III Stars. Such kind of stellar residues would have masses of a few. Another fascinating group of these black holes might have been emerged in the early infant Universe, among the presence of messy conditions that appeared after the Big Bang. According to Carr and Hawking [4], the primordial black holes were formed as a result of energy fields and high pressure those were implanted by the presence of inhomogeneous quantum fluctuations that might have stimulated the sufficient compression of overly dense sectors. Once the black hole has emerged it might grow from the surrounding medium by accreting gas. For a particle to be captured, it must fall within the null hypersurface or the event horizon, i.e. the Schwarzschild radius ($R_S = 2GM/c^2$) [3]. Also, the Eddington limit is provided as a recognized principle for black hole's growth inhibition. The Eddington limit can be defined as the highest luminosity that an accreting body could attain before the hurling of an intense radiation-driven wind [4].

2 Black Holes Under the Theory of General Relativity

The most dense recipients that were predicted by the theory of general relativity were the black holes and are made of pure gravitation [5]. The history of black holes starts with the discovery of the Schwarzschild solution in 1916, following Einstein's formulation of the famous general relativity theory. The Schwarzschild solution is the interpretation of the vacuum Einstein equation which is symmetric spherically and is dependent on the positive parameter M (mass) along with $r = 2M$, and the area of the spheres which are orbits of rotation group $4\pi r^2$. However, according to a coordinate change made by Eddington transformed the Schwarzschild solution into a configuration that is not singular at $r = 2M$, but he failed to explain. Later, through Lemaitre's work, it was understood that $r = 2M$ is not truly a singularity but a non-success of the original system of coordinates. Much after these propositions, Eddington's transformation was re-discovered by the scientist Finkelstein who noticed that $r = 2M$ is an event horizon; the limit of space-time which is connected to infinity and validates the dynamic nature of the region r is less than $2M$. It has been assumed that singularities that occur in the gravitational collapse are not visible from outside yet are hidden behind the boundary [6]. A black hole situated on a space-like surface can be described as a connecting element of the area of the surface that is being bounded by the event horizon [7].

Einstein equation of general relativity is given by

$$RC_{\mu\nu} - {}^1\!/_2 S_c g_{\mu\nu} + \Lambda g_{\mu\nu} = 0 \tag{1}$$

where $R_{\mu\nu}$ is defined as Ricci curvature tensor, R is the scalar curvature, $g_{\mu\nu}$ is the metric tensor while Λ is the constant of cosmology [8] (see also [9]). The general relativity theory is established based on several principles as in the principle of covariance, the principle of equivalence, the principle of general relativity, the principle of correspondence and on the principle of minimal coupling [5]. Here is the Schwarzschild solution to the above equation

$$ds^2 = -(1 - 2m/r_c)dt^2 + (1 - 2mb/r_c)dr_c^2 + r_c^2\left(d\theta_x^2 + \sin x dy^2\right) \qquad (2)$$

where t is the Schwarzschild time coordinate, r is radial coordinate, θ and ϕ are the angular coordinates and m corresponding to the mass of black hole (for more details Ref. [10]). The Schwarzschild solution defines the space-time of spherically symmetric and an electroneutral black hole as he was finding a solution to Einstein's vacuum equation explaining the extrinsic gravitational field of a spherically symmetric and neutral body. Earlier the Schwarzschild solution was believed to have zero physical effect being it is the radial coordinate to be much lesser than the actual physical radius of any astronomical object, where the outer solution must be postulated with the internal vacuum solution which describes the gravitational field in the interiors of the object [11]. Therefore, in the 4-D general relativity, the black holes are defined as lucid objects that are being identified by a few constants [12].

3 The Astrophysical Black Holes

In the general relativity thesis, there are no limits on the mass rate of black holes that could be instinctively small as well as large [12]. The astrophysical black holes are identified by two parameters, which are spin and mass [13]. Through astronomical observations we get sturdy evidence of two types of astrophysical black holes: the super-massive (SM) black holes and the stellar (STM) mass black holes [12]; however, the presence of intermediate-mass (IM) black holes is also speculated.

3.1 The SM Black Holes

As per astronomical findings, there exists a huge amount of a mass in a relatively small magnitude centred at a large number of galaxies. Corresponding to standard interpretation these massive objects are the SM black holes with a mass ~ 10^5–10^{10} $M_\odot$ [12]. Even our galaxy of total mass $M \sim 10^{12}$ $M_\odot$ [14] is centred by an SM black hole of mass estimation $M \approx 4 \times 10^6$ $M_\odot$ [15]. It is predicted through models that there is a presence of SM black holes at the centre of light galactic clusters. The SM black holes can also be formed by the fainting heavy primeval clouds or may be formed as the result of the merging of several black holes [16].

3.2 STM Black Holes

It is expected that there is an existence of 10^8–10^9 black holes that are created at the edge of heavy star evolutions and a similar number is also expected in other similar galaxies as well [12]. Stellar black holes should have an initial mass which depends upon the properties of a progenitor star's evolution, mass and super-nova explosion [17]. Whenever a star is exhausted its nuclear fuel and the gas pressure are unable to stabilize the gravitational force anymore and the body gradually collapses and a significant fraction of matter gets expelled away by this vigorous process [11]. STM black holes might possess a mass estimation of about 3–100 $M_\odot$ [12] while at this present moment all-known STM black holes in the X-ray binaries would be having a mass of $M \approx 3$–$20\ M_\odot$ [13]. Also, there are pieces of evidence of black holes with a mass estimation ranging to be ~ 10^2–$10^4\ M_\odot$ which means filling up the vacancy between both the SM and the STM black holes [18]. While the STM black holes are believed to be the end products of the heavy stars, the origin of SM black holes is still a partial puzzle. Since heavy objects tend to move towards the centre of a multi-body system, it becomes feasible to think that an actual black hole could be developed by swallowing the surrounding matter.

3.3 IM Black Holes

As discussed above, there exists another type of black holes with an estimated mass lying between that of the STM black holes and the SM black holes, to be ~10^2–10^4 $M_\odot$ [12]. These are the black holes which are speculated to be a link between the fore-mentioned black holes. IM black holes are considered as the seedling which could lead towards the development of SM black holes and also define the existence of quasars possessing black hole masses about $10^{10}\ M_\odot$ at the time when the Universe was only 0.8 Galactic years old. The black holes have an impact not only on the development of the SM black holes but also on the development of a galaxy. Thus, for understanding their impact the analysis of their emissive properties along with their accretion physics plays an important part [19]. However, the IM black holes can also be found in the accretion discs of active galactic nuclei, if they could develop similarly as the planets do in the proto-planetary discs [20] (see also [21]) since the accretion discs surrounding the active galactic nuclei are not stable enough to the star formation at huge radii [22] as massive objects like stellar left-outs and stars are present in these discs wherein there is a possibility of collision accretion and growth [21]. In fact, with the availability of a mechanism through which the compact objects could be captured with the possibility of colliding, the huge mass build-up might direct to the formation of IM black holes in active galactic nuclei discs [21].

4 Kerr Black Holes

Kerr black hole is defined as an electroneutral black hole having a spin angular momentum ω and mass M [12]. When their angular momentum vanquishes, they transform into the Schwarzschild black holes. However, it was proposed by Janis and Newman that the complex deformation ($z \rightarrow z + ia$) of Schwarzschild solution results in the Kerr solution [23]. Kerr's solution is known to be symmetric under the time reversal if the sign of angular momentum is also reversed as well [6]. The region of the Kerr black hole boundary which captures light differs from its non-spinning limit and is called a photon sphere. For understanding this, one should begin by considering the light rays to be incident on the black hole's equatorial plane approaching at the light-catching region. External to the equatorial plane there exists photons which orbit black hole but their respective orbits do not confine to a plane. This rotation of black hole hauls the space time and makes an initial non-equatorial orbital plane [5]. In the context of no-hair hypothesis, the black holes are featured only by their angular momentum, charge and mass, while implying that it behaves like a point particle [23] (see also [24]).

5 Testing the Fundamental Physics with a Black Hole

5.1 Motivations

Even though the strong gravitational system remains mostly unexplored the general-relativity theory is comprehensively scrutinized among the frail gravitational fields [25]. The system of black holes is determined as an ideal lab for the scrutiny of fundamental physics in the strata of strong gravitational fields [11]. It can be executed using the electromagnetic or gravitational wave methodology, as these two methods explore different sections of the theory. While the electromagnetic tests can test the motion of gigantic and massless matter around a black hole in the strong gravitational field and scrutinize if the projections of the general relativity theory are met or not, whereas the gravitational wave tests can directly inspect the projections of Einstein's equations. The modified versions of the gravitational theories still defined by Kerr solution might speculate the general relativity as an electromagnetic spectrum and yet possess a varying gravitational wave sign as the metric of space-time remains same but field equations differ [11].

5.2 Electromagnetic Tests

Any astrophysical model that requires Kerr metric for describing a component of an accreting black hole's electromagnetic spectrum can be amplified for scrutinizing the fundamental physics with black holes [11]. At present, the most effective electromagnetic technique is X-ray reflection spectroscopy containing quantitative observational controls [26]. Generally, in these tests, geodesic motion is assumed and then analogy of the theoretical model with the observational data for restricting any possible deviations from the Kerr metric [27]. Just like X-ray reflection spectroscopy, the continuum, the fitting procedure can also be expanded for testing the nature of astrophysical black holes [28], though this method can only be utilized for testing stellar-mass black holes since independent measurements of the mass, distance and thermal component of the disc along with the angle of inclination of the disc are required [29]. Also, further tests include the study of pulsars which orbit the black holes [30] and many more.

5.3 Gravitational Wave Tests

These tests are believed to be the only technique for probing the regime of a dynamically strong field. The independent model tests the consistency of various measurements of the system constants considering general relativity [11]. The tests which depend upon the model and the observational data are compared with the theoretical predictions to frame a model [31]. However, there are drawbacks in the model-dependent tests in terms of ease of working [11]. The present findings of the gravitational waves coming from the black holes are constrained [32]. But with further advancements, it will be possible through laser interferometer [11].

6 Concluding Remarks

Through this comprehensive review, we have tried to explain the basics of black hole physics. The contribution of the noble laureate Albert Einstein in the field of physics is unfathomable, mainly his theory of general relativity which laid the foundation of modern-day physics as well as the latest advancements in the field of research. Through general relativity, it can be understood that an abundantly compact mass is capable of deforming the space-time and lead towards the creation of black holes. Also, it is well explained that the null hypersurface or the region from where it is impossible to even for the light to escape to infinity is termed as the event horizon. Although there are some drawbacks to the Schwarzschild solution it has been quite successful in describing the singularities. The existence and subtypes of astrophysical black holes have also been postulated. However, the presence of IM black holes has also been speculated which could fill up the aperture between the SM and STM black holes, but their disposition is not yet clear. Kerr black holes, on the other hand, possess no electrical charge but do possess an angular momentum and mass, but can transform into Schwarzschild black holes in the absence of angular momentum. Through the testing of fundamental physics with black holes it has been tried to make clear that though the general relativity theory is being challenged at every step, it still remains to be an irreplaceable key to understand various aspects of physics.

Acknowledgements We acknowledge Lovely Professional University, Phagwara, Punjab, India.

References

1. Hawking SW (1975) Particle creation by black holes. Commun Math Phys 43(3):199–220
2. Hawking SW (1976) Black holes and thermodynamics. Phys Rev D 13(2):191
3. Smith A, Bromm V (2019) Supermassive black holes in the early universe. Contemp Phys 60(2):111–126
4. Carr BJ, Hawking SW (1974) Black holes in the early Universe. Mon Not R Astron Soc 168(2):399–415
5. Lemos JP, Herdeiro CA, Cardoso V (2019) Einstein and Eddington and the consequences of general relativity: black holes and gravitational waves. arXiv preprint arXiv:1911.01959
6. Christodoulou D (2012) The formation of black holes in general relativity. In: The twelfth marcel grossmann meeting: on recent developments in theoretical and experimental general relativity, astrophysics and relativistic field theories (In 3 volumes), pp 24–34
7. Hawking SW (1972) Black holes in general relativity. Commun Math Phys 25(2):152–166
8. Arnowitt R, Deser S, Misner CW (2008) Republication of: the dynamics of general relativity. Gen Relativ Gravit 40(9):1997–2027
9. Aoki S, Onogi T, Yokoyama S (2020) Conserved charges in general relativity. arXiv preprint arXiv:2005.13233
10. Kunz J (2019) Introduction to black holes. J Phys Conf Ser 1263(1):012007 (IOP Publishing)
11. Bambi C (2019) Astrophysical black holes: a review. arXiv preprint arXiv:1906.03871
12. Bambi C (2018) Astrophysical black holes: a compact pedagogical review. Ann Phys 530(6):1700430
13. Casares J, Jonker PG (2014) Mass measurements of stellar and intermediate-mass black holes. Space Sci Rev 183(1–4):223–252
14. McMillan PJ (2016) The mass distribution and gravitational potential of the Milky Way. Mon Not R Astron Soc. stw2759
15. Boehle A, Ghez AM, Schödel R, Meyer L, Yelda S, Albers S, Matthews K (2016) An improved distance and mass estimate for Sgr A* from a multi-star orbit analysis. Astrophys J 830(1):17
16. Volonteri M (2010) Formation of supermassive black holes. Astron Astrophys Rev 18(3):279–315
17. Belczynski K, Bulik T, Fryer CL, Ruiter A, Valsecchi F, Vink JS, Hurley JR (2010) On the maximum mass of stellar black holes. Astrophys J 714(2):1217
18. Coleman Miller M, Colbert EJ (2004) Intermediate-mass black holes. Int J Mod Phys D 13(01):1–64
19. Mezcua M (2017) Observational evidence for intermediate-mass black holes. Int J Mod Phys D 26(11):1730021
20. McKernan B, Ford KES, Lyra W, Perets HB (2012) Intermediate mass black holes in AGN discs–I. Production and growth. Mon Not R Astron Soc 425(1):460–469
21. Bellovary JM, Mac Low MM, McKernan B, Ford KS (2016) Migration traps in disks around supermassive black holes. Astrophys J Lett 819(2):L17
22. Dittmann AJ, Miller MC (2020) Star formation in accretion discs and SMBH growth. Mon Not R Astron Soc 493(3):3732–3743
23. Arkani-Hamed N, Huang YT, O'Connell D (2020) Kerr black holes as elementary particles. J High Energy Phys 2020(1):46

24. Maybee B, O'Connell D, Vines J (2019) Observables and amplitudes for spinning particles and black holes. J High Energy Phys 2019(12):1–32
25. Will CM (2014) The confrontation between general relativity and experiment. Living Rev Relativ 17(1):4
26. Bambi C, Cárdenas-Avendaño A, Dauser T, García JA, Nampalliwar S (2017) Testing the Kerr black hole hypothesis using X-ray reflection spectroscopy. Astrophys J 842(2):76
27. Bambi C (2014) Constraining possible variations of the fine structure constant in strong gravitational fields with the Kα iron line. J Cosmol Astropart Phys 2014(03):034
28. Bambi C, Barausse E (2011) Constraining the quadrupole moment of stellar-mass black hole candidates with the continuum fitting method. Astrophys J 731(2):121
29. Kong L, Li Z, Bambi C (2014) Constraints on the spacetime geometry around 10 stellar-mass black hole candidates from the disk's thermal spectrum. Astrophys J 797(2):78
30. Wex N, Kopeikin SM (1999) Frame dragging and other precessional effects in black hole pulsar binaries. Astrophys J 514(1):388
31. Yunes N, Yagi K, Pretorius F (2016) Theoretical physics implications of the binary black-hole mergers GW150914 and GW151226. Phys Rev D 94(8):
32. Berti E, Yagi K, Yunes N (2018) Extreme gravity tests with gravitational waves from compact binary coalescences:(I) inspiral–merger. Gen Relativ Gravit 50(4):46

Kenaf-Fiber-Based Bio-materials: A Review on Processing and Mechanical Properties

Jai Inder Preet Singh, Sehijpal Singh, Vikas Dhawan, Amrinder Singh Pannu, Ankur Bahl, Piyush Gulati, Rajeev Kumar, and Manpreet Singh

1 Introduction

Because of its optimized properties such as high strength and less weight, more precise strength, etc., natural fiber-reinforced composites play an important role in advanced technology. Natural fiber composites are used in most applications, starting from household products, interior decoration, and automotive applications, as well as in medical applications, due to global environmental concerns. Lot of car companies are beginning to develop the cars' interior products from natural fiber-reinforced composites. Natural fibers are mainly classified according to their source. Figure 1 shows the types of natural fibers.

The whole stalk and external bast fibers of Kenaf have many possible specific uses, containing as paper and composite. The use of kenaf fibers is also particularly important from the point of view of environmental approachability. A number of new uses for kenaf products are currently emerging including paper products, building materials, absorbents, and bedding for feed and livestock. Kenaf is one of the major seed crops grown in temperate climates by synthetic oils. Figure 2 shows the kenaf fiber structure.

J. I. P. Singh (✉) · A. Bahl · P. Gulati · R. Kumar · M. Singh
School of Mechanical Engineering, Lovely Professional University, Phagwara, India

S. Singh · A. S. Pannu
Guru Nanak Dev Engineering College, Ludhiana, India

V. Dhawan
Pro Vice Chancellor, SGT University, Gurugram, India

A. Parey et al. (eds.), *Recent Trends in Engineering Design*, Lecture Notes in Mechanical Engineering, https://doi.org/10.1007/978-981-16-1079-0_10

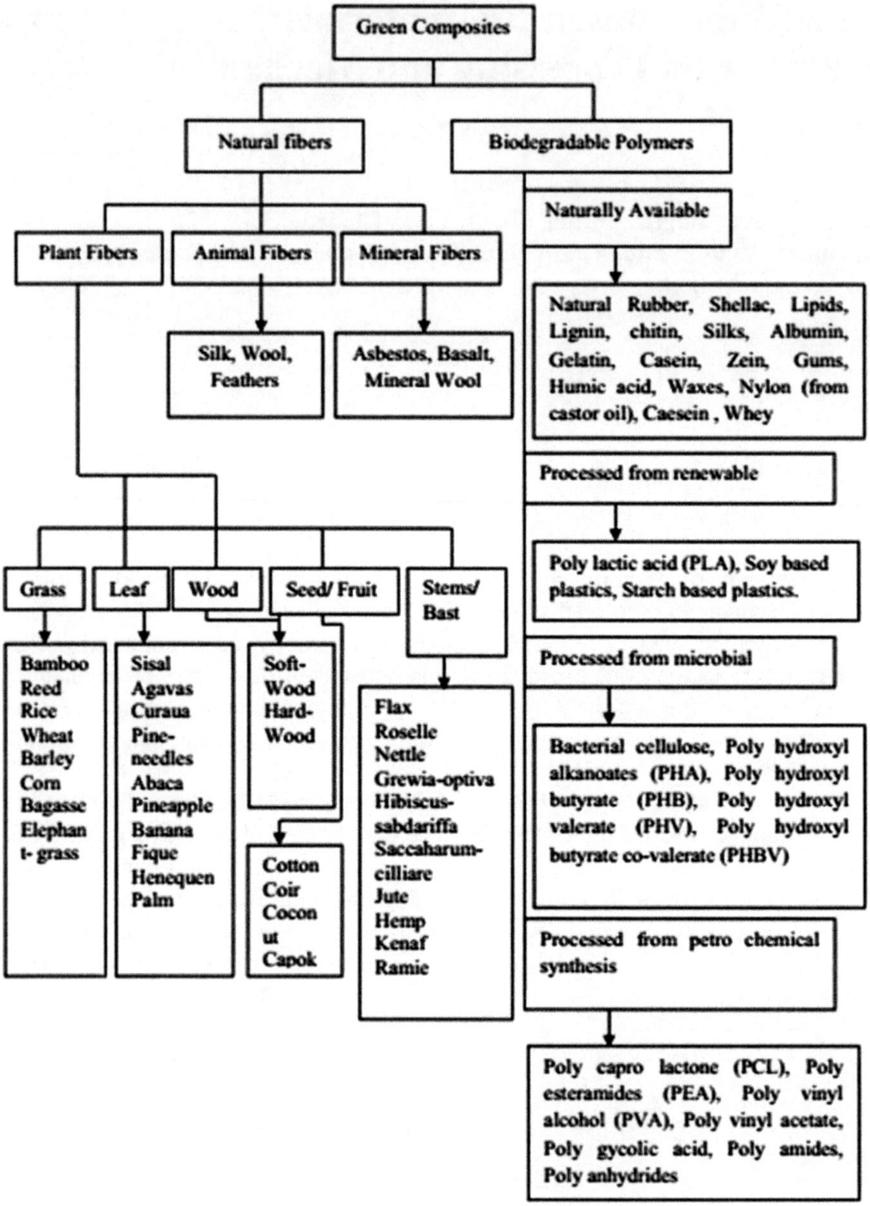

Fig. 1 Constituents of green composites [1]

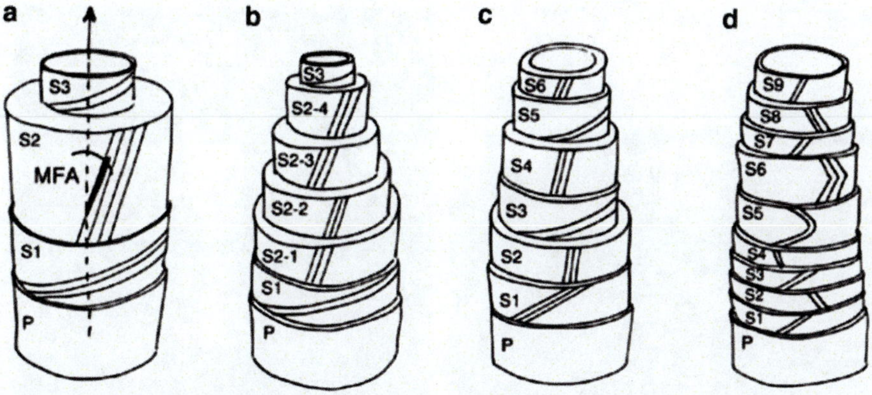

Fig. 2 Schematics of possible cell wall organization in **a** wood fibers, **b** bast fibers, **c** plant fibers, and **d** seed fibers. Black lines indicate orientation of cellulose microfibrils

2 Kenaf Fiber

The kenaf plant has a lifecycle of 90–125 days including growth, flowering, and maturation periods. The kenaf stem has a diameter of 1–2 mm in size, with a height of about 80 cm. The fundamental fibers which are both cellulose material are the primary and secondary cell walls. Cellulose fibrils (diameter between 0.2 and 0.25 μm) are introduced into a compact lamella consisting of approximately 3% pectins and 14% hemicellulose, resulting in thermal fiber and water degradation. The kenaf plant grows when the temperate climate suffices for humidity. There are two kinds of plants of kenaf seed and fiber keanf. Keanf is cultivated for both seeds and fibers. Fiber keanf is designed for small, strong fiber growth. The kenaf plant is an annual plant that reaches a height of approximately 100 cm. The plants are pulled by hand or machine and all the fibers from the top to the root are intact. Figure 3 shows the kenaf plant.

The stem part is taken for the after pulling retting operation. The fleshy portion of the stem is rotted almost for a week due to contact with water. Finally, the kenaf stem is pushed and pounded on a sharp edge, where fibers are loosened.

3 Kenaf Fiber Physical and Mechanical Properties

Kenaf is the best among organic cellulosic fibers. Kenaf is two to three times more wear- and abrasion-resistant than cotton fiber. It absorbs the moisture well, and is a very breathable fabric. Because of the lack of elasticity of kenaf fibers, fabrics made from kenaf do not seem to lose their shape quickly. In recent years, the use of kenaf fibers in the automotive industry has increased dramatically [2]. As regard to properties, kenaf fiber has the potential to replace the synthetic fibers. Some of the

Fig. 3 Kenaf plant

investigators had researched the properties of kenaf fibers. According to Pan et al. [3], kenaf fiber reinforced with PLLA improves the mechanical properties as well as accelerates the crystallization of PLLA. According to Bernard et al. [4], the effect of processing parameters on the mechanical properties of the kenaf/PP composites was investigated and it reveals that the processing parameters have a great impact on properties of the composites. Nitta et al. [5] explored the influence of alkali treatment on tensile properties of kenaf fibers and results reveal that tensile strength is greatly improved with 15 wt% NaOH for 1 h as compared to the untreated fiber. Figure 4 shows the SEM image raw fibers and surface-treated kenaf fibers.

Bajpai et al. [7] developed the jute/hemp/flax-reinforced hybrid composite and investigated its mechanical properties. Mechanical characterization results reveal that the ductile properties are more in hybrid composite as compared to the jute fiber-reinforced epoxy, flax fiber-reinforced epoxy, and hemp fiber-reinforced epoxy composites. Figure 5a shows the SEM images of jute/hemp/flax-reinforced epoxy composites.

Interfacial bonding among the fiber and matrix shows a vital part for defining the mechanical properties of the composite material [8]. Some of the findings are tabulated in Table 1.

4 Kenaf Fiber-Reinforced Polymer Composites

Nowadays, lots of researchers are working in the area of hybrid composites especially for the structural applications. Srinivasan et al. [14] evaluate the tensile and thermal

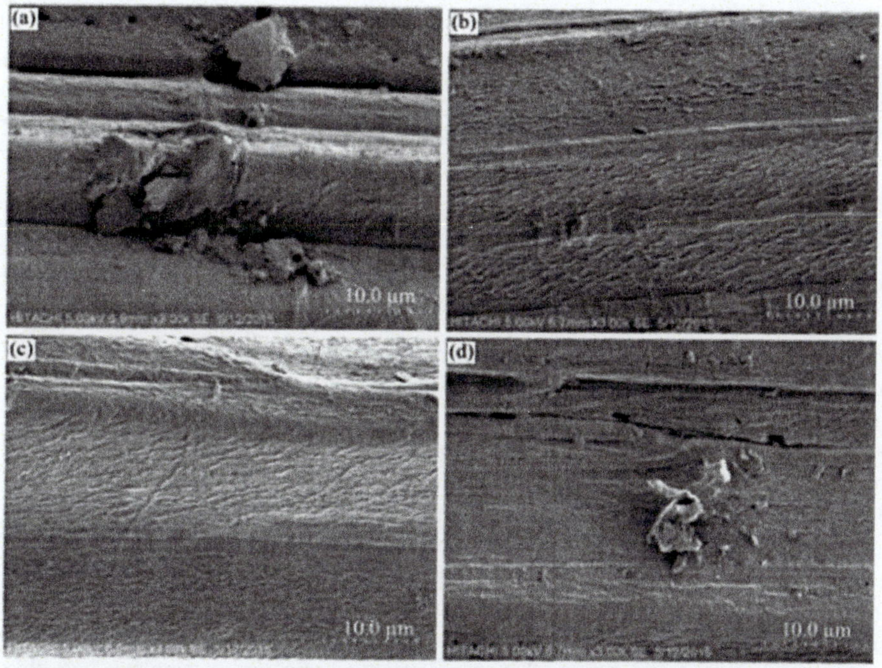

Fig. 4 SEM of **a** untreated, **b** alkali solution, **c** silane, and **d** alkali-silane-treated Kenaf fiber [6]

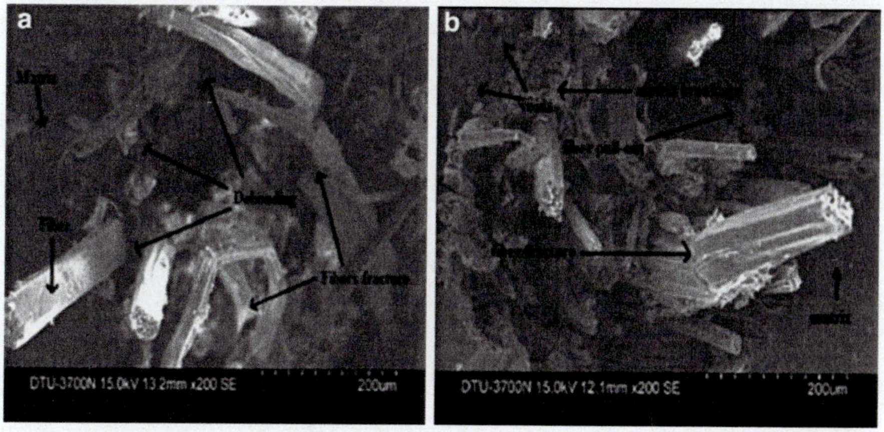

Fig. 5 SEM image of **a** jute/hemp/flax-reinforced epoxy composites, **b** hemp/flax/epoxy composite [7]

Table 1 Physical and mechanical properties of Kenaf fibers

Diameter (mm)	Ultimate stress (MPa)	Density (kg/m^3)	Young's modulus (GPa)	Water absorption (%)	Source
0.16–0.40	370–630	1510	13–51	0.85	[9]
	229–970	1200–14,500	15.7–57	13–17	[10]
0.13–0.29	950	1460	57	11–18	[11]
0.85	953	1500	56	12–18	[12]
–	800	1500	15.5–58	12–15	[13]

properties of banana-kenaf/epoxy hybrid composites with the help of compression molding machine. Mechanical properties such as tensile strength, flexural strength, and impact strength have been investigated and structural morphology has been tested using SEM. Results found that the mechanical properties of glass-banana-kenaf-based hybrid composites possess higher values as compared to the glass/epoxy, banana/epoxy, and keanf/epoxy composites. Ghani et al. [15] deliberated the mechanical behavior of kenaf/PLA composites and record that 35% fiber volume fraction leads to better mechanical properties. It was also found that at less fiber volume fraction, mechanical properties get deteriorated. Poor bonding between the matrix and reinforcement was observed at higher fiber loading. Apart from the mechanical properties of the kenaf-reinforced polymer composites, researchers had also tried to study the effect of aging on the ductile properties. In one of the studies, immersion of the composites into the water has been shown to have a degrading effect on the mechanical properties. Figure 6 shows the SEM images of fractured biocomposites. Atiqah et al. [16] deliberate the impact of hygrothermal aging on the ductile properties of kenaf/epoxy-based composites. Results indicate that the

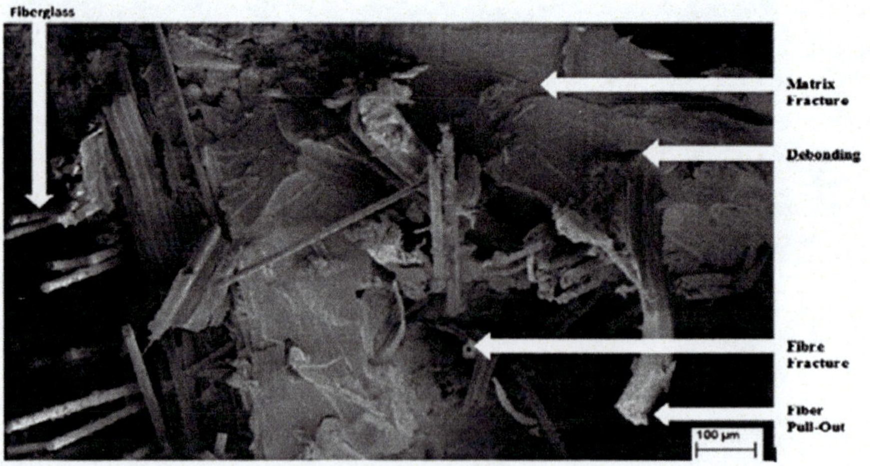

Fig. 6 The SEM images of fractured biocomposites [15]

mechanical properties were affected significantly due to the aging process. Jacob et al. [17] studied the environmental influences on the fabrication of kenaf textile yarn. Reducing environmental influences associated with the production of hemp yarn should give priority to reducing energy consumption during the fiber processing and yarn production phases and reducing eutrophication during the crop production phase. Yang et al. [18] premeditated the effect of thermal treatment on kenaf fibers for the improvement of wetting and swelling properties. The thermal treatment has been shown to make fibers less hydrophilic. Jai Inder et al. [19] studied the effect of three different surface treatment chemicals and molding temperature on mechanical properties of jute/PLA green composites and results revealed that the alkali-treated jute/PLA composite has 170 °C molding temperature, maximum tensile, and flexural strength.

5 Conclusions

The complete study focuses on the critical review for the reinforcement of kenaf fibers in the polymer composites. Natural fiber-reinforced polymer composites have increased wide interest because of ecofriendly and sustainable content. Specific polymer composites of kenaf fiber were intentionally the possible material for substituting synthetic fibers for polymer composites. Kenaf fiber is one of the natural fibers having more mechanical properties as compared to other fibers. The complete review also touched upon the different mechanical properties of kenaf fibers and its applications.

6 Future Work

From the various studies mentioned in the above sections, cited by the various researchers on the mechanical properties of polymer composites reinforced by kenaf fiber. The outcome of these research studies leads to provide the future scopes in the following areas:

1. Most researches involve the enhancement of mechanical properties with the use of fiber surface treatments.
2. The area of development in the surface finish of natural fiber-reinforced polymer composites has been less explored.

References

1. Singh H, Jai Inder Preet S, Dhawan V, Kumar Tiwari S (2018) A brief review of jute fibre and its composites. Mater Today Proc 5(14):28427–28437
2. Inder J, Singh P, Singh S, Dhawan V (2019) Influence of fiber volume fraction and curing temperature on mechanical properties of jute/PLA green composites. Polym Polym Compos 1–12
3. Pan P, Zhu B, Kai W, Serizawa S, Iji M, Inoue Y (2007) Crystallization behavior and mechanical properties of bio-based green composites based on Poly(L-lactide) and Kenaf Fiber. J Appl Polym Sci 105:1511–1520
4. Bernard M et al (2011) The effect of processing parameters on the mechanical properties of kenaf fibre plastic composite. Mater Des 32(2):1039–1043
5. Nitta Y, Goda K, Noda J, Il Lee W (2013) Cross-sectional area evaluation and tensile properties of alkali-treated kenaf fibres. Compos Part A Appl Sci Manuf 49:132–138
6. Bulota M, Budtova T (2015) Highly porous and light-weight flax/PLA composites. Ind Crop Prod 74:132–138
7. Chaudhary V, Bajpai PK, Maheshwari S (2018) Studies on mechanical and morphological characterization of developed jute/hemp/flax reinforced hybrid composites for structural applications. J Nat Fibers 15(1):80–97
8. Inder J, Singh P, Singh S, Dhawan V (2019) Effect of alkali treatment on mechanical properties of jute fiber-reinforced partially biodegradable green composites using epoxy resin matrix. Polym Polym Compos 1715:1–10
9. Akil HM, Omar MF, Mazuki AAM, Safiee S, Ishak ZAM, Bakar AA (2011) Kenaf fiber reinforced composites: a review. Mater Des 32(8–9):4107–4121
10. Shah AUM (2019) Effects of fibre treatment on mechanical properties of kenaf fibre reinforced composites: a review. Integr Med Res 8(3):3327–3337
11. Nor W et al (2012) Thermal properties of PLA/Kenaf green nanocomposite: effect of chemi-mechanical treatment. Adv Mater Res 576:342–344
12. Akhtar MN et al (2016) Influence of alkaline treatment and fiber loading on the physical and mechanical properties of kenaf/polypropylene composites for variety of applications. Prog Nat Sci Mater Int 26(6):657–664
13. Gardner L, Munro T, Villarreal E, Harris K, Fronk T, Ban H (2018) Thermal characterization of alkali treated kenaf fibers and Kenaf-epoxy composites. Fibers Polym 19(2):393–402
14. Srinivasan VS, Rajendra Boopathy S, Sangeetha D, Vijaya Ramnath B (2014) Evaluation of mechanical and thermal properties of banana–flax based natural fibre composite. Mater Des 60:620–627
15. Ghani MAA, Salleh Z, Hyie KM, Berhan MN, Taib YMD, Bakri MAI (2012) Mechanical properties of Kenaf/fiberglass polyester hybrid composite 41:1654–1659
16. Atiqah A, Maleque MA, Jawaid M, Iqbal M (2014) Development of kenaf-glass reinforced unsaturated polyester hybrid composite for structural applications. Compos Part B Eng 56:68–73
17. Jacob M, Bellmann C, Anandjiwala RD (2010) Kenaf—polypropylene composites: effect of amphiphilic coupling agent on surface properties of fibres and composites. Carbohydr Polym 82(3):549–554
18. Yang B et al (2014) Effects of chemical versus enzymatic processing of kenaf fibers on poly(hydroxybutyrate-co-valerate)/poly(butylene adipate-co-terephthalate) composite properties. Compos Part B Eng 56:926–933
19. Jai Inder Preet S, Sehijpal S, Dhawan V, Gurjot S, Piyush G, Rajeev K, Manpreet S (2020) Influence of surface treatment and molding temperature on mechanical properties of jute/PLA-based green composites. Adv Mater Sci Eng 149–158

Numerical Simulation of Bi-material Interfacial Adiabatic Crack Using EFGM and XFEM

Nalla Suresh, Ishan Wath, and Sahil Garg

1 Introduction

Materials have become the part of human life since the very existence and also there has been various developments that took place in metal industry. The materials with numerous properties are most preferred by the manufacturers for making automotive parts as they need to withstand to a variety of loading conditions during their life time. Properties of materials including high strength, less weight, resistance to corrosion, etc. are likely to be best for applications of automotive and aerospace industries [1]. Then the researchers show some attention to develop layered materials which possess the potentiality of being used for various applications mostly in mechanical and aerospace industries [2]. Layered materials consist of a substrate coating of any other material on the substrate material, for example, aluminium coating on steel, in which aluminium provides the corrosion resistance in the top of substrate steel material [3]. Coatings are widely used on automotive body panels, engine parts, exhaust to increase resistance to corrosion, as thermal barriers, wear resistance, anti-chip protection, etc. Similarly, there are a variety of substrates available in the market which can be applied as a coating on different materials using various processes [4]. Finite Element Method (FEM) has been widely used for layered domain analysis [5, 6]. FEM has inherent problems such as non-conformal mesh and locking of elements. These gaps were circumvented by the use of MMs such as EFGM [7] and XFEM [8].

Both EFGM and XFEM have been extensively used in the area of fracture mechanics [9–14]. Owing to their dexterity and wide range of applications to solid mechanics this work employs a comparative study of these methods on interfacial crack problem under adiabatic conditions. Since these methods are well established

N. Suresh (✉) · I. Wath · S. Garg
School of Mechanical Engineering, Lovely Professional University, Phagwara, Punjab 144411, India

A. Parey et al. (eds.), *Recent Trends in Engineering Design*, Lecture Notes in Mechanical Engineering, https://doi.org/10.1007/978-981-16-1079-0_11

and provide cognizable results the concurrency of simulations presented in this paper is accepted.

2 EFGM and XFEM Formulation

The variable u can be estimated by MLS approximation for EFGM, $u^h(x)$ can be written as [7]

$$u^h(\boldsymbol{x}) = \sum_{j=1}^{m} p_j(\boldsymbol{x})a_j(\boldsymbol{x}) \equiv \boldsymbol{p}^T(x)\boldsymbol{a}(x) \tag{1}$$

where $p(x)$ is the basis function, $a(x)$ can be obtained at any random point x in the domain by minimization of weighted least square sum of the difference between local approximation, $u^h(x)$ and field function nodal parameters, u_I. The approximation function is obtained as

$$u^h(\boldsymbol{x}) = \sum_{I=1}^{n} \Phi_I(\boldsymbol{x})u_I \tag{2}$$

For modelling cracks in XFEM, the approximation function [15] takes the following form:

$$\boldsymbol{u}^h(\boldsymbol{x}) = \sum_{i=1}^{n} \boldsymbol{N}_i(\boldsymbol{x})\left[\overline{\boldsymbol{u}_i} + \underbrace{\mathbf{H}(\boldsymbol{x})\boldsymbol{a}_i}_{i\in \boldsymbol{n}_r} + \underbrace{\sum_{j=1}^{4} \boldsymbol{\beta}_j(\boldsymbol{x})\boldsymbol{b}_i^j}_{i\in \boldsymbol{n}_A}\right] \tag{3}$$

The terms used in these equations are represented by their usual meanings. A fully enriched basis function approach [16] for crack modelling is used, which is given as

$$\beta_j(x) = \left[\sqrt{r}\sin\frac{\theta}{2}, \sqrt{r}\cos\frac{\theta}{2}, \sqrt{r}\cos\frac{\theta}{2}\sin\theta, \sqrt{r}\sin\frac{\theta}{2}\sin\theta\right] \tag{4}$$

where r is the radial distance of the evaluation point from the crack tip and θ being the angular orientation of the line joining the point to crack tip.

For modelling weak discontinuity in the EFG domain jump function [17] approach has been used and in case of XFEM level sets [18] are used.

3 Numerical Implementation

The work is performed using MATLAB software for coding the mathematical modelling related to XFEM and EFGM. The bi-material domain is assumed with Young's modulus ratio of (1/2, 1/4, 1/6, 1/8 and 1/10). Thermal conductivity of both domains is kept to be 1:1 initially. Poisson's ratio is assumed to be 0.3 for booth materials. Centre crack is modelled at the interface and adiabatic conditions [14, 19] are assumed. The dimensions of the domain are 160 by 160. Discretization is done using 1600 nodes for both EFGM and XFEM as shown in Fig. 1a, b. Modified thermal interaction integral [20] has been used to find SIFs. The concurrency in the results of both the methods validates the simulation.

3.1 *Results and Discussions*

The results of the work are elaborated by comparison of distribution of temperature and stresses obtained by both EFGM and XFEM. Temperature and stress distribution by EFGM are shown in Fig. 2a, b and by XFEM are shown in Fig. 3a, b. The SIFs obtained by both the methods are −4.5106 (EFGM) and −5.1274 (XFEM) which shows the sync in simulations. The negative value of SIF shows compressive nature of stresses which is also confirmed by deformed configurations shown in Fig. 4a, b. Variation in ratio of Young's modulus does not impact the values of SIFs obtained.

When different ratios of thermal conductivities are taken (1/100, 75/100, 1/50, 1/25, 1/20 and 1/10) there arises a complete change in the behaviour of domain and results are plotted in Fig. 5. The plot shows normalized mixed mode SIFs variation w.r.t. thermal conductivity. The simulation for this plot has been done at Young's modulus ratio of 1:4.

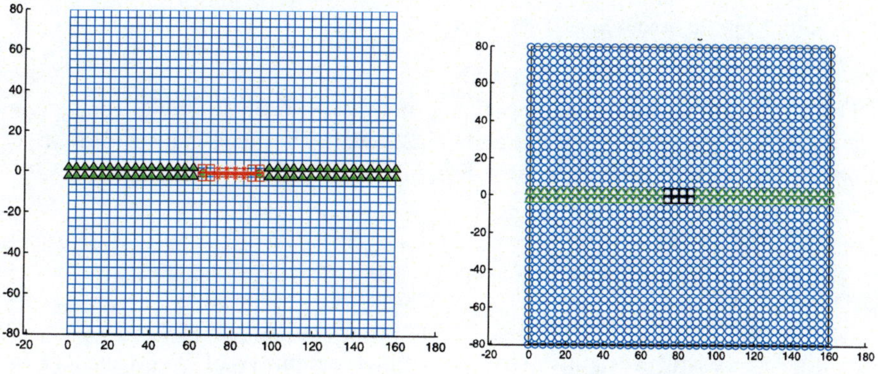

(a) (b)

Fig. 1 **a** XFEM domain, **b** EFGM domain

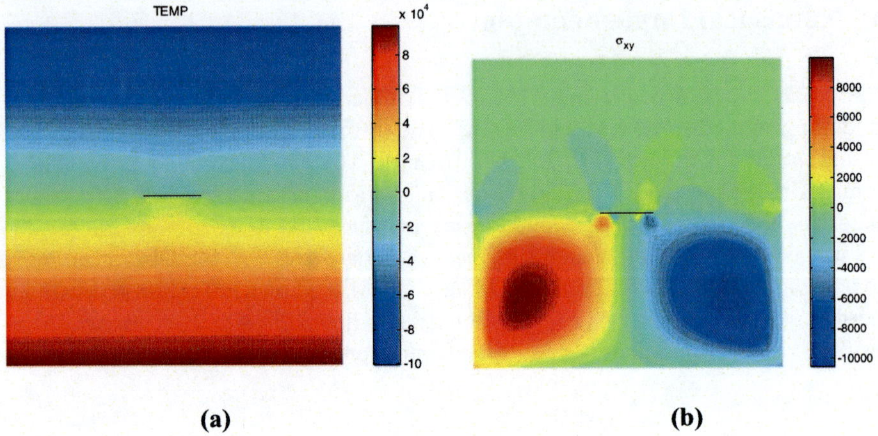

Fig. 2 EFGM domain **a** temperature field, **b** stress distribution

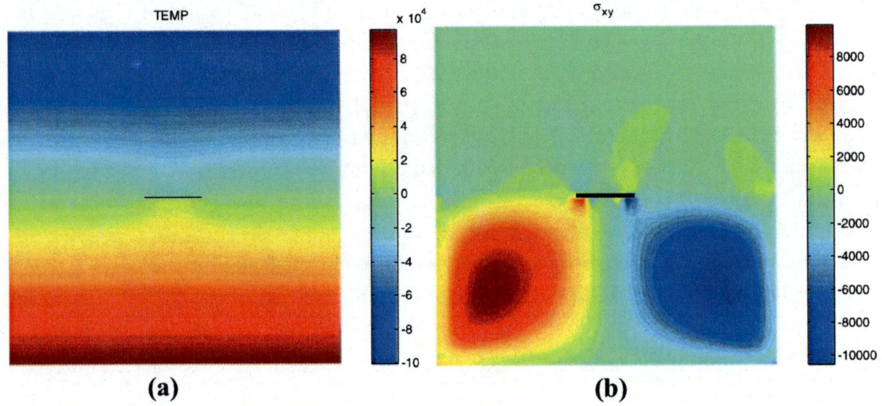

Fig. 3 XFEM domain **a** temperature field, **b** stress distribution

These simulations provide an important proof that XFEM and EFGM can handle wide changes in properties in complex domains as well. The plot is made with average values of mixed mode SIFs taken from EFGM and XFEM.

4 Conclusion

The work is successful in establishing the modelling capabilities of EFGM and XFEM in simulation of interfacial centre cracks under thermal loads. The work establishes the robustness of these methods in analysing complex domains with intricate stress and displacement fields. The analysis can be extended to simulations of welded or

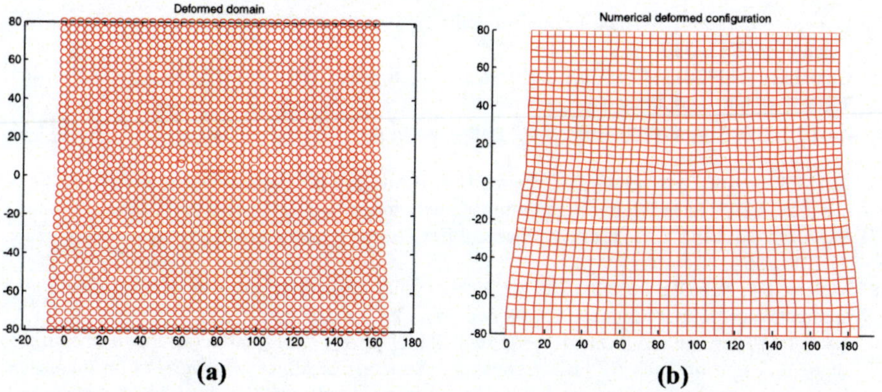

Fig. 4 Deformed domain **a** EFGM, **b** XFEM

Fig. 5 Plot of thermal conductivity variation with normalized SIFs

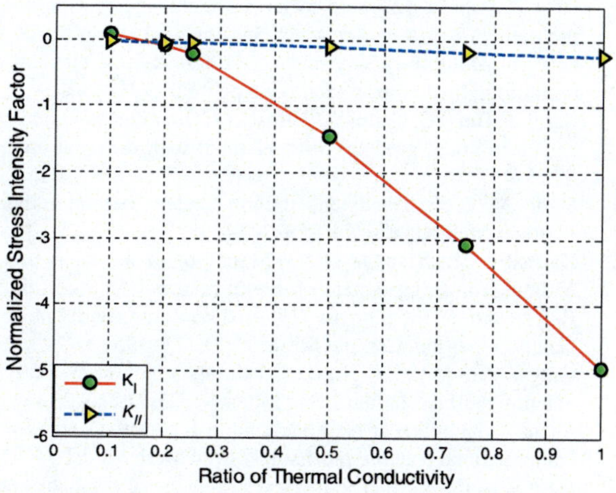

brazed structures which is difficult experimentally and even by FEM. Moreover, centre crack modelling is difficult due to the presence of two crack tips which is also successfully achieved in this case.

References

1. Bhardwaj G, Singh IV, Mishra BK (2015) Stochastic fatigue crack growth simulation of interfacial crack in bi-layered FGMs using XIGA. Comput Methods Appl Mech Eng 284:186–229
2. Bouhala L et al (2013) An XFEM crack-tip enrichment for a crack terminating at a bi-material interface. Eng Fract Mech 102:51–64

3. Dai-Heng C (1994) A crack normal to and terminating at a bimaterial interface. Eng Fract Mech 49(4):517–532
4. Chamat A et al (2013) Crack behaviour in zinc coating and at the interface zinc-hot galvanised TRIP steel 800. Eng Fract Mech 114:12–25
5. Śliwa A et al (2016) Prediction of the properties of PVD/CVD coatings with the use of FEM analysis. Appl Surf Sci 388:281–287
6. Kot M, Rakowski W, Lackner JM, Major Ł (2013) Analysis of spherical indentations of coating-substrate systems: experiments and finite element modeling. Mater Des 43:99–111
7. Belytschko T, Lu YY, Gu L (1994) Element free Galerkin methods. Int J Numer Methods Eng 37:229–256
8. Melenk JM, Babuška I (1996) The partition of unity finite element method: basic theory and applications. Comput Methods Appl Mech Eng 139(1–4):289–314
9. Jameel A, Harmain GA, Anand Y, Masoodi JH, Najar FA (2017) Effect of inclusions on the shape and size of crack tip plastic zones by element free effect of inclusions on the shape and size of crack tip plastic zones by element free galerkin method. Int J Mech Aerosp Ind Mechatron Manuf Eng 1:414–419
10. Garg S, Pant M (2018) Meshfree methods: a comprehensive review of applications. Int J Comput Methods 15(4):1830001-1–1830001-85
11. Garg S, Pant M (2018) Accelerated element-free Galerkin method for analysis of fracture problems. J Brazilian Soc Mech Sci Eng 40(11):541
12. Moës N, Dolbow J, Belytschko T (1999) A finite element method for crack growth without remeshing. Int J Numer Methods Eng 46(1):131–150
13. Kang Z, Bui TQ, Saitoh T, Hirose S (2016) Quasi-static crack propagation simulation by an enhanced nodal gradient finite element with different enrichments. Theor Appl Fract Mech 87:61–77
14. Duflot M (2008) The extended finite element method in thermoelastic fracture mechanics. Int J Numer Methods Eng 74(5):827–847
15. Karihaloo BL, Xiao QZ (2003) Modelling of stationary and growing cracks in FE framework without remeshing: a state-of-the-art review. Comput Struct 81(3):119–129
16. Belytschko T, Fleming M (1999) Smoothing, enrichment and contact in the element-free Galerkin method. Comput Struct 71(2):173–195
17. Garg S, Pant M (2018) Numerical simulation of thermal fracture in coatings using element free galerkin method. Indian J Eng Mater Sci 25(3):217–232
18. Yazid A, Abdelkader N, Abdelmadjid H (2009) A state-of-the-art review of the X-FEM for computational fracture mechanics. Appl Math Model 33(12):4269–4282
19. Garg S, Pant M (2017) Numerical simulation of adiabatic and isothermal cracks in functionally graded materials using optimized element-free Galerkin method. J Therm Stress 40(7):846–865
20. Garg S, Pant M (2018) Numerical simulation of thermal fracture in coatings using element free Galerkin method. Indian J Eng Mater Sci 25:217–232

Structural and Modal Analysis of PEEK-CF Composite for Aircraft Wing

Kanishkha Jha, I. V. S. Yeswanth, Desai Manish, and Y. K. Tyagi

1 Introduction

The aircraft industry is one of the rapidly growing industries in recent years. The objective of the industry is to reduce the weight of the aircraft and improve the fuel efficiency by increasing the number of flight hours from 40,000 to 60,000 h [1]. One way to achieve this objective is through weight reduction by developing the materials with low density and high strength. The development of lightweight materials for aeronautical structural components like wings, fuselage can be made of composite materials which further can reduce the weight of aircraft components. Lightweight materials offer the advantage of improved fuel efficiency and increases the aircraft's payload and flight range which helps in reduction of aircraft's operating cost. The recent researchers focus on developing materials with optimized properties to reduce weight, improve damage tolerance, fatigue, and corrosion resistance. The structural requirement of airframe materials includes a life of 60,000 flight hours for static weight, service loads, damage tolerance, and resistance to UV radiation [1]. The history of aircraft materials development dates back to 1900 when conventional wood frame materials are first replaced with metals. However, with introduction of metals, the weight of aircraft increased thereby increasing the consumptions of fuel. Since then the aircraft industry is looking for the development of materials with lightweight and aluminum alloys were introduced to replace the metal structures and dominated the industry for 80 years [2]. The common Al

K. Jha · I. V. S. Yeswanth (✉)
School of Mechanical Engineering, Lovely Professional University, Phagwara, Punjab, India
e-mail: venkata.11815942@lpu.in

D. Manish
School of Aeronautical Sciences, Hindustan Institute of Technology and Science, Padur, Chennai, India

Y. K. Tyagi
School of Mechanical Engineering, DIT University, Makka Wala, Uttarakhand, India

A. Parey et al. (eds.), *Recent Trends in Engineering Design*, Lecture Notes in Mechanical Engineering, https://doi.org/10.1007/978-981-16-1079-0_12

alloys used in airframe materials include Al2024, Al6065, and Al7075 which are some of the common airframe materials. Al2024 possesses better damage tolerance; however, the lower yield strength limits it usage. Al2024 is common material used in lower panels of wing where it is subjected to tension and damage tolerance of high priority [3]. Al7075 alloy is commonly used in upper wing surface where it is subjected to compression which requires high strength. Though Al7075 provides high strength its poor damage tolerance and corrosion resistance limit its usage [4]. The use of composite materials is rapidly increasing in the structural parts of the aircraft. Polymer composites are known for their low density, high specific strength, and modulus. The density of carbon fiber (CF)-reinforced epoxy composite is only half that of Al alloys, while the tensile strength and elastic modulus are three times and two times higher, respectively [5]. Thermoset and thermoplastic polymers are widely used in aeronautical structure applications such as ailerons, flaps, and landing-gear doors. Around 50% weight of Boeing 777 and 787 aircrafts are made of carbon fiber-reinforced polymer [6]. Although carbon fiber offers a great improvement in the strength of polymer composites, it is likely to suffer stress concentration due to its brittleness [7]. However, there are some limitations for the epoxy-carbon fiber apart from the advantages. Epoxy resins exhibit poor moisture absorption which degrades mechanical properties. The structural materials degrade with time due to radiation effects [8]. Recent developments in high-performance polymers like Polyether Ether Ketone (PEEK) has gained importance because of its high strength, stiffness, and densities very close to epoxy making it a suitable replacement for conventional materials used in aircraft industry. Recent investigations carried out on PEEK showed that it is "70% lighter than steel, 40% lighter than Aluminum alloys. PEEK exhibits 20% high strength and stiffness compared to Al7075 T6 under same loading conditions with 100 times longer fatigue life" says Victrex, a leading manufacturer of thermoplastic polymers for aircraft industry. The present study focuses on investigating the suitability of PEEK-CF as aircraft wing's structural material and compares the results with Al7075 T6 and epoxy-CF composites.

2 Methodology and FEA Simulation

2.1 Problem Description

Conventional airframe materials used in aircraft wing applications are made of aluminum alloys which offers the advantage of high strength, stiffness but their density is a bit high and aircraft industry demands materials with low density, high strength, and stiffness with improved mechanical properties. The development of epoxy-based polymers with carbon fiber as reinforcements has improved the strength of the materials further. However, the epoxy-based polymers absorb moisture which degrades the mechanical properties.

The present work focuses on investigating the possibility of replacement of Al alloys and epoxy-CF used in aircraft wing materials with PEEK-CF composite as PEEK-based composites are resistant to chemical and moisture absorption.

2.2 *Selection and Properties of Materials*

The present materials used by aircraft wings are Al 7075 and epoxy-CF composites.

- **Al7075T6**: Al7075T6 is an alloy consisting of aluminum-zinc. Al7075T6 offers the advantage of high strength, hardness, wear resistance, and corrosion-resistant properties; however, Al7075T6 is subjected to embrittlement because of micro-segregation of $MgZn_2$ precipitates leading to catastrophic failure of components produced from it.
- **Epoxy-CF composites**: The epoxy-carbon fiber is widely popular material in aircraft industry as it offers high strength, stiffness, and improved mechanical properties at low density as compared to Al alloys.

Material chosen for study:

- **PEEK-CF composites**: The PEEK-CF is recently developed high-performance materials which offers the advantages of good chemical resistance, low flammability, and low moisture absorption compared to the conventional materials used in the industry. The material properties chosen for study are tabulated in Table 1.

Table 1 Comparison of material properties

Material property	Al7075T6	Epoxy-CF	PEEK-CF
Density (kg/m^3)	2804	1490	1900
Poisson's ratio	0.33	0.27	0.470
Young's modulus of elasticity (GPa)	71	121	172
Ultimate tensile strength (Mpa)	572	1030	2900
Tensile yield strength (Mpa)	503	945	2070

2.3 Design Methodology

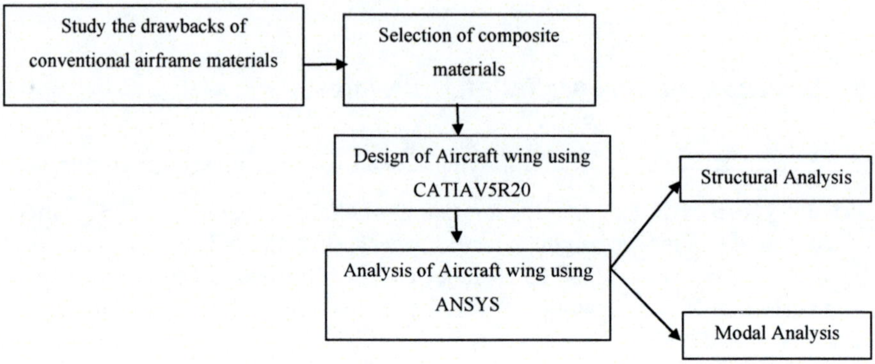

2.4 Design of Aircraft Wing Using CATIAV5R20

The wing that was designed based on Ikarus C42.Ikarus C42 is a two seater, fixed tricycle landing-gear, micro-light aviation aircraft. The wing span, wing area, and chord length are taken as per the dimensions of the aircraft. The construction of the wing is done according to the procedure mentioned in "Aircraft design—by Jan Ruskom". The location of the front spar is at 25% of the chord from the leading edge and the rear spar is located at 75% of the chord. The thickness of each of the rib is 20 mm and the spacing between the ribs is 500 mm. I- section is selected for the main and rear spars because top flange supports the compression and the bottom flange supports the tension in the ribs. Stringers of 10 mm square cross section and a skin of 1 mm thickness are used as the cover. The aircraft specifications chosen for study are tabulated in Table 2.

Table 2 Wing design parameters

Parameter	Dimensions
Wing span	9.45 m
Wing area	12.5 m^2
Max. takeoff weight	450 kg
Half wing	4.520 m
Chord length	1.4 m
Area of half wing	6.328 m^2
Wing loading on half wing	35.55625 kg/m^2
Selected airfoil	Clark y

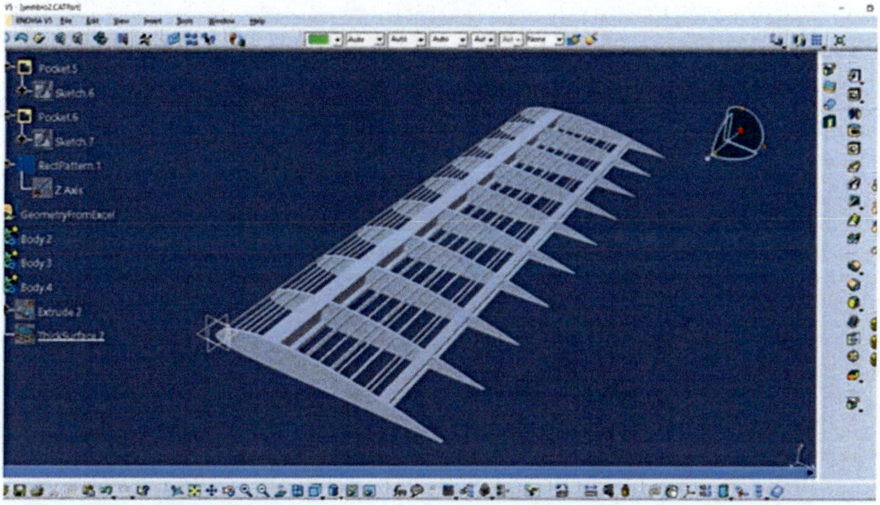

Fig. 1 Wing skeleton

2.5 Procedure

The airfoil of 1400 mm chord length is plotted using data obtained from arifoil-stools.com [9] and the plot is extruded using Pad command in CATIA V5 and is shaped for fitting spars and stringers. Pattern command is used for creating similar ribs. Spar and stringers are also sketched and extruded. The skin is designed using extrude command in generative shape design and thickness is given using thick surface command in part design. Each part is designed as a separate body. The skeleton of aircraft wing with ribs, spars, stringers, and airfoil is shown in Fig. 1 and the wing with skin is shown in Fig. 2.

2.6 Analysis of Aircraft Wing

2.6.1 Structural Analysis

A static structural analysis is performed to determine the deformation, stresses, and strains that are produced in a structural member on application of load.

Meshing: Meshing is done in ANSYS workbench for aircraft wing structure by applying sweep and automatic method. In swept method, quad mesh element is chosen and automatic method is used for generating mesh on wing body and as a result 160785 nodes are formed. The meshed view is show in Fig. 3.

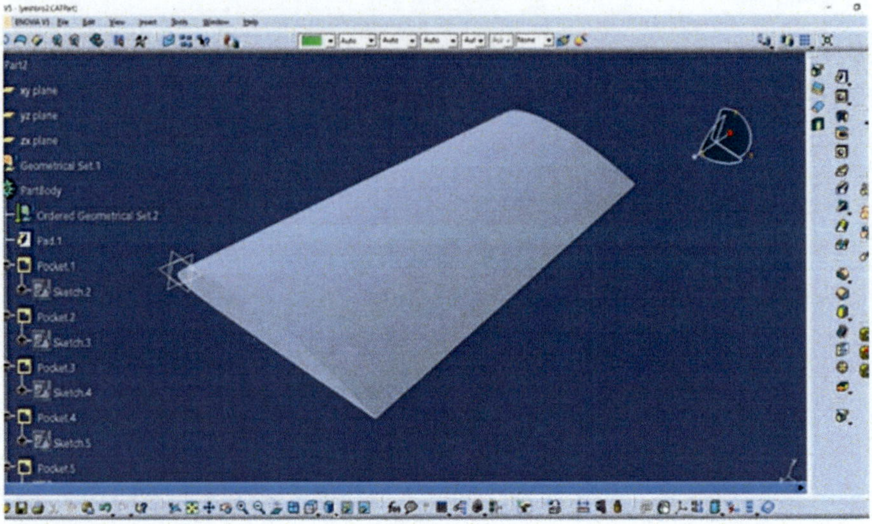

Fig. 2 Aircraft wing

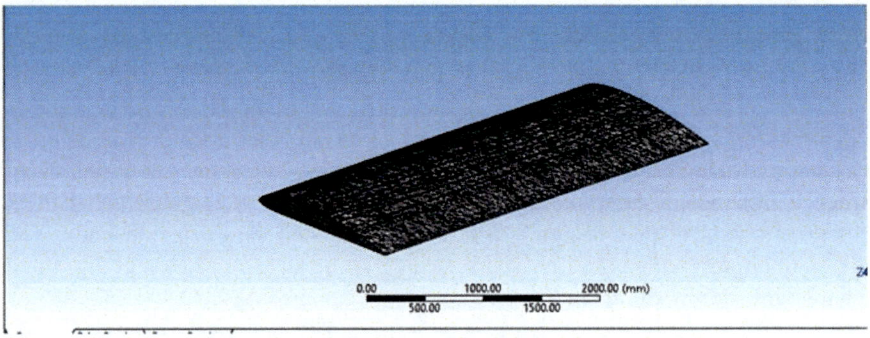

Fig. 3 Meshing in ANSYS workbench

Boundary condition: The boundary condition for an aircraft wing structure is a cantilever beam condition. The end of the wing which connects fuselage is fixed and bottom of the wing is subjected to a pressure which gives lift for the wing. Based on design of wing and aircraft conditions the pressure applied on the bottom of the aircraft is 318 Pa at speed of 180 km/h as shown in Fig. 4.

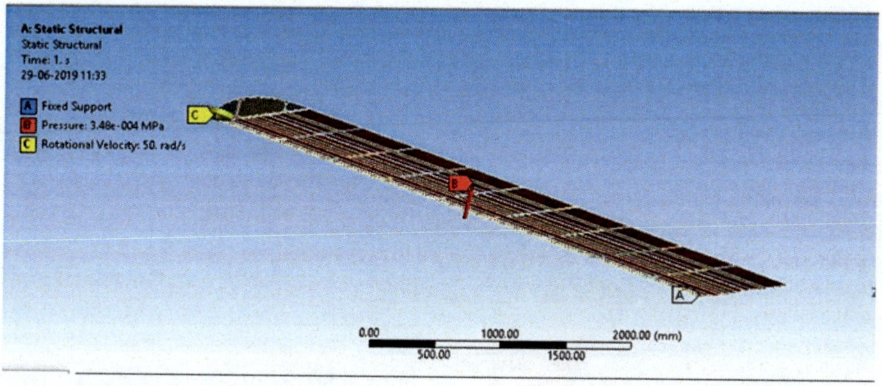

Fig. 4 Boundary conditions for aircraft wing

2.6.2 Modal Analysis

To understand the dynamic properties of the vibrating structures, modal analysis is performed. Modal analysis determines the natural frequency of the structural member. The structure having high natural frequency delays the resonance.

3 Results and Discussion

3.1 Structural Analysis

From the results obtained, it is evident PEEK-CF with 6.61 mm shown in Fig. 5 showed better deformation values compared to its counterparts Epoxy-CF with 9.33 mm in Fig. 6 and Al7075T6 with 17.75 mm as shown in Fig. 7. The distribution of stresses developed is found to be uniformly distributed and a maximum stress of 274 MPa for Al 7075T6, 275MPA fir Epoxy-CF, and 273 MPa for PEEK-CF composites is found at root of the wing. The results of equivalent stress also make it clear that structure is safe as the values obtained are less than the yield strength of the materials. Also the stress developed in PEEK-CF is less when compared to epoxy-CF and Al 7075T6 and moreover the results obtained show that stress values obtained for the three materials are more or less same; however, when it comes to deformation, PEEK-CF composites exhibited far better values[10–12]. The results are given in Table 3.

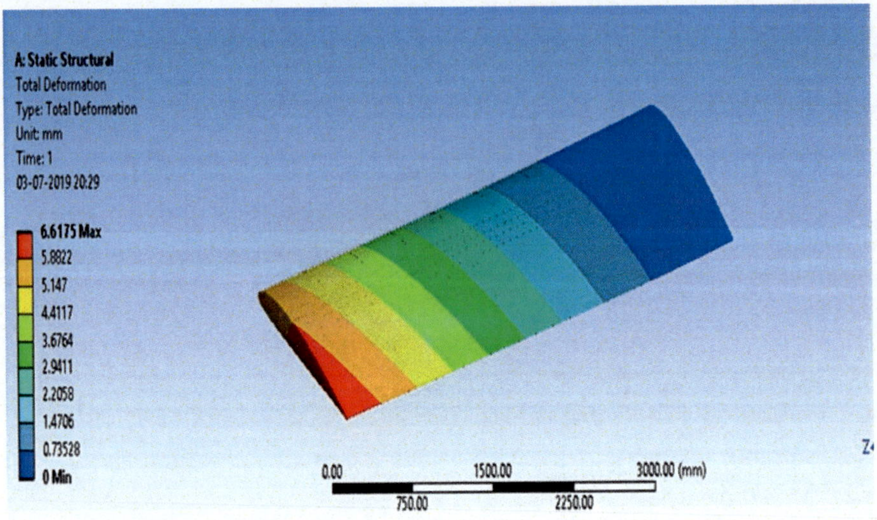

Fig. 5 Deformation of PEEK-CF max 6.617 mm

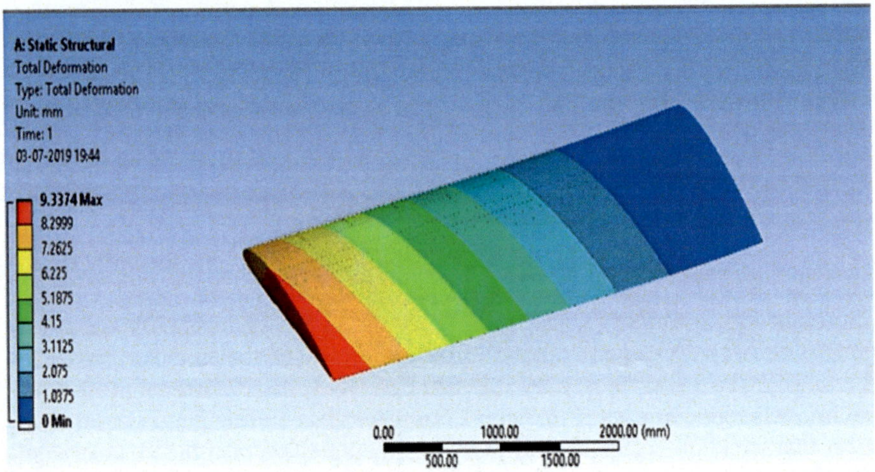

Fig. 6 Deformation of epoxy-CF max 9.337 mm

3.2 Results of Modal Analysis

Modal analysis is used to determine the dynamic properties of the structure under vibration. In aircraft wing structure, wing provides the necessary lift and carries the load of the engine which causes the vibrations. Modal analysis provides sufficient information to modify the structural design to reduce noise and vibration. Modal

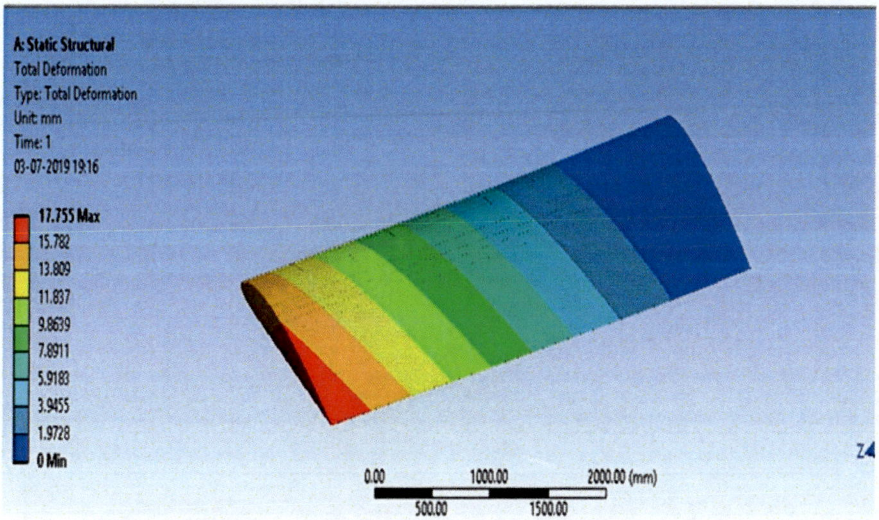

Fig. 7 Deformation of Al7075 T6-17.755 mm

Table 3 Results of structural analysis

Material	Deformation (mm)	Equivalent stress (MPa)	Equivalent strain
Al7075 T6	17.755	274.15	0.0054
Epoxy-carbon fiber	9.337	275.88	0.00316
PEEK-carbon fiber	6.617	273.38	0.00211

analysis uses mass and stiffness to determine natural frequency at which it resonates. In present analysis, five mode shapes are created. The natural frequency of Al7075 T6, epoxy-CF, and PEEK-CF has been analyzed. The mode shapes, frequency, and deformation are tabulated in Table 4 and figures corresponding highest frequency and deformation are shown in Figs. 8, 9 and 10. At mode 1, latitude type of deformation is observed, at mode 2 torsion, at mode 3 latitude, mode 4 bending, and at mode 5 torsion types of deformations are observed. From the results, it can be observed that PEEK-CF composite has high natural frequency as compared to epoxy-CF and Al7075T6 materials which shows that resonance is delayed in PEEK-CF composites. The results are given in Table 4.

Table 4 Results of modal analysis

Al7075 T6		Epoxy-CF		PEEK-CF		Type of deformation observed
Frequency (Hz)	Deformation (mm)	Frequency (Hz)	Deformation (mm)	Frequency (Hz)	Deformation (mm)	
6.22	3.93	11.49	5.44	12.385	4.84	Latitude
31.54	6.08	57.83	8.27	59.351	7.83	Torsion
33.378	3.97	61.27	5.62	64.659	4.74	Latitude
39.255	5.82	70.78	7.83	76.117	6.49	Bending
95.42	7.73	174.57	10.57	185.99	9.14	Twisting

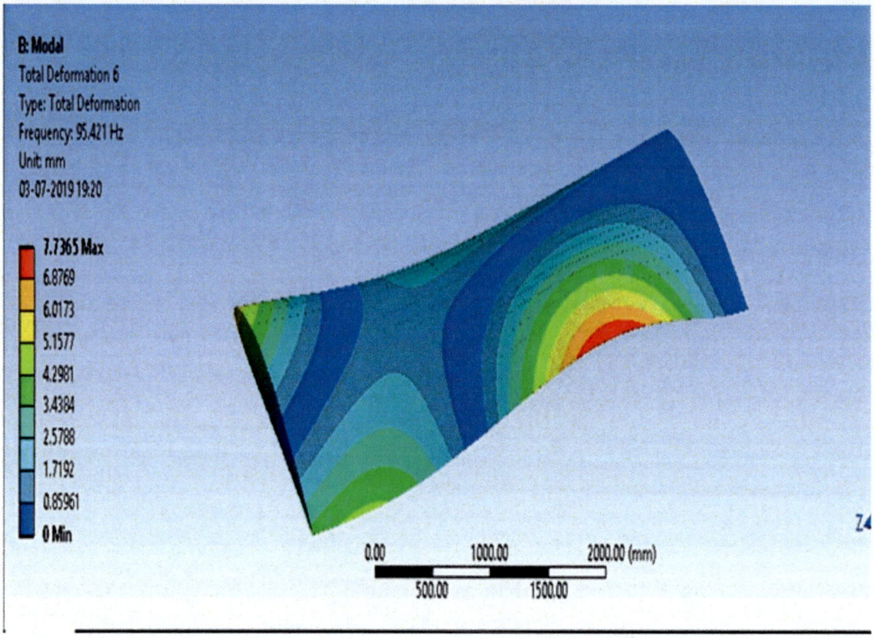

Fig. 8 Al7075T6 frequency-95.42 Hz, Def-7.73

4 Conclusions

The results show that PEEK-CF can be further investigated as an alternative material for aeronautical structural elements. PEEK offers the advantage of better corrosive, chemical, and moisture absorption properties. The results for structural and modal analysis show that the deformation, equivalent stress, and strain obtained are close to the values of epoxy-CF and better than Al alloys. Also the density of PEEK is close to the density of epoxy which makes it an attractive alternative for the metals and epoxy-based composite materials. Moreover, thermoplastic polymers like PEEK are high-performance engineering materials that are recyclable which offers the advantage

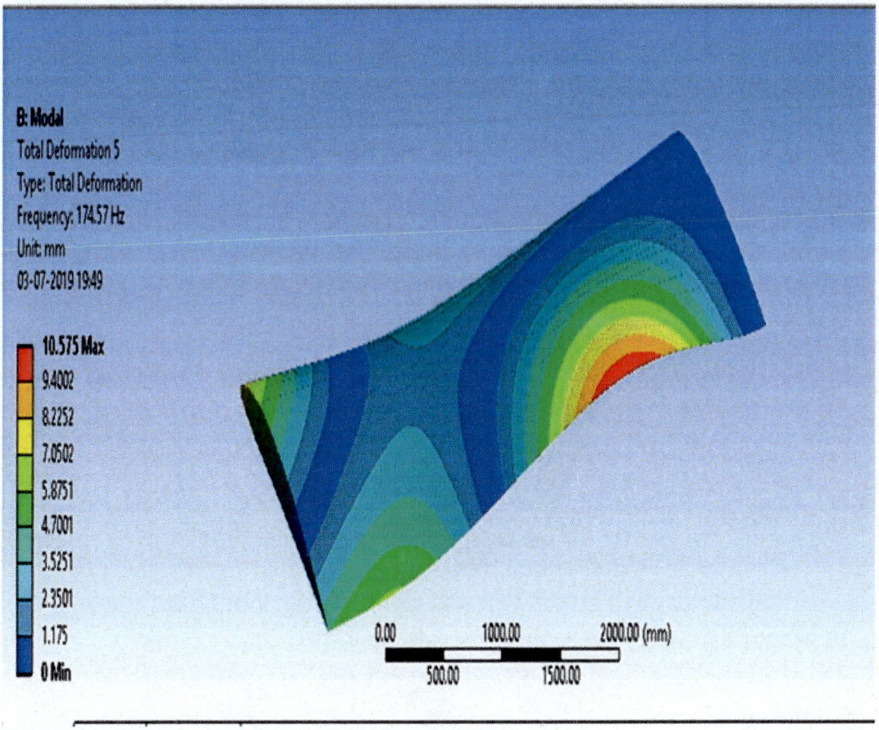

Fig. 9 Epoxy-CF frequency 174.573 Hz, Def-10.575

of reducing the usage of plastic materials as plastic wastage nowadays is becoming a serious threat to global environment. PEEK also offers less toxicity with almost zero flammable properties. PEEK shows better moisture-absorbing abilities (0.02%) when compared to epoxy's 4–5% which can help in avoiding the degradation of mechanical properties which is one of the drawbacks of epoxy-based composite materials used in aircraft industry.

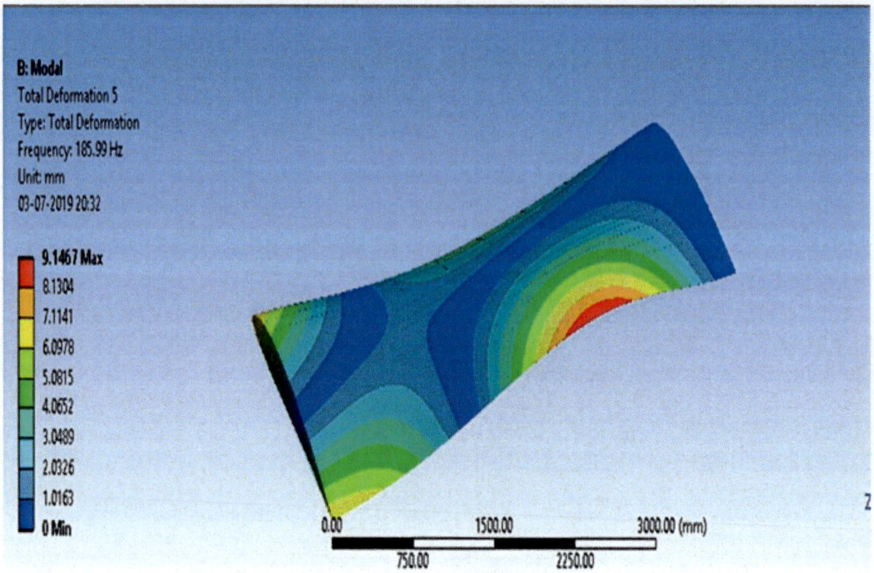

Fig. 10 PEEK-CF frequency-185.99 Hz, Def-9.1467

References

1. Dursun T, Soutis C (2014) Recent developments in advanced aircraft aluminium alloys. Mater Des 56:862–871
2. Starke E, Staley J (1996) Application of modern aluminum alloys to aircraft. Prog Aero Sci 32(2–3):131–172
3. Banerjee S et al (2010) Effect of trace additions of Sn on microstructure and mechanical properties of Al–Cu–Mg alloys. Mater Des 31(8):4007–4015
4. ELING. Aluminium Alloys. Available from: https://www.eling.rs/AluminiumAlloys.html
5. Wang R-M, Zheng S-R, Zheng YG (2011) Polymer matrix composites and technology. Elsevier
6. Pimenta S, Pinho ST (2011) Recycling carbon fibre reinforced polymers for structural applications: technology review and market outlook. Waste Manag 31(2):378–392
7. Paiva JMFD, Santos ADND, Rezende MC (2009) Mechanical and morphological characterizations of carbon fiber fabric reinforced epoxy composites used in aeronautical field. Mater Res 12(3):367–374
8. Mangalgiri PD (1999) Composite materials for aerospace applications. Bull Mater Sci 22(3):657–664
9. https://airfoiltools.com/plotter/index?airfoil=clarky-il
10. Kumar Das S, Roy S (2018) Finite element analysis of aircraft wing using carbon fiber reinforced polymer and glass fiber reinforced polymer. IOP Conf Ser Mater Sci Eng 402:012077
11. Ramesh Kumar A, Balakrishnan SR, Balaji S (2013) Design of an aircraft wing structure for static analysis and fatigue life prediction. Int J Eng Res Technol (IJERT) 2(5)
12. Sarath S, Issac JC, Girish KE (2013) Analysis of the wingbox with spliced skin and estimation of the fatigue life for the wing box. Int J Mech Eng Rob Res 2(2):155–163

Formulation of Stress Concentration Factor of a Finite Plate with an Elliptical Hole of High Eccentricity Ratio

Prafull Agarwal, Dhruv Mathur, and S. S. Ghosh

1 Introduction

The presence of sudden changes in the cross section and irregularities is what that sets off the phenomena of stress concentration in a body. Practically changes in the cross section in the form of threads, keyways, and holes become unavoidable. Because of the changes in the cross sections, the equations in the standard form become redundant if provisions for the prevalent stress concentration are not made.

For accommodating the effect of the stress concentration in the prevalent standard equations, a unique factor known as the stress concentration factor (SCF) is used. SCF can be defined as the ratio of the highest value of stress about the discontinuity to the nominal stress obtained without the discontinuity. The points on the body with maximum stress concentration are the points most susceptible to early failure and stress concentration factors of those points are of utmost importance for the safety of component.

Rectangular plates of finite width find full applications in various fields such as aerospace, automobiles, and many others. It is since the early 1960s that the analysis of stress concentration and the stress concentration factors has intrigued researchers and scientists alike. Earlier, the main focus of the researchers was to determine SCFs experimentally or analytically. Advances in the field of finite element modeling allowed many researchers to take the finite element route for their studies. Stresses induced due to the presence of sharp corners and cracks were presented by Inglis [1]. In the domain of stress concentration analysis, a significant amount of work has been done by Peterson [2], who compiled the works of numerous researchers on rectangular plates with some sort of notches and discontinuities, subjected to torsion, bending, or tension. Young et al. [3] have compiled the works of R J Roark in the

P. Agarwal · D. Mathur (✉) · S. S. Ghosh
Department of Mechanical Engineering, Manipal University Jaipur, Jaipur, India

S. S. Ghosh
e-mail: sasankas.ghosh@jaipur.manipal.edu

A. Parey et al. (eds.), *Recent Trends in Engineering Design*, Lecture Notes in Mechanical Engineering, https://doi.org/10.1007/978-981-16-1079-0_13

field of stress concentration. Many researchers [4–11] used analytical and numerical methods to find stress concentration factor of elliptical, circular holes on isotropic and orthotropic plates.

Even though R. J. Roark has given SCF equation for an elliptical hole in a rectangular plate subjected to a tensile force, the equation is valid for eccentricity factor varying from 0.5 to 10. For the eccentricity ratios greater than 10, the calculations of SCF using the equation given by Roark will not provide right result. The main aim of this paper is to find out the SCF for eccentricity ratio of elliptical holes more than 10 using finite element software ANSYS. Then an empirical relationship has been established for SCF, which links the width of the plate and the major and minor axes of the elliptical hole similar to Roark equation.

2 Problem Description

A finite flat rectangular plate of width *D* with a central elliptical hole is subjected to in-plane force *P*, as shown in Fig. 1. The SCF (K_t) can be calculated using the following formula:

$$K_t = \frac{\text{Maximum Principal Stress}}{\text{Nominal Stress}}$$

where the maximum principal stress is calculated using the following formula:

$$\sigma_{\max} = \sigma_a = \sigma_{\text{nom}}$$

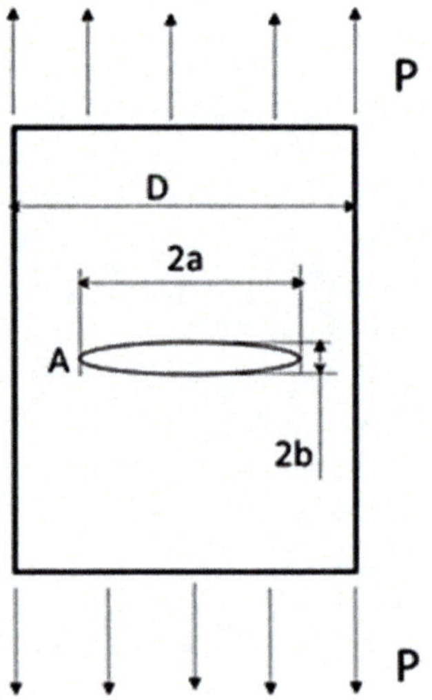

Fig. 1 Finite rectangular plate with a central elliptical hole under in-plane loading

The nominal stress, i.e., σ_{nom} is calculated from the following relation:

$$\sigma_{\text{nom}} = \frac{P}{t(D - 2a)}$$

where

P = applied force, as shown in Fig. 1,
t = thickness of the plate,
D = width of the plate, and
a = semi-major axis of the elliptical hole.

SCF equations for an elliptical hole in a rectangular plate have been given by Roark [3]:

$$K_t = C1 + C2\left(\frac{2a}{D}\right) + C3\left(\frac{2a}{D}\right)^2 + C4\left(\frac{2a}{D}\right)^3$$

where

$C_1 = 1+2\left(\frac{a}{b}\right), C_2 = -0.351-0.021\sqrt{\frac{a}{b}}-2.483\left(\frac{a}{b}\right), C_3 = +3.621-5.183\sqrt{\frac{a}{b}}+ 4.494\left(\frac{a}{b}\right), C_4 = -2.270 + 5.204\sqrt{\frac{a}{b}} - 4.011\left(\frac{a}{b}\right)$.and
a = semi-major axis of the elliptical hole,
b = semi-minor axis of the elliptical hole,
D = width of the rectangular plate, and
K_t = SCF for an elliptical hole in a flat plate.

The SCF for elliptical holes has been analyzed for different sizes by ANSYS and validated using Roark's formula. Since the above formula is valid only in the range $0.5 < \frac{a}{b} < 10$, the SCF for elliptical holes with high eccentricity ratio is obtained by ANSYS and a new formulation is derived.

2.1 Validation of the Process (Roark's Formula)

The primary task at hand is to validate Roark's formula so that the method of performing the analysis is verified. A plate with dimensions of 150 × 200 × 5 mm in length, width, and thickness has been considered. The two varying parameters include the eccentricity ratio, the values for which are 2, 2.5, 3.33, and 5 and the ratio $2a/D$, for which the values are 0.1, 0.133, and 0.2. For validating the formula, simulations are run to calculate the values of the maximum principal stress which are then used to calculate the SCF. The theoretical calculation values for each case have been calculated by substituting the values of the two ratios in Roark's formula, and error is calculated based on theoretical values. Table 1 represents the values obtained.

The error obtained in most of the cases is within ±6%, with a maximum variation of −8.58%. So, the process used above is valid for the next steps. The SCFs on ANSYS for eccentricity ratio higher than 10 can be calculated using the same procedure and the results are formulated.

Table 1 Validation of Roark's formula

S. No.	$2a/D$	a/b	K_t (Theoretical)	K_t (ANSYS)	% error
1	0.1	2	4.51519	4.70477048	4.19875
2	0.1	2.5	5.40337	5.798468	−7.31
3	0.1	3.33	6.87938	7.17	−4.22
4	0.1	5	9.85303	10.3954995	−5.505
5	0.133	2	4.37401	4.410	−0.822
6	0.133	2.5	5.22988	5.52	−5.54
7	0.133	3.33	6.65338	6.97	−4.758
8	0.133	5	9.52407	9.09	+4.55
9	0.2	2	4.11837	4.372	−8.58
10	0.2	2.5	4.91554	5.2588	−6.98
11	0.2	3.33	6.24453	6.7572	−8.2
12	0.2	5	8.93196	9.186	−2.844

3 Methodology

For deriving the relations, primarily three programs have been used: the package ANSYS Workbench 19.2 premium, for simulating the model, SolidWorks 2019 for modeling the part and MATLAB 2020 Ra for calculations and the final derivation of the empirical correlation. As with the case for validation of Roark's formula, the experiment primarily involved the variation in two parameters, namely, $2a/D$ and the eccentricity ratio a/b. The values of $2a/D$ considered were 0.1, 0.15, 0.2, 0.3, and 0.4. For each value of $2a/D$, three formulations were done, on values of a/b equal to 11, 15, and 20. The values of the semi-major axis a, the thickness, and length of the plate were kept constant at 20 mm, 5 mm, and 150 mm, respectively. Applying a force of 20 N on the top face of the plate while fixing the bottom face of the plate completed the model.

3.1 Geometry

The part, i.e., the plate with an elliptical hole was modeled in SolidWorks (Fig. 2) which enabled us to parameterize the model which allowed us to change the value of the major and the minor axes for every simulation. The part models in SolidWorks were saved as a step file format and imported onto ANSYS static structural to analyze the models.

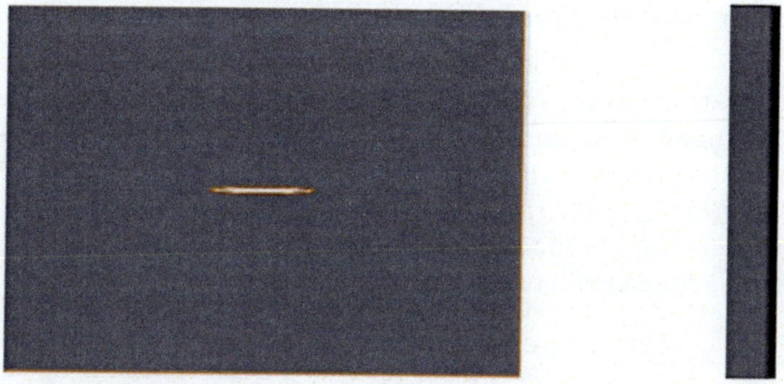

Fig. 2 Front and side views of the plate

3.2 Meshing

After importing the model onto the static structural module, meshing was performed on the ANSYS mechanical. The meshing process involved the generation of mesh with the default size to analyze the model better. Following a threshold 300,000 nodes with a maximum tolerance of $\pm 10,000$ nodes, the element size was varied concerning the size of the plate. After performing the refinement on the inner surface of the elliptical cavity with refinement set at 3, a good enough mesh was achieved (Fig. 3).

Fig. 3 The meshing of the plate

3.3 Loads and Constraints

In the ANSYS static structural module, after a good enough mesh was achieved, the necessary constraints were applied to the model. The bottom face of the plate was fixed in all the three directions to prevent any sort of movement. With the application of a uniaxial tensile force of 20 N on the top face of the plate, the loading of the model was deemed complete. On simulating, the model, the values of maximum principal stress, and the deformations produced in the plate were obtained.

4 Results and Discussion

For the analysis, the plate considered is of the uniform cross section with the only discontinuity being at the center of the plate in the form of an elliptical cavity with varying dimensions to analyze different situations. The material of the plate is structural steel with the material properties as follows: density = 7850 kg/m^3; Young's modulus = 200 GPa; bulk modulus = 166 GPa; shear modulus = 76.9 GPa; Poisson's ratio = 0.3; and tensile and compressive yield strength = 250 MPa.

By varying the eccentricity ratio, new values of stress concentration factor were derived. The value of the major axis has been kept constant at 20 mm. For values of the eccentricity ratio greater than 10, the size of the elliptical cavity for every subsequent case that is talked about here is petite.

The two constant dimensions of the plate include the length and the thickness of the plate, which are 150 mm and 5 mm in magnitude, respectively. The variation in the value of width is by the ratio $2a/D$, which implies that the width is constant only for a specific value of $2a/D$ and is changing on changing the ratio.

A rectangular coordinate system has made use of with the planes being the *XY*-plane, *YZ*-plane, and the *ZX*-plane. The origin is at the left bottom edge of the plate with the major axis of the elliptical cavity along the *x*-axis and the minor axis along the *y*-axis.

4.1 Results for Eccentricity Ratio More Than 10

Figure 4 represents a sample result obtained after performing the simulation on ANSYS, while Table 2 represents the values of the SCF obtained for various cases.

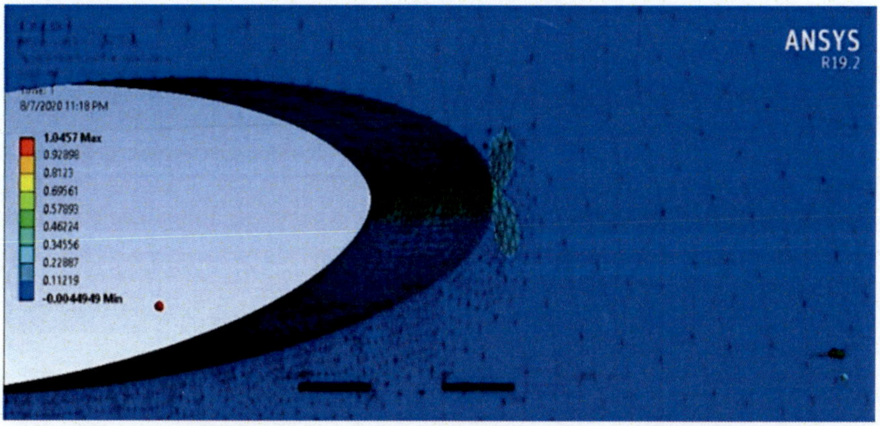

Fig. 4 Stress distribution around the elliptical hole

Table 2 Results obtained by ANSYS with high eccentricity ratio

S. No	$2a/D$	a/b	K_t (ANSYS)
1	0.1	10.5	21.0152
2	0.1	15	22.8321
3	0.1	20	27.9
4	0.15	10.5	20.6283
5	0.15	15	22.6059
6	0.15	20	27.8
7	0.2	10.5	19.42
8	0.2	15	22.3836
9	0.2	20	27.6
10	0.3	10.5	18.834
11	0.3	15	19.6
12	0.3	20	27.00027
13	0.4	10.5	16.574
14	0.4	15	18.9
15	0.4	20	24.538

4.2 Empirical Formulation

Using the value of the maximum principal stress and the nominal stress, the stress concentration factor for each case has been calculated. Same form of equation as given by Roark is assumed for high eccentricity ratio. Twelve constants are obtained to specify the equation for SCF as

$$K_t = \left(f_1 + x\, f_2 + x^2 f_3\right) + \left(f_4 + x\, f_5 + x^2 f_6\right)y$$

$$+ (f_7 + x f_8 + x^2 f_9) y^2 + (f_{10} + x f_{11} + x^2 f_{12}) y^3$$

where $f_1, f_2, \ldots, f_{12}$ are the constants to be found out, $x = \sqrt{\frac{a}{b}}$ and $y = \frac{2a}{D}$.

The first 12 values of the SCFs have been used to find out the values of the 12 constants in the general equation. Substituting the values of the constants derived in the above general equation gives us the empirical formula for calculating the SCF for the plate with an elliptical hole having a high eccentricity ratio. The equation derived is as

$$K_t = (64.0967 - 25.9154x + 4.1015x^2) + (31.2039 - 36.6164x + 5.0120x^2)y + (154.5237 + 0.7372x - 0.6972x^2)y^2 + (-154.3764 - 84.9123x + 14.4825x^2)y^3$$

which is of the form: $K_t = C_1 + C_2\left(\frac{2a}{D}\right) + C_3\left(\frac{2a}{D}\right)^2 + C_4\left(\frac{2a}{D}\right)^3$.

where

$$C_1 = 64.0967 - 25.9154\sqrt{\frac{a}{b}} + 4.1015\left(\frac{a}{b}\right)$$

$$C_2 = 31.2039 - 36.6164\sqrt{\frac{a}{b}} + 5.012\left(\frac{a}{b}\right)$$

$$C_3 = 154.5237 + 0.7372\sqrt{\frac{a}{b}} - 0.6972\left(\frac{a}{b}\right)$$

$$C_4 = -154.3764 - 84.9123\sqrt{\frac{a}{b}} + 14.4825\left(\frac{a}{b}\right)$$

On substituting values of 15 set of parameters into the formula, it is found out that the results are within ±6% of the value obtained from ANSYS. Figure 5 shows

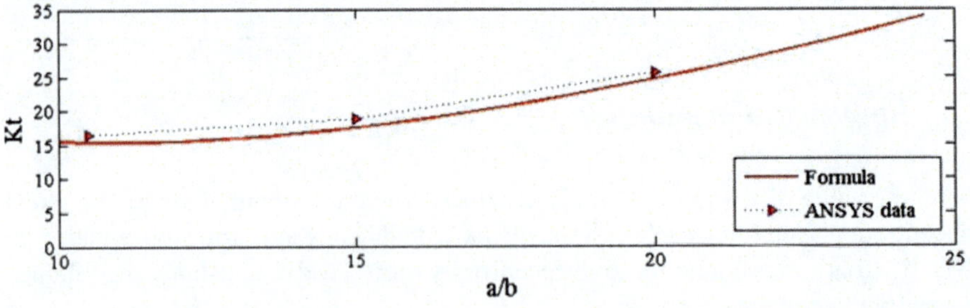

Fig. 5 K_t versus *a/b* for $2a/D = 0.4$

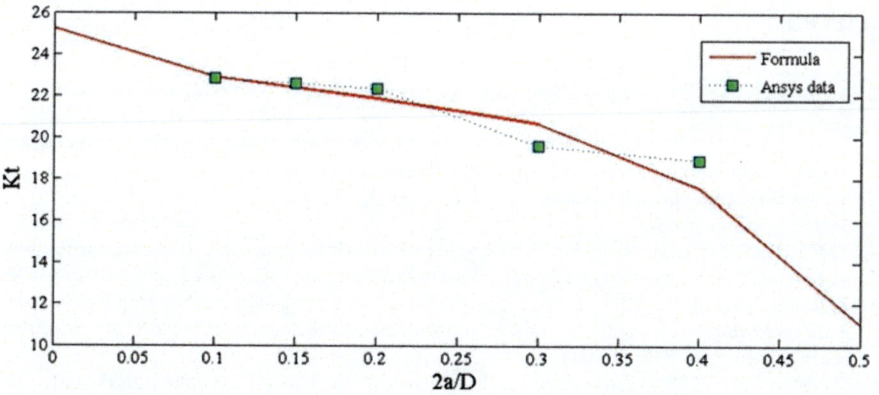

Fig. 6 K_t versus $2a/D$ for $a/b = 15$

K_t versus a/b obtained from the formulation and simulation for $\frac{2a}{D} = 0.4$, while Fig. 6 shows K_t versus $2a/D$ for $a/b = 15$. So, the above formula is valid in the range $11 < \frac{a}{b} < 20$ and $0.1 < \frac{2a}{D} < 0.4$. For $\frac{a}{b} > 20$, the trend in the change in the values of K_t becomes different, and thus it is difficult to obtain the convergence and thus formulate the results.

5 Conclusions

For flat plates with elliptical hole subjected to a static uniaxial tensile force, SCFs can be calculated from Roark's formula. Roark formula is valid when the eccentricity ratio of the elliptical hole lies between 0.5 and 10. For eccentricity ratios greater than 10, the availability of such formulation is rare. In this paper, an empirical relation is established for calculating the SCF of plates with elliptical cavities of eccentricity ratio greater than 10. FEM analysis on ANSYS is used first to validate Roark's formula wherein the values obtained using FEM were compared to those calculated using Roark's formula. From the results, the procedure is extended for deriving a relation for SCF of a flat plate with an elliptical hole with eccentricity ratios greater than 10 and subjected to uniaxial tension. The new empirical relation for SCF is valid for $11 < \frac{a}{b} < 20$ and $0.1 < \frac{2a}{D} < 0.4$. The empirical relations thus derived can be used easily to obtain significantly accurate results which can ultimately lead to the safer design of the specimens.

References

1. Inglis CE (1913) Stresses in a plate due to the presence of cracks and sharp corners. Trans Inst Naval Archit 55:219–241
2. Peterson RE (1953) Stress concentration design factors: charts and relations useful in making strength calculations for machine parts and structural elements
3. Young WC, Budynas RG, Sadegh AM (2002) Roark's formulas for stress and strain, vol 7. McGraw-Hill, New York
4. Li F, He Y, Fan C, Li H, Zhang H (2008) Investigation on three-dimensional stress concentration of LY12-CZ plate with two equal circular holes under tension. Mater Sci Eng A 483:474–476
5. Mittal N, Jain N (2007) The optimise design of a square simply supported isotropic plate with central circular hole for reduction of stress concentration subjected to transverse static loading. In: Proceedings of ICTACEM
6. Growney W (2008) Using FEA results to determine Stress Concentration Factors. In: International ANSYS conference
7. Snowberger D (2008) Stress concentration factor convergence study of a flat plate with an elliptic hole under elastic loading conditions. Master thesis, Rensselaer Polytechnic Institute, Connecticut, USA
8. Patle B, Bhope D (2012) Evaluation of stress concentration factors in plate with oblique hole. IOSR J Mech Civ Eng 2:28–32
9. Pan Z, Cheng Y, Liu J (2013) Stress analysis of a finite plate with a rectangular hole subjected to uniaxial tension using modified stress functions. Int J Mech Sci 75:265–277
10. Zappalorto M, Carraro PA (2015) An engineering formula for the stress concentration factor of orthotropic composite plates. Compos Part B 68:51–58
11. Chawla K, Ray-Chaudhuri S (2018) Stress and strain concentration factors in orthotropic composites with hole under uniaxial tension. Curv Layer Struct 5:213–231

Design and Development of the Automated Multimedia Device for Memorials

Sunil Sharma

1 Introduction

Commemorating the dearly departed members of family and society at graveyards is a social custom. This custom has been primarily affected by technological prowess such as video screens, laser shows, and other technological inputs. The use of memorial technologies is not new [1–6]. Currently, research is being done for digitization of the memorials [7–11]. In this paper, we propose a novel way of funeral tribute by a new product. This automated gravesite multimedia device can recreate the memories and allow the bereaved family to relive a few moments with the deceased ones. The product can rise from the ground and reach the waist level to display the recorded multimedia. Other applications can be in pet memorials, libraries, plant houses, and graveyards of army men. Casket and Funeral Supply Association of America predicted a sale of 1.7 million coffins in the US (2006) with a price range of $1000–10,000. Harris Interactive conducted a nationwide online survey (2004) sampling 2158 US adults showed that 83% (1793) wanted a funeral and 62% wanted some form of personalization by displaying photos, playing special music, or video tributes of the deceased, or integrating the person's hobby or interests into the funeral.

1.1 Background

Serenity Panel [12], a patented multimedia player, is used in conjunction with a grave marker to pay tribute to a deceased person. It is mounted on the gravesite and covered with a solar panel. The solar panel charges the battery that powers the multimedia device for 10 min, as shown in Fig. 1. Ziegler [13] has patented an audiovisual system for use at the memorial site, as shown in Fig. 1. It includes an audiovisual means

S. Sharma (✉)
IIT Ropar, Punjab, India

A. Parey et al. (eds.), *Recent Trends in Engineering Design*, Lecture Notes in Mechanical Engineering, https://doi.org/10.1007/978-981-16-1079-0_14

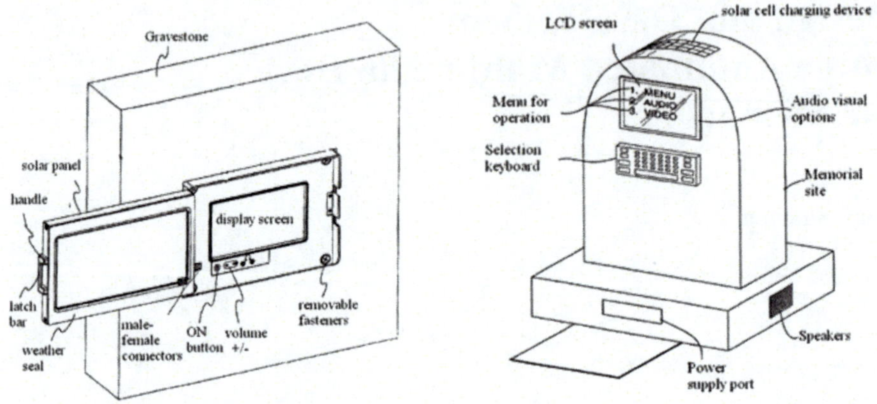

Fig. 1 Serenity panel [12] and Zeigler's audiovisual system [13]

with LCD. The audio is played using a speaker and can be stored on flash memory, magnetic tapes, and DVD. A keyboard or touch screen can be used as input selection means. In order to protect from weather and extreme temperature, this product has been hermetically sealed. It also has a power supply medium, such as a battery. A solar cell, exposed to solar radiation, has been provided to charge the battery and maintain the full charge.

Issues in the current commercial product. Serenity panel is a similar commercial product that is solar-powered, installed on the gravestone, and has an LCD. When the viewer opens the solar panel, the power supply is disconnected, and the video is played. However, the height of the installation on the gravestone is 12″, which is insufficient to properly watch the tribute. For a good view, the screen's position must be slanted (if the screen is not up to eye level, which it cannot be) and up to the person's waist level, i.e., 36″. Protection from vandalism is another problem existing in other products. Damage protection of solar panel, which acts as a cover, and the LCD screen has not been provided. Therefore, considering the need and lacunae in the existing product, we propose an automated gravesite multimedia device to be installed near the grave that rises to waist level (36″) and will be vandalproof and accident resistant. This product will be embedded in the ground at the foot of the grave. Once it is opened with a remote, it will rise to waist level to view multimedia records.

2 Methodology and Development

The conventional product design approach [14, 15] is followed for designing. This method consists of the following steps. Step 1 is the specification and translation of problems. Step 2 is the selection of components based on specifications. Step 3 is

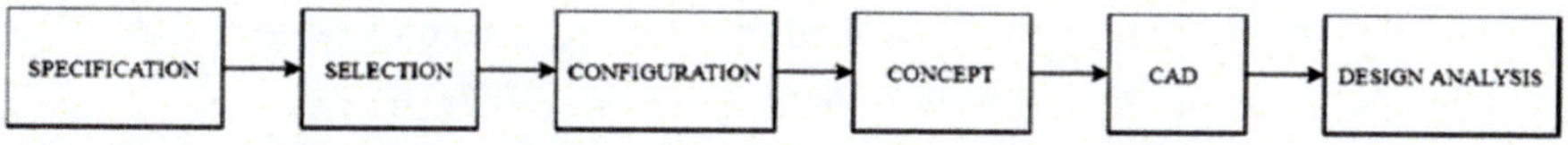

Fig. 2 A product design process [15]

to configure components according to the functionality and specifications. Step 4 is to generate conceptual sketches. Step 5 is CAD design and Step 6 is design analysis (Fig. 2).

Step 1: Specification of design. A product design specification list has been formulated to elaborate on the functionality of the automated gravesite multimedia device, which is as follows:

1. A video screen displays recorded information about memorials (person, pet animal), plants, books, and others.
2. A mechanism that can bring the multimedia device to the user interface level, i.e., 36 in. above from the ground.
3. A solar-powered battery that can power the mechanism.
4. An IR remote can be used to open and close the automated gravesite multimedia device. It can also control the lifting mechanism, volume output, and can be used to turn off the power of the automated gravesite multimedia device.
5. All the components must be secured in an enclosure to prevent possible damage from surroundings and vandalism.
6. Lawnmowers should not damage it, i.e., 2.5 in. ground clearance.
7. The product should pump out water that may accumulate in it due to rain.
8. The product should not corrode over some time.

Working of the automated multimedia device. The solar panel will convert solar energy into electrical energy that will charge the battery. The charged battery will run the multimedia device, lifting mechanism, and pump. When discharged, the battery will be automatically recharged from the solar panel. The lifting mechanism will be operated by the infrared (IR) remote. It will activate the automated gravesite multimedia device to open. The lid will be open due to the lifting mechanism in the enclosure. The enclosure will rise to the waist level and stop. Another push of the remote button will activate the multimedia device. The video can then be played. After the video has been played, the power to the multimedia device will be shut off. The interface will ask the user to play the video again or switch off. The user can thus select the lifting mechanism to go underground or play the video again. If the user selects to switch off the device, the lifting mechanism will go underground. The top lid will be closed automatically through a linkage mechanism. The device will be locked by pressing the lock button on the remote and again be opened with the remote.

Step 2: Selection of components. As per the working mechanism, some of the components from specifications have been selected. These components should be 12 V compatible. The details of the off-the-shelf components are given below:

1. Solar panel: The solar panel is needed for charging a rechargeable battery. The solar panel should be waterproof, rustproof, and corrosionproof.
2. Solar charge controller: A charge controller (also called charge regulator) regulates the voltage and current coming from the solar panels going to the battery.
3. Electric pump: A pump can remove the accumulated water due to heavy rain inside the automated gravesite multimedia device.
4. Battery: The battery will provide power to the multimedia device, linear actuator, and pump. A comparison of different types of batteries has been made to select a suitable battery and AGM battery can be the right choice.
5. Linear actuator: For lifting the automated gravesite multimedia device, a linear actuator can be used. Linear actuators have motors that provide push and pull forces in a straight line. For the automated gravesite multimedia device, a linear actuator is needed that can rise to 36 in. from the ground. The up and down motion will be controlled by remote control. The linear actuator should be able to withstand moisture and dust.
6. Multimedia device: A high contrast video screen (10.4 in. digital photo frame) is needed, displaying the pictures and videos, which is readable in sunlight. The video screen will be powered by the battery and can be activated by the remote control. The input voltage for the multimedia device is 12 V. A 3.2 Ah battery can run 1.3 A multimedia device for 2.46 h.Before designing the other required parts of the automated gravesite multimedia device for housing the off-the-shelf components and supporting the structure, a list of possible damage by unprecedented accidents has been made. This list has been illustrated in Table 1.

Step 3: Configuration of components. The components which are to be purchased and those that are to be custom modeled are sketched to decide the orientation and location of various components. This configuration is shown in Fig. 3. Once the components are selected in the configuration phase, the next task is how these components will be configured so that a complete assembly of a gravesite multimedia device is formed. The dimensions and technical specifications are mandatory to know about.

Step 4. Concept generation. This phase presents concept sketches about the problems the current design poses or the specification a new design has to meet. In this design, the main challenge is to mechanize the lifting mechanism. Conceptual solutions of design have been provided for the lifting mechanism of the device. Mechanizing the multimedia device to lift automatically is a challenge. For solving, these two solutions are sought. Firstly, a gas spring has been used in a lifting mechanism. Secondly, the lifting mechanism is supported by a linear actuator. A comparison of the lifting mechanism is made based on different parameters, and the linear actuator is selected as the finalized one. Refer to Fig. 4.

Step 5: CAD Design. In order to develop the device, manufactured components need to be designed first. Modeled in ProE, these components form sub-assemblies with selected components. These sub-assemblies, when assembled, form a final

Table 1 Possible damage and protection to the automated gravesite multimedia device

Damage from	Damage and relevant variables	Protection
1. Hail balls falling from the sky	Damage top enclosure and solar panel by hail size 5–150 mm (0.20–5.9 in) and average weight 0.29 lbs/in	Lexan sheet to protect the solar panel
2. Rain	During rain, the inner enclosure is filled with water which may damage components	Electric pump to pump out water
3. Lightning strike	High voltages/current can damage parts	Lightning arrestor
4. Fire	High temperature can melt material	The enclosure should withstand high temperatures
5. Walk-behind Lawnmowers	Accidently, the tire can damage the product by 100 lbs force	Withstand the load on the structure
6. Riding mowers	Force exerted by a wheel on the product will be 1070 lbs	Protect solar panel by providing Lexan sheet
7. Stepping foot by a 250 lbs man	Sufficient to break the solar panel	Protect the solar panel
8. Ten-year kid weighing 70 lbs	Can damage the solar panel, but the structure may not be damaged	Protect the solar panel
9. Falling tree	Severe damage by 2983 kg (6562 lbs) tree	Protect the structure
10. Vandalism	Solar panel and multimedia device can be stolen	Vandalproof fasteners and Lexan sheet Multimedia device mounted in structure
11. Dust	Reduce the transmission of light to the solar panel	Maintenance and nano-layered coatings
12. Intruders	Protect from the third person using the device	Remote control key
13. Water accumulated	Water accumulated during rain can damage all parts	Electric pump to pump out water

product. Refer to Fig. 5. CAD design has been done for the following components: (1) outer enclosure, (2) inner enclosure, (3) top plate for pump, (4) base enclosure for pump, (5) top support plate for linear actuator, (6) side support plates for linear actuator, (7) enclosure for bushing, (8) bushing, (9) rubber seal between mount and bushing, (10) mount, (11) Lexan for multimedia device, (12) frame for multimedia device, (13) base hinge, (14) enclosure for solar charge controller, (15) bottom enclosure for solar assembly, (16) top enclosure for solar assembly, (17) Lexan for solar panel, (18) rubber cushion on frame, and (19) enclosure for pump control panel.

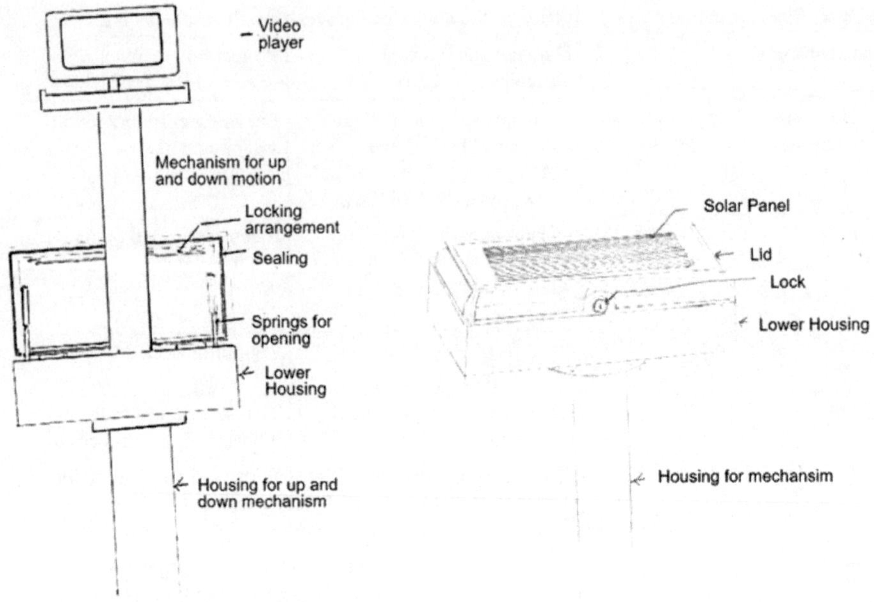

Fig. 3 Initial configuration sketches

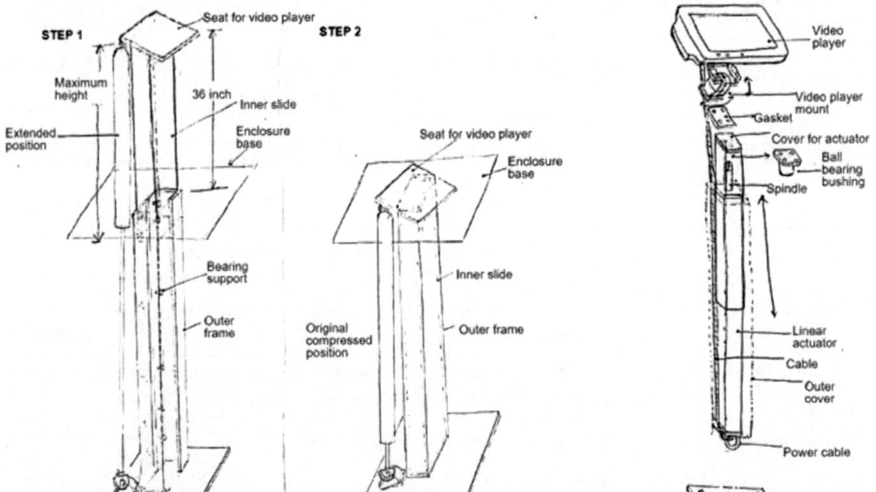

Fig. 4 First and second sketches show the gas spring-based lifting mechanism. Third sketch shows the linear actuator-based lifting

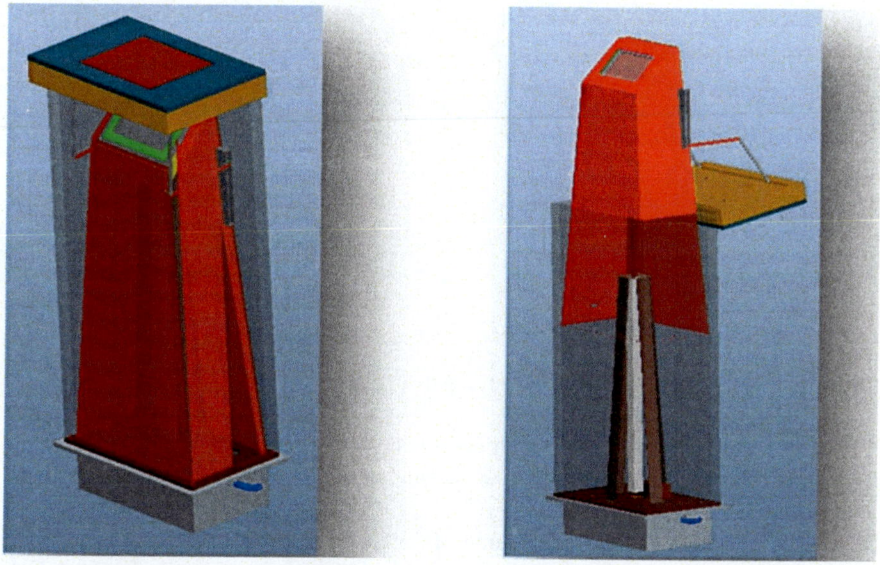

Fig. 5 CAD assembly in the open and closed positions

Step 6: Analysis of design. Different part components of automated gravesite multimedia device assembly are analyzed for failure, as described next. Refer to Table 1. Three such cases of product failure have been considered. Firstly, a hail ball falls and collides with the top of the device. Secondly, a heavy man accidentally steps on a top of the Lexan sheet, the Lexan bends, and breaks. Thirdly, a lawnmower's wheel is accidentally run on the Lexan sheet's top during the lawn-mowing process. This action causes the Lexan sheet to bend and ultimately break.

1. *Calculations for force created by hail ball impact.* Manual calculations are performed using the formula of drag force and flow velocity. The kinetic energy of the hail ball is then predicted using research data. The force generated is 11 kN, which can damage the Lexan sheet.
2. *Calculations for deflection of the Lexan sheet.* Two cases are described. *Case I:* A 250 lbs man steps on the top Lexan sheet of the automated gravesite multimedia device. Refer to Table 2. *Case II:* A 2140 lbs lawnmower runs its wheel on the Lexan sheet of the automated gravesite multimedia device. Refer to Table 3.

Table 2 Case I: theoretical and FEA deflection results for different thicknesses of Lexan sheet

Thickness (in.)	Theoretical deflection (in.)	FEA deflection (in.)
0.125	23.93	23.988
0.25	2.99	3.002
0.5	0.374	0.374

Table 3 Case II: theoretical and FEA deflection results for different thicknesses of Lexan sheet

Thickness (in.)	Theoretical deflection (in.)	FEA deflection (in.)
0.25	12.8	12.868
0.5	1.6	1.607
0.75	0.474	0.475

2.1 Other Design Concerns

Crucial design decisions have also been taken and described as follows:

- Spring fit on protrusions of the enclosure to prevent the automated gravesite multimedia device to wobble.
- To ensure the safe working of components, the enclosure can have a thickness of 1 in. with a slot on the enclosure's inner side with a depth of 0.5 in. This slot will support the inner enclosure.
- Calculation of average discharge time of the battery.
- Installation of the automated gravesite multimedia device.

3 Conclusions

An automated gravesite multimedia device is laid at the feet of the grave. The system will require a power source to operate, a multimedia device, a pump, a lifting mechanism, and an enclosure to hold and protect the product from surroundings. Initial selections for off-the-shelf and to be manufactured components are made. Design calculations have been done for off-the-shelf and to be manufactured components. The material has been selected based on environmental conditions where the automated gravesite video player will work. CAD models and drawings are made. The cost of making this automated gravesite video player as calculated from the bill of materials comes to be $2300. An interdisciplinary approach to the problem can improve a second version of the product. More tests and calculations can be done on the automated gravesite video player. We summarize the conclusions as follows:

- The product design process has been followed to design automated multimedia device from specification to design analysis.
- Conceptual sketches sketched, and CAD and analysis are developed for this device.
- Design calculations and considerations with manufacturability and actual products have been taken into consideration.
- Detailed drawings have been prepared with assemblies and sub-assemblies.

References

1. Casey C, Maresco P (2006) Stories in stone, stories on screen: an examination of increased personalization of cemetery memorials. In: Proceedings of the media ecology association
2. Maresco PA (2006) Personalized gravestones: your life's passion for all to see and hear. J Relig Popular Cult 13:4–4
3. Nansen B, Arnold M, Gibbs M, Kohn T The restless dead in the digital cemetery, 14
4. Sabra JB, Andersen HJ, Rodil K (2015) Hybrid cemetery culture: making death matter in cultural heritage using smart mobile technologies. In: 2015 international conference on culture and computing. IEEE, Kyoto, pp 167–174
5. Walter T, Hourizi R, Moncur W, Pitsillides S (2012) Does the internet change how we die and mourn? Overview Anal 64:275–302. https://doi.org/10.2190/OM.64.4.a
6. Moncur W, Bikker J, Kasket E, Troyer J (2012) From death to final disposition: roles of technology in the post-mortem interval. In: Proceedings of the 2012 ACM annual conference on human factors in computing systems—CHI'12. ACM Press, Austin, Texas, USA, p 531
7. Häkkilä J, Colley A, Kalving M, Forsman M-T (2020) Exploring pervasive displays for cemeteries and memorial sites. Pers Ubiquit Comput. https://doi.org/10.1007/s00779-019-01359-1
8. Häkkilä J, Colley A, Kalving M (2019) Designing an interactive gravestone display. In: Proceedings of the 8th ACM international symposium on pervasive displays. ACM, Palermo, Italy, pp 1–7
9. Gotved S (2015) Privacy with public access: digital memorials on quick response codes. Inf Commun Soc 18:269–280. https://doi.org/10.1080/1369118X.2014.989250
10. Montogomery D, Lintz J (2018) Interactive device for memorial information
11. Refslund Christensen D, Gotved S (2015) Online memorial culture: an introduction. New Rev Hyperm Multim 21:1–9. https://doi.org/10.1080/13614568.2015.988455
12. Aguirre SE (2006) Multimedia player for grave marker
13. Ziegler M (2005) Audio visual display system for use at interment or memorial site
14. Retooling manufacturing: bridging design, materials, and production. National Academies Press, Washington, D.C. (2004)
15. Ulrich KT, Eppinger SD (2016) Product design and development. McGraw-Hill Education, New York, NY

Statistical Investigation on Reliability of Single-Hole Pin Joints Strength and Failure Mechanisms in Glass–Epoxy Composite Laminates

Akash Gupta and Manjeet Singh

1 Introduction

The selection of the right material seems to be a primary challenge for the engineers, which should have been high in strength with the reduction of weight and as per the perspective of demand. The incorporation of composite material during the last few decades has offered numerous benefits and substitutes the metals mostly with composites [1]. Regardless of the benefit of excellent mechanical properties, fiber-reinforced polymer composites provide the most durable engineering materials throughout the automotive, aerospace, shipbuilding applications, etc. Due to the high costs associated with geometric constraints, it becomes almost difficult to construct an entire structure in one unit. The choice is the development of smaller sections that will later be assembled. The assembly approach requires the utilization of joints [2]. Generally, these approaches involve the joining of composite materials to other composites or even to metallic materials. However, such joints are sensitive to large stress concentrations inside and outside the circular cavity, and frequently tend to cause unexpected failure in structural components made of composites. Due to the extreme flaw established in the material which induces local stress accumulation all across the cavity, it can lead to component failure. And subsequently, the strength of such a structural system relies on the joint's strength [3]. The entire nature of pinned joints gained the attention of the researchers who have been interested in investigating the composite joints. Okutan et al. [4] examined the behavior of the glass–epoxy composite pin-loaded joints by considering the E/D and W/D ratios differed between 1 and 5 and between 2 and 5 with six distinct stacking configurations and with the surge in W/D and E/D proportions, superior bearing strength of the joint observed. Aktas et al. [5] examined experimentally the influence of geometric variables on the strength and failure states of the pin-loaded joint made from carbon epoxy pin joint.

A. Gupta · M. Singh (✉)
School of Mechanical Engineering, Lovely Professional University, Phagwara, Punjab, India
e-mail: manjeet.21545@lpu.co.in

A. Parey et al. (eds.), *Recent Trends in Engineering Design*, Lecture Notes in Mechanical Engineering, https://doi.org/10.1007/978-981-16-1079-0_15

And it was claimed that the high bearing strength witnessed at E/D and W/D equals 4. Because composite materials are anisotropic, and in various directions have dissimilar mechanical properties. By comparison, they show varying strengths due to its internal composition, and therefore it is hard to express the strength of the mechanical behavior of the material by a single magnitude. This results in the need to incorporate statistical analyses for their effective use during the design and manufacturing [6, 7]. The Weibull distribution tool has indeed been proven effective as well as a flexible statistical tool for assessing composite structural reliability. Weibull distribution has the potential to model different characteristics of experimental results. That is one of the prime motivations presently for broad utilization [8, 9]. The incorporation of Weibull distribution methodology has frequently been employed to evaluate the dynamic and static mechanical performance of polymer matrix (PMC), ceramics matrix (CMC), and metal matrix composites (MMC), respectively [10]. Weibull's distribution function has proved to be a valuable and flexible way to characterize composite material properties when evaluating the reliability of composite structures [11].

This research deals with the preparation of specimens of pinned joints made from glass–epoxy composite. Experimentally and statistically, these pinned joints were analyzed. By utilizing Weibull distribution, the variations in bearing strength intensity were validated. To every specimen configuration, four bearing strength tests were performed. Finally, the reliability of composite laminated pin joints in terms of bearing strength was evaluated at a 99% reliability level compared with the average bearing strength of similar specimens during the experimental test.

2 Sample Formation and Testing Method

In this study, the composite laminate's geometrical configuration consists of a single circular opening with diameter "D" (4 mm) positioned at a distance from one end "E" along the midline of the plate and the stacking sequence of lamina was considered [0/45/90]. The layout of composite laminate is shown in Fig. 1. A constant force "P" was introduced to the plate and therefore the force was parallel to the length of the plate and there is no bending existed.

Bearing stress (σ_b) can be defined as

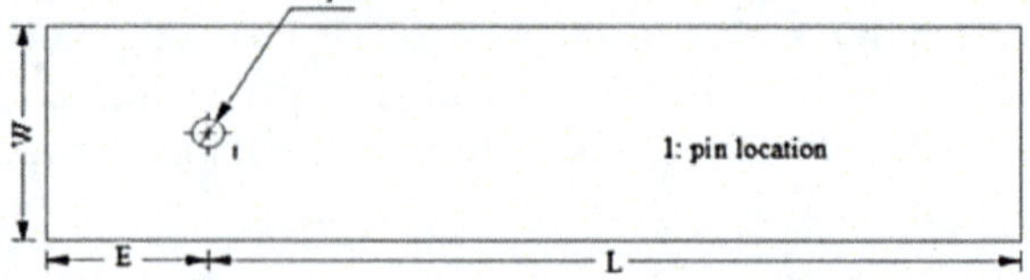

Fig. 1 Geometrical configuration of composite laminate

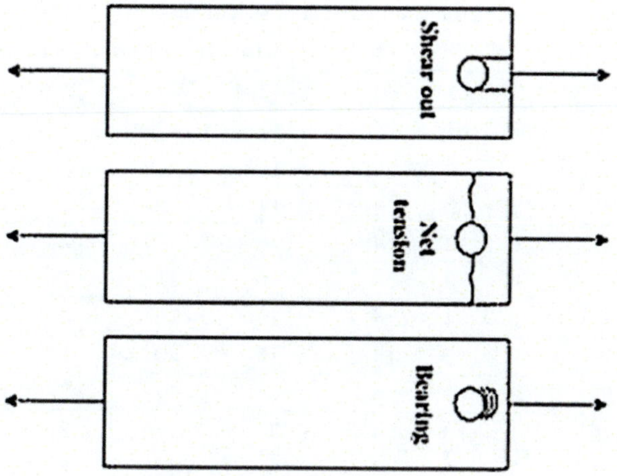

Fig. 2 Mechanically fastened pin joint failure modes [3]

$$\sigma_b = \frac{P_{max}}{D.t} \quad (1)$$

Generally, the mechanically fastened pin joint composite laminate comprises three main possible failure manners, i.e., bearing, shear-out, and net tension. Among all, bearing failure mechanism is the suggested failure manner that has high strength as per the design point of view due to non-catastrophic. And rest two failure modes are severe and not recommended type of failure mode having the least strength [12, 13] as presented in Fig. 2. The various parameters that can influence the joint strength are material, geometry, stacking sequence, fasteners parameters, and design parameters [6, 14]. Unidirectional glass fiber was used as a reinforcement of 1200 gsm fabric and epoxy "L-12" was used as a matrix agent. Preparation of laminates includes the curing process at room temperature for 24 h and then after compressing the laminates use a compression molding technique at a steady load of 120 KN with curing at 150 °C for 2 h and were finally cooled down at room temperature by keeping constant pressure.

All the specimens were cut into a rectangular shape for tensile testing as per ASTM D3039 and specimen was prepared from the laminates. So a circular cavity of diameter 4 mm and length $L = 100$ mm along the middle line is shown in Fig. 1. In total, 12 geometric configurations have been selected by keeping E/D proportion as 3 and 4 and W/D proportion altered 2, 3, 4, and 5, respectively, and maintaining W/D proportion as 3 and 4 and E/D proportions altered 2, 3, 4, and 5, respectively. Each specimen of selected configuration has been tested four times to get the mean bearing strength value. The specimens made of distinct proportions of edge/diameter and width/diameter have been presented in Table 1.

The tensile test was performed using Zwick-Roell Universal Testing Machine—Model Z010 at a temperature of 25 ± 2 °C. For all the specimens, a crosshead speed of 2 mm/min was being kept.

Table 1 Geometric configuration of specimen

S. No.	*E*/*D*	*W*/*D*	*D* (mm)	*E* (mm)	*W* (mm)	Thickness (mm)	*L* (mm)	Layup
1	3	2	4	12	8	2	100	[0/45/90]
2	3	3	4	12	12	2	100	[0/45/90]
3	3	4	4	12	16	2	100	[0/45/90]
4	3	5	4	12	20	2	100	[0/45/90]
5	4	2	4	16	8	2	100	[0/45/90]
6	4	3	4	16	12	2	100	[0/45/90]
7	4	4	4	16	16	2	100	[0/45/90]
8	4	5	4	16	20	2	100	[0/45/90]
9	2	3	4	8	12	2	100	[0/45/90]
10	5	3	4	20	12	2	100	[0/45/90]
11	2	4	4	8	16	2	100	[0/45/90]
12	5	4	4	20	16	2	100	[0/45/90]

3 Weibull Distribution

Weibull distribution tool has been deployed to model the extreme values of bearing strength at different geometric configurations. Two common arrangements of the Weibull distribution approach can be used, i.e., two- and three-parameter-based Weibull distribution. The general expression of three-parameter-based Weibull distribution can be written as [8]

$$F(x;\gamma,n,\beta)=1-\exp\left[-\left(\frac{x-\gamma}{n}\right)^{\beta}\right] \tag{2}$$

$$\gamma\geq 0, n\geq 0 \text{ and } \beta\geq 0$$

where γ, n, and β are variables of position, scale, and shape, respectively. When $\gamma = 0$ in Eq. (2), then it becomes two-parameter-based Weibull distribution. For this analysis, the Weibull two-parameter-based distribution function was presented to predict the load-carrying capacity of glass–epoxy composite laminates at a 99% reliability level. The two-parameter Weibull function can be written as [6, 7, 11]

$$F(x;n,\beta)=1-\exp\left[-\left(\frac{x}{n}\right)^{\beta}\right] \tag{3}$$

$$n\geq 0 \text{ and } \beta\geq 0$$

In this approach, probability density function can be represented by $F(x;\, n,\, \beta)$ which means the probability of bearing strength would be equal or less than x.

$$F(x; n, \beta) + R(x; n, \beta) = 1 \tag{4}$$

where $R(x; n, \beta)$ represents the reliability of the factor, which means the probability of bearing strength would be equal to or more than the value of x.

$$R(x;\ n,\ \beta) = \exp\left[-\left(\frac{x}{n}\right)^{\beta}\right] \tag{5}$$

The linear regression approach was adopted to evaluate the scale and shape variables. Since this is the simplest technique that can be used to assess the Weibull variables, this approach allows transforming Eq. (3) into linear equation manner, i.e., $Y = mx + c$. by taking double natural logarithmic on both sides of the expression.

$$\ln\left[\ln\left(\frac{1}{1 - F(x, n, \beta)}\right)\right] = \beta \ln(x) - \beta \ln(n) \tag{6}$$

To determine the $F(x, n, \beta)$, Benard's approximation method is employed, and with the help of Median Rank ($Q(t)$), n and β will be evaluated.

Benard's approximation (Median Rank) [15]

$$F(x, n, \beta) = Q(t) = \frac{i - 0.3}{j + 0.4} \tag{7}$$

where i is the order of the specimen and j is the total size of the number.

4 Result and Discussion

The strengths were obtained experimentally for single-hole joints glass fiber-reinforced epoxy composite (GFREC) and validate the data with a statistical approach, and the influence of geometrical parameters, i.e., edge/diameter and width/diameter proportions were addressed. Table 2 presents the mechanical properties of unidirectional fiberglass epoxy (GFREC) laminate.

Table 2 Mechanical properties of GFREC laminate

$E1$ (MPA)	$E2$ (MPA)	$G12$ (MPA)	ν_{12}	Fiber %
17,400	6200	3100	0.3	62

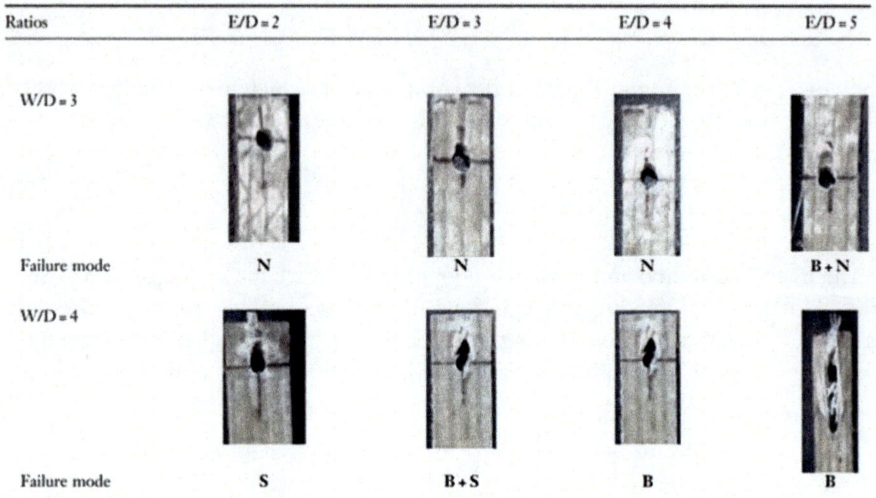

Fig. 3 Schematic of failure modes observed experimentally at width/diameter = 3 and 4 and edge/diameter = 2, 3, 4, and 5, respectively

4.1 Experimental Results of Unidirectional GFREC

The pin joint strength of a single cavity GFREC was reported, and it has been identified that the bearing failure mechanism was induced when the proportions of E/D and W/D exceed 4. Apart from the bearing failure mode, combination of failure mechanism was also addressed. At E/D equals 5 and W/D equals 3, combined net tension and bearing failure mechanism detected. As well as at $E/D = 3$ and $W/D = 4$ combined bearing and shear failure mode revealed. And at $W/D = 3$ and $E/D = 2$, 3, and 4, respectively, catastrophic net tension failure is noticed as shown in Fig. 3. Bearing mode was observed at $E/D = 4$ and $W/D = 4$ and 5. At lower edge to diameter and width to diameter proportions, catastrophic failure, i.e., shear-out and net-tension failure mechanism, has been noticed as represented in Fig. 4.

4.2 Probability Study

Weibull reliability functions have been analyzed by considering different sets of arrangements with four specimens each. The influence of geometric parameters has been examined at a 99% reliability level of composite laminate joint in terms of the normalized bearing stress ratio, i.e., ratio of σ_{bi} to σ_{avg}, where σ_{bi} is the bearing stress for *i*th specimen at given set of geometrical parameters and σavg is the average bearing strength of all four specimens at each set of geometrical configuration. Table 3 shows the bearing strength for each specimen at a given set of parameters along with

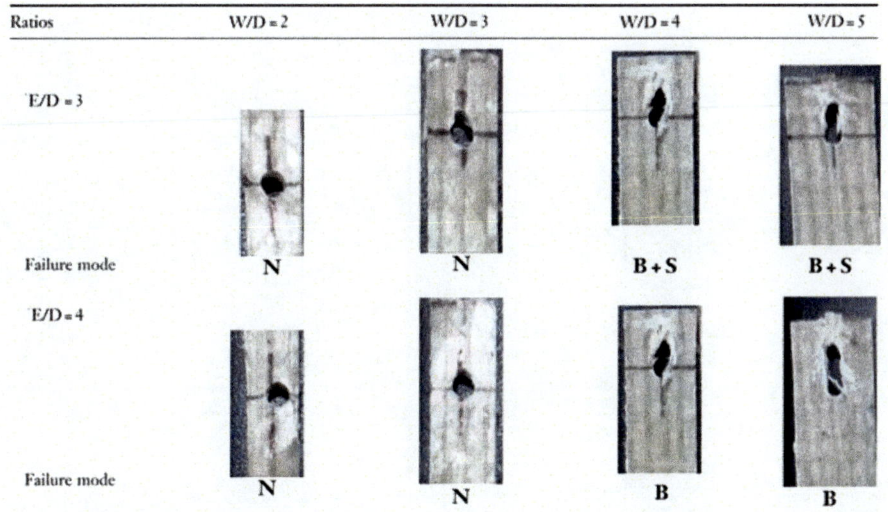

Fig. 4 Schematic of failure modes observed experimentally at edge/diameter = 3 and 4 and edge/diameter = 2, 3, 4, and 5, respectively

the average bearing strength from the experimentally recorded values and theoretically calculated Weibull parameters at each geometrical configuration. The Weibull reliability plots concerning normalized stress ratio of data obtained from bearing test at different geometric parameters offer identical test plots of bearing stress. The abscissa represents the normalized stress ratio (σ_{bi}/σ_{avg}) where σ_{bi} is the bearing stress for *i*th specimen and σ_{avg} is the average bearing stress of the joint. The ordinate represents the Weibull reliability distribution as shown in Fig. 5.

The Weibull reliability was taken into account from the study point of view with 99% reliability for all configuration sets. Figure 5a curve depicts the bearing strength value of 193.7 MPA noticed at 99% reliability point, which offers a high level of reliability at E/D equals 3 and *W*/*D* equals 2 with net-tension failure mechanism. Similarly, in Fig. 5b, the Weibull reliability curve gives the bearing strength of 153.76 MPA at *E*/*D* and *W*/*D* proportions equal to 3 with again catastrophic net-tension failure. A similar plot has been observed at $E/D = 3$ and $W/D = 4$ with a magnitude of bearing stress of 282.35 MPA as shown in Fig. 5c but the dual failure mode has been noticed, i.e., bearing and shear-out failure mechanism. Similar failure modes were also observed at $E/D = 3$ and $W/D = 5$ with the magnitude of stress 252.48 MPA as shown in Fig. 5d. Therefore, those curves support the finding that can be used to approximate the assurance of joint strength.

Figure 6 shows that edge/diameter proportion equals 4 and width/diameter ranged between 2 and 5, respectively, offered bearing stress magnitude 189.72, 297.58, 460.91, and 450.89 MPA with identical plots at 99% reliability level. It represents that increasing the *W*/*D* ratio bearing failure mode can be achieved at $W/D = 4$ and 5 and at lower ratio catastrophic failure obtained as confirmed from Fig. 4.

Table 3 Tensile test data and Weibull parameters at different geometric configurations

S. No.	*E/D*	*W/D*	*S*1 (MPA)	*S*2 (MPA	*S*3 (MPA)	*S*4 (MPA)	Average bearing Strength (MPA)	Shape parameter (*β*)	Scale parameter (*η*)
1	3	2	250	240	278	272	260	14.08	268.54
2	3	3	213	245	229	273	240	9.348	251.52
3	3	4	358	324	332	346	340	22.32	346.98
4	3	5	332	374	396	418	380	10.17	396.89
5	4	2	220	230	235	215	225	24.27	229.38
6	4	3	332	348	364	356	350	25.52	356.36
7	4	4	494	490	510	506	500	50.78	504.61
8	4	5	484	480	499	497	490	49.65	494.66
9	2	3	178	191	182	209	190	13.11	196.68
10	5	3	420	444	421	415	425	28.31	432.05
11	2	4	200	209	218	221	212	22.09	118.78
12	5	4	396	418	405	421	410	34.65	415.50

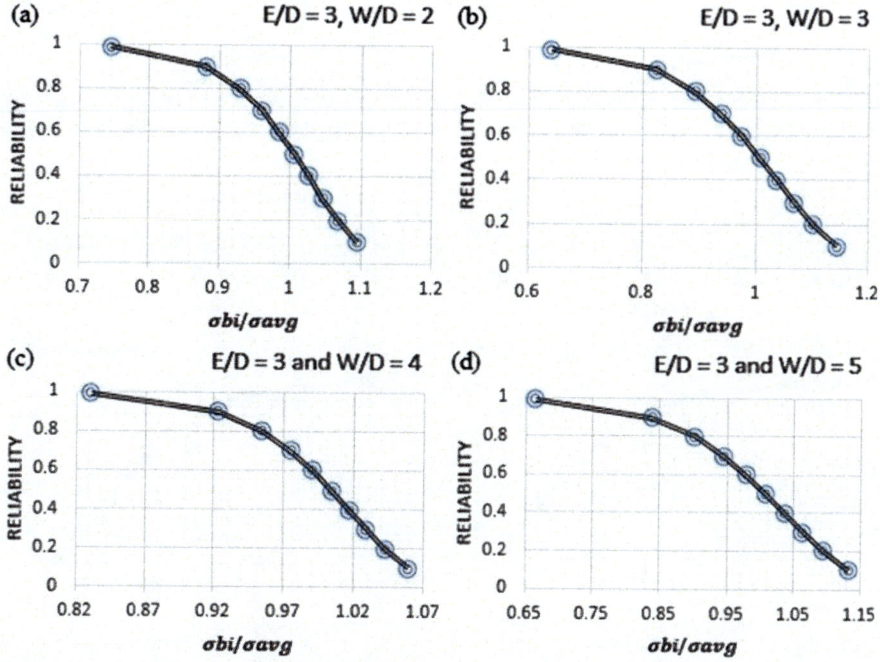

Fig. 5 Weibull reliability distribution at $E/D = 3$ and $W/D = 2, 3, 4,$ and 5

The bearing strength noticed from Weibull reliability at $E/D = 2$ and $W/D = 3$ was 138.47 MPA, which was 27% lower than the average bearing strength with net-tension failure mechanism as represented in Fig. 7a. A similar plot was also observed at $E/D = 2$ and $W/D = 4$ with high reliability bearing strength of 96.45 MPA and the shear-out failure mode was detected as shown in Fig. 7b. Similar plots are also observed in Fig. 8 with greater bearing strength at $E/D = 5$ and W/D varied 3 and 4, respectively, i.e., 367.24 MPA and 363.85 MPA as the selected ratios are in the high range which gives bearing failure modes.

The average pin joint strength attained from experimentation and Weibull probability distribution function at constant E/D equal 3 and W/D proportion altered between 2 and 5, respectively, as shown in Fig. 9a.

The lowest bearing strength noted at E/D equals 3 and W/D equals to 2 and 3, respectively, for both statistical and experimental outcomes. However, pin joint strength surges with the surging of W/D proportion above or equal to 4. The strength obtained from the Weibull distribution function at 99% reliability gives 28% lower than the average bearing strength. As an assurance aspect, if the theoretical value with the geometric parameters was preferred, then the component failure life would significantly reduce as well as the composite structure reliability would enhance. The average bearing strength obtained from experimentation and Weibull probability distribution function at E/D equals 4 concerning W/D as shown in Fig. 9b. It can be shown that, in an experimental and statistical study, the deviation in the bearing

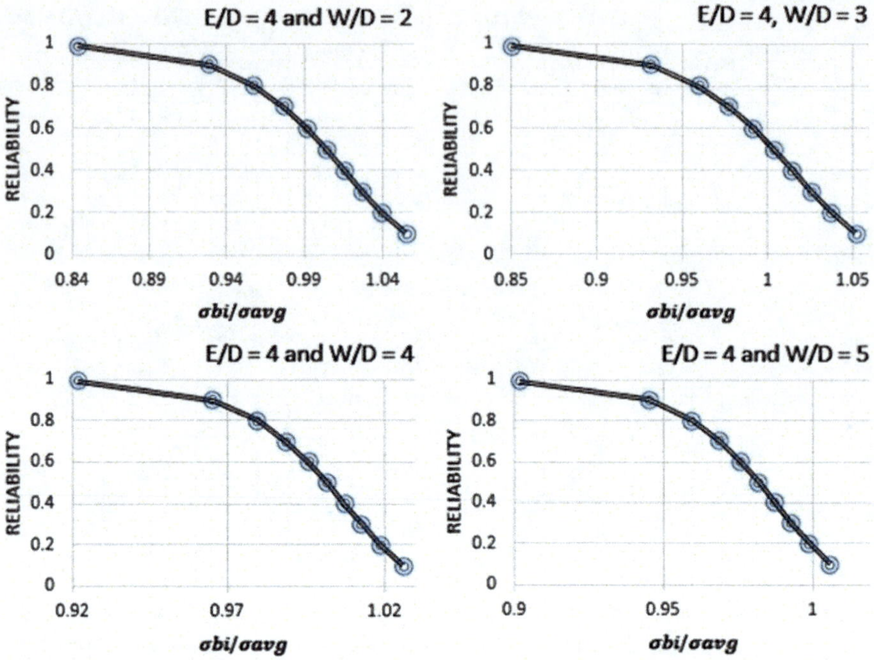

Fig. 6 Weibull reliability distribution At *E*/*D* equals 4 and *W*/*D* ranged between 2 and 5, respectively

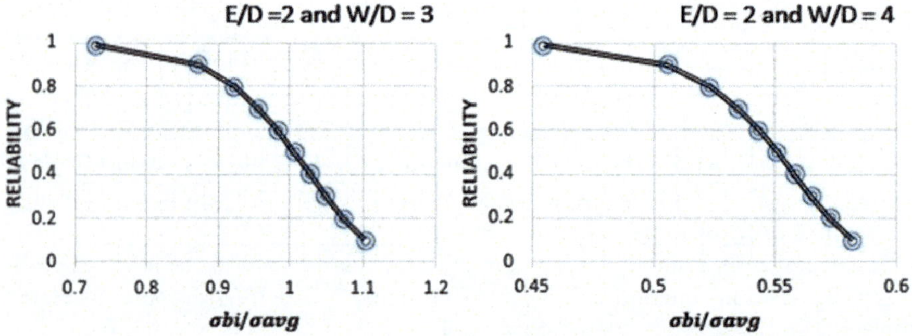

Fig. 7 Weibull reliability distribution at $E/D = 2$ and $W/D = 3$ and 4

strength seems to be an identical pattern. And the maximum bearing strength was noticed along with bearing failure mode at *E*/*D* and *W*/*D* ratio equal to 4. The strength obtained from the Weibull distribution function at 99% reliability gives 11% lower than the average bearing strength.

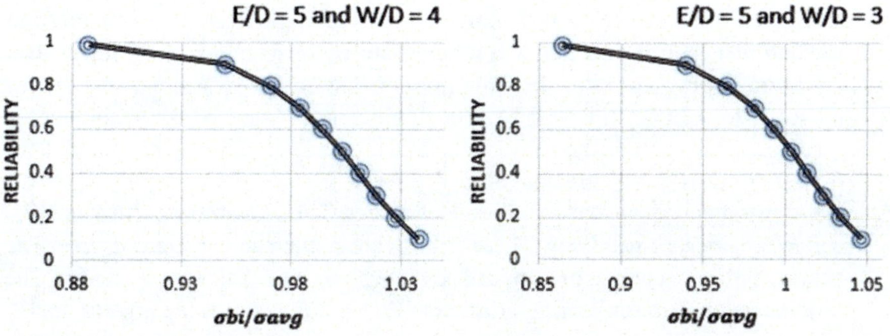

Fig. 8 Weibull reliability distribution at $E/D = 5$ and $W/D = 3$ and 4

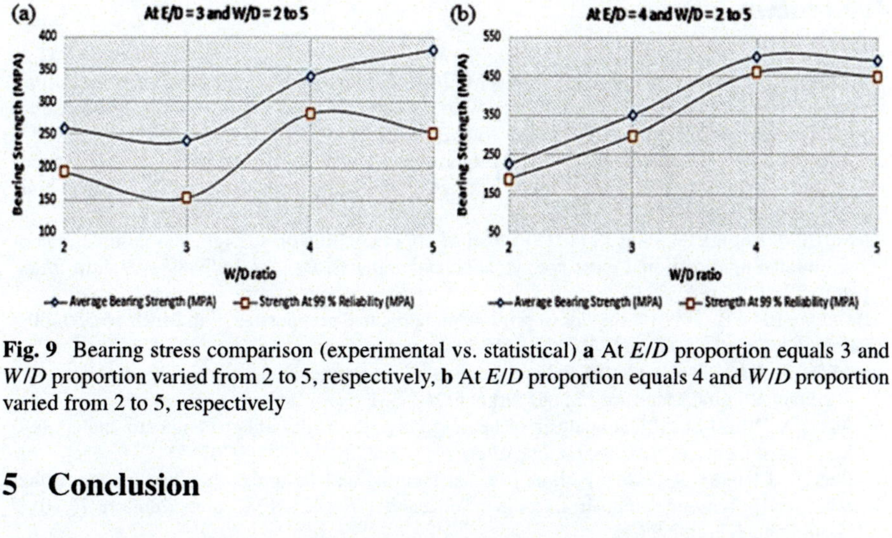

Fig. 9 Bearing stress comparison (experimental vs. statistical) **a** At E/D proportion equals 3 and W/D proportion varied from 2 to 5, respectively, **b** At E/D proportion equals 4 and W/D proportion varied from 2 to 5, respectively

5 Conclusion

In this analysis, both experimentally and statistically, the pin joint strength of the single-hole unidirectional GFREC laminate has been explored. The research's primary objective was to identify the triggered failure mechanism during the applied load and also to study the impact of dimensional configuration on the pin joint strength of the laminated composite at constant tensile loading. The following findings are taken based on both experimental and statistical outcomes which are as follows:

1. In an experimental study, bearing strength has been recorded at different sets of geometrical configurations for each specimen and the same data has been employed to conduct the statistical analysis.
2. In statistical analysis, the Weibull distribution function has been applied to predict the reliability of the composite laminate at 99% of the reliability level.
3. The pin joint strengths extracted from the Weibull study were almost 22% lower than that obtained from experimental tests by maintaining the value of E/D equals 3 as well as the W/D spanning from 2 to 5.

4. At E/D equals 4 involving W/D ranged from 2 to 5, the strength extracted from Weibull distribution function at a 99% reliability gives almost 11% lower than the average pin joint strength. This implies that designers account the fact of the average values, so it will be less significant in terms of the assurance. If assurance is required, then statistical bearing strength can be considered for further analysis with a great significance level.
5. Net-tension and shear-out failure modes detected at a lower proportion of W/D and E/D values, respectively. These failures are sudden and precaution-free. The bearing failure mechanism is indeed advantageous since this is progressive. The failure mechanism for bearing arises for W/D and E/D ratios are greater than 4.

References

1. Karakuzu R, Taylak N, Içten BM, Aktaş M (2008) Effects of geometric parameters on failure behavior in laminated composite plates with two parallel pin-loaded holes. Compos Struct 85:1–9. https://doi.org/10.1016/j.compstruct.2007.10.003
2. Aktaş A, Imrek H, Cunedioğlu Y (2009) Experimental and numerical failure analysis of pinned-joints in composite materials. Compos Struct 89:459–466. https://doi.org/10.1016/j.compstruct.2008.09.009
3. Singh M, Bhunia H, Saini JS (2015) Effect of ply orientation on strength and failure mode of pin jointed unidirectional glass-epoxy nanoclay laminates. Def Sci J 65:489–499. https://doi.org/10.14429/dsj.65.8917
4. Okutan Baba B (2006) Behavior of pin-loaded laminated composites. Exp Mech 46:589–600. https://doi.org/10.1007/s11340-006-8735-z
5. Aktas A, Karakuzu R (2010) Failure analysis of two-dimensional carbon-epoxy composite plate pinned joint. Mech Adv Mater Struct 6:347–361
6. Aktaş A (2007) Statistical analysis of bearing strength of glass-fiber composite materials. J Reinf Plast Compos 26:555–564. https://doi.org/10.1177/0731684407075531
7. Bedi R, Chandra R (2009) Fatigue-life distributions and failure probability for glass-fiber reinforced polymeric composites. Compos Sci Technol 69:1381–1387. https://doi.org/10.1016/j.compscitech.2008.09.016
8. MCCool JI (2012) Using the weibull distribution. Wiley
9. Sakin R (2008) Materials and design statistical analysis of bending fatigue life data using Weibull distribution in glass-fiber reinforced polyester composites 29:1170–1181. https://doi.org/10.1016/j.matdes.2007.05.005
10. Dirikolu MH, Aktaş A, Birgören B (2002) Statistical analysis of fracture strength of composite materials using Weibull distribution. Turkish J Eng Environ Sci 26:45–48
11. Barbero E, Fernández-Sáez J, Navarro C (2000) Statistical analysis of the mechanical properties of composite materials. Compos Part B Eng 31:375–381. https://doi.org/10.1016/S1359-8368(00)00027-5
12. Hassan J, Higgins RO, Waimer M, Mccarthy CT, Feser T, Toso N Investigation of geometrical and composite material parameters for tension-absorbing bolted joints 1–9
13. Singh M, Saini JS, Bhunia H (2018) To study the contribution of different geometric parameters on the failure load for multi holes pin joints prepared from glass / epoxy nanoclay laminates. J Compos Mater 52:629–644. https://doi.org/10.1177/0021998317712572
14. Okutan B (2002) The effects of geometric parameters on the failure strength for pin-loaded multi-directional fiber-glass reinforced epoxy laminate. Compos Part B Eng 33:567–578

15. Khashaba UA, Sebaey TA, Alnefaie KA (2013) Failure and reliability analysis of pinned-joints composite laminates: effects of stacking sequences. Compos Part B Eng 45:1694–1703. https://doi.org/10.1016/j.compositesb.2012.09.066

Three-Dimensional Non-linear Transient Analysis for Predicting the Defect Propagation Rate of Taper Roller Bearing Using Finite Element Method

Sumit Shoor and Manpreet Singh

1 Introduction

Mechanical systems such as motors, pumps, engines, and turbines are operating on shafts which are rotating at different speeds. The bearing is at the heart of all rotating equipment, and the condition of the bearing often reflects how well a machine is running [1]. Due to unexpected operating and loading conditions, results in high stresses and amplitude. Factors which influence reaction forces generated in taper roller bearing are curvature, misalignment, contact angle, clearance, diameter, lubrication, and material properties, etc. In most of the rotating equipment bearing is one of the major causes of failure, so more attention is paid by researchers on bearing failure in past two decades [2–4]. Main reason of failure in bearing is contact stresses [5], Temperature field [6] fatigue, crack propagation, misalignment, unbalance, friction, etc. Different types of non-destructive techniques are available to detect the bearing defects and such as ultrasonic testing, wear analysis, thermal analysis, finite element analysis, acoustic signal analysis, and vibration signal analysis [7–10]. State of bearing can be predicted by contact analysis using FEM and vibration analysis is also a mostly adopted for condition monitoring which gives information about stresses, displacement, and penetration which helps to evaluate bearing performance under different loading conditions. Contact between surface can be generated by using ANSYS software and friction heat generated can be used to predict the state of system. Taper roller bearing is designed to carry radial and thrust loads so they are used under severe vibration conditions [11]. The major parameters in taper roller bearing are the contact between moving element and stationary element. The friction generated due to contact between elements causes friction and rise in temperature and hence produces contact stresses [12] which may further lead to failure of bearing

S. Shoor · M. Singh (✉)
School of Mechanical Engineering, Lovely Professional University, Phagwara, Punjab 144411, India
e-mail: Manpreet.20360@lpu.co.in

A. Parey et al. (eds.), *Recent Trends in Engineering Design*, Lecture Notes in Mechanical Engineering, https://doi.org/10.1007/978-981-16-1079-0_16

if it increases a certain limit. Very less study has been carried out on rolling element bearing by using FEM. The main aim of the study is to develop a CAD model of taper roller bearing and to obtain a simulated vibration signal using FEM analysis.

2 Mathematical Model

Hertzian tells the contact stress produce when parts rub each other [13]. In Bearing contact is generally a higher pair or lower pair when a sphere came into contact with surface make a line contact and cylinder with surface make a surface contact. Increase in pressure cause yielding of two surfaces and creates a very small indentation through yielding. Contact stresses thus created are also called hertzian contact stresses [14]. Hertz stress also known as stress closest to this area of contact between two spheres. The contact angle between the roller and inner raceway is α_{i}, the angle between the center line of the roller and the center line of the bearing is α_{m}, the contact angle between the roller and outer raceway is α_{o}. Taper roller bearing geometry is shown d: cone bore diameter, D: outside cup diameter, T: Bearing width, D_{pw}: Pitch diameter, L_{Re}: Length of rolling element bearing, D_{m}: Mean diameter of rolling element bearing at its center Ore Figs. 1 and 2.

$$\text{Stiffness } k = \frac{\mathrm{d}f}{\mathrm{d}u} = \left(E^2 R F\right)^{1/3} \tag{1}$$

$$\text{Depth of indentation } \mu = \left(\frac{2F^2}{E^2 R}\right)^{1/3} \tag{2}$$

$$\frac{1}{E} = \frac{1}{2}\left(\frac{1-\mu^2}{E1} + \frac{1-\mu^2}{E2}\right) \tag{3}$$

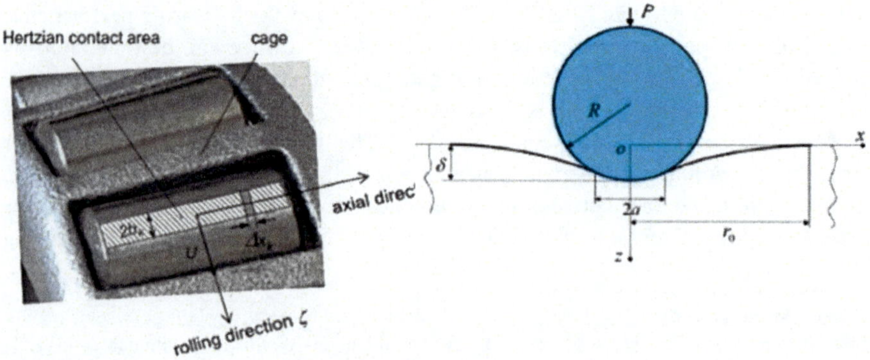

Fig. 1 Hertzian contact zone and indentation of a roller on an elastic surface

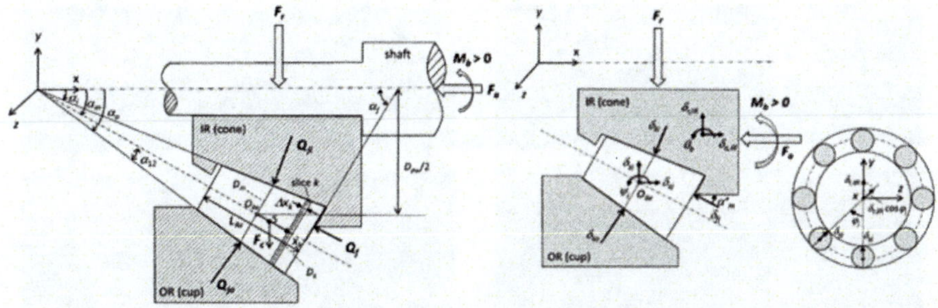

Fig. 2 Loading and degree of freedom of taper roller bearing [15]

$$\text{Stress } \sigma_{\max} = \frac{3F}{2\pi a^2} = 0.4\left(\frac{E^2 F}{R^2}\right) = 0.4\frac{K}{R} \tag{4}$$

3 Finite Element Analysis

Contact between roller and outer race, roller and inner race is line contact before loading. Line contact converted into surface contact after loading. Size and shape depends upon load applied on bearing. In this 30205 taper roller bearing was modeled using CREO with stationary outer race is considered. All degree of freedom is constrained in outer race and radial load is applied on inner ring as shown in Fig. 5. Contact is simulated by finite element method.

3.1 Building of Bearing Finite Element Model

ANSYS model once built can be changed in terms of material properties, loading conditions, boundary conditions, etc. ANSYS model helps in creating and optimizing design and enhancing reliability. Taper roller bearing mainly composed of outer race, inner race, roller, and cage [16]. Figure 3 shows outer race, inner race, taper roller bearing, three-dimensional CAD model, and meshed model of taper roller bearing.

3.2 Defining Material Properties

It is assumed that taper roller bearing material is isotropic with having following Material properties: Density: 7750 kg/m^3: Young's modulus: 200,000 Mpa Poisson's Ratio: 0.29 [17]

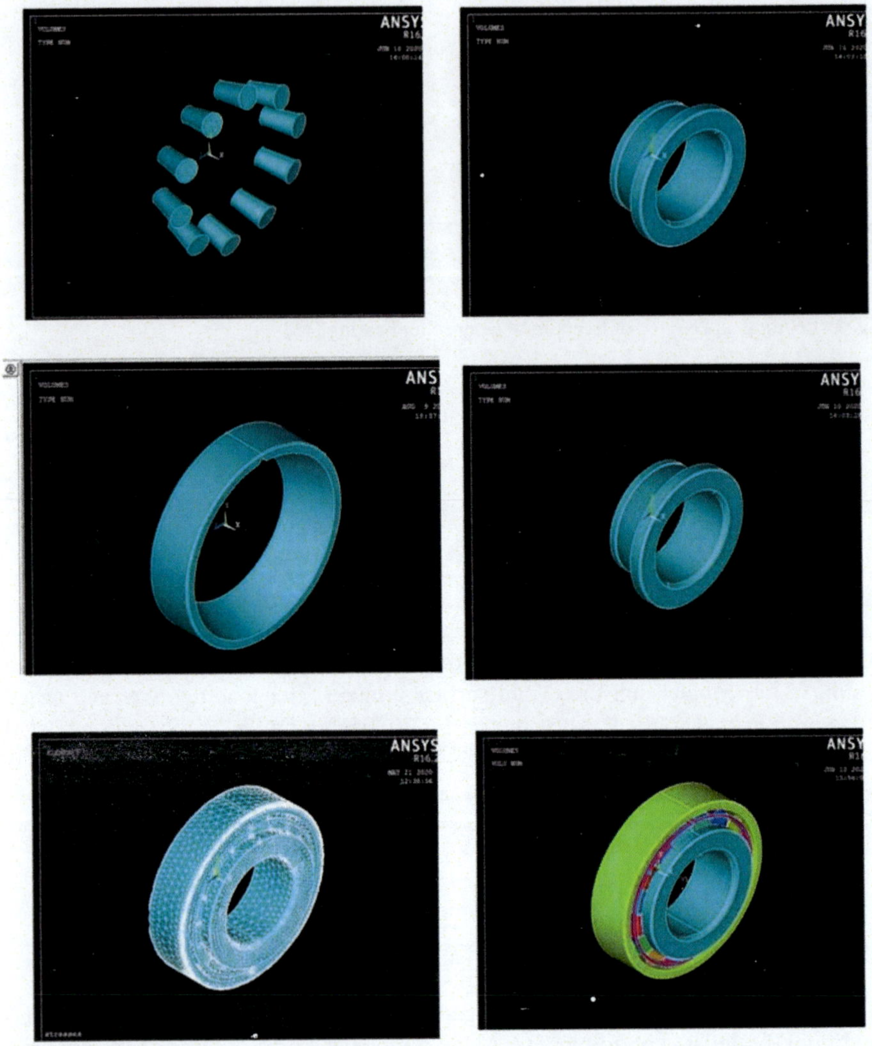

Fig. 3 Inner race, defective outer race, cage, roller, assembled model and meshed model of taper roller

3.3 *Define Contact and Check Contact Element*

As shown contact has been created between roller and outer race, roller and inner race. Friction coefficient taken is 0.29. Contact in ANSYS was created by pick target, Pick contact and create Fig. 4.

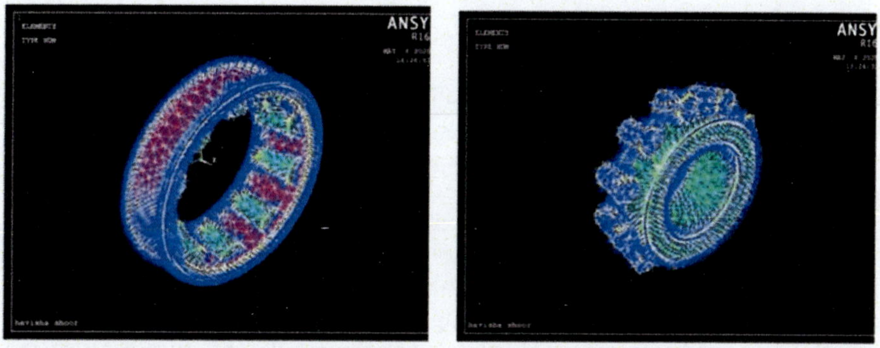

Fig. 4 Contact between roller and inner race and roller and outer race

3.4 *Selecting Element Type and Meshing*

Brick 8 node 185 was used for meshing. Target 170 and contact 174 element was used for creating contact between roller and outer race and roller and inner race. The finite element model was composed of 24,734 nodes and 22,722 elements.

3.5 *Boundary Conditions and Apply Loads*

Boundary conditions were applied as shown in Fig. 5 which outer race is fixed or all degree of freedom is constrained and radial loads are applied to inner part of ring. In analysis type transient analysis was selected.

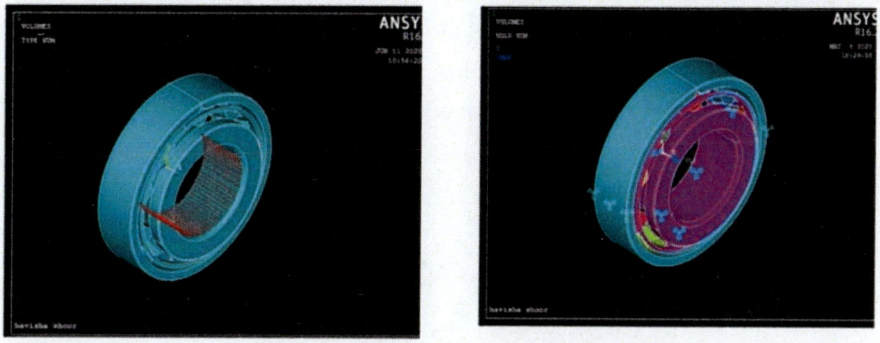

Fig. 5 Boundary conditions and normal load

4 Result and Discussion

Bearings are mechanical assemblies that allow machinery to carry heavy loads with efficiency and ease and move at extremely high speed in linear or rotational motion while reducing friction. By using finite element method contact stresses, strain, von misses stress, displacement can be get once the process simulated. In this three dimensional model was created by CAD and non-linear analysis was carried out by using APDL (Ansys parametric design language) using ANSYS software. Nonlinearity is further classified as material nonlinearity and geometric nonlinearity. Where stress–strain relationship is not linear. By applying the boundary and loading conditions the contact stresses and deformation is simulated by using ANSYS software. Simulated result helps in design, optimization, and predict the contact stresses and deformation. The maximum stress generated in the taper roller having outer race defect was analyzed by transient analysis using the finite element method Figs. 6 and 7.

Taper roller bearing axial load was applied radial load of 150 N and angular velocity of 150 rad/s was applied along *x*-direction. Transient analysis is carried out to predict the behavior of taper roller bearing under loading. As shown from simulated results maximum displacement is 36.5 mm at 50 s. Figure 8 shows the

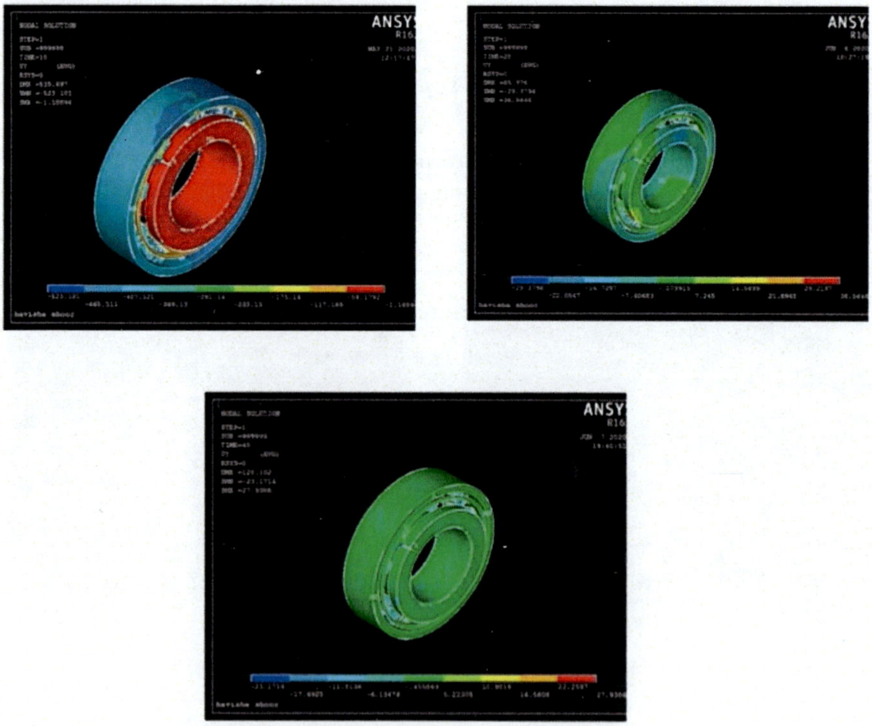

Fig. 6 Transient analysis of displacement

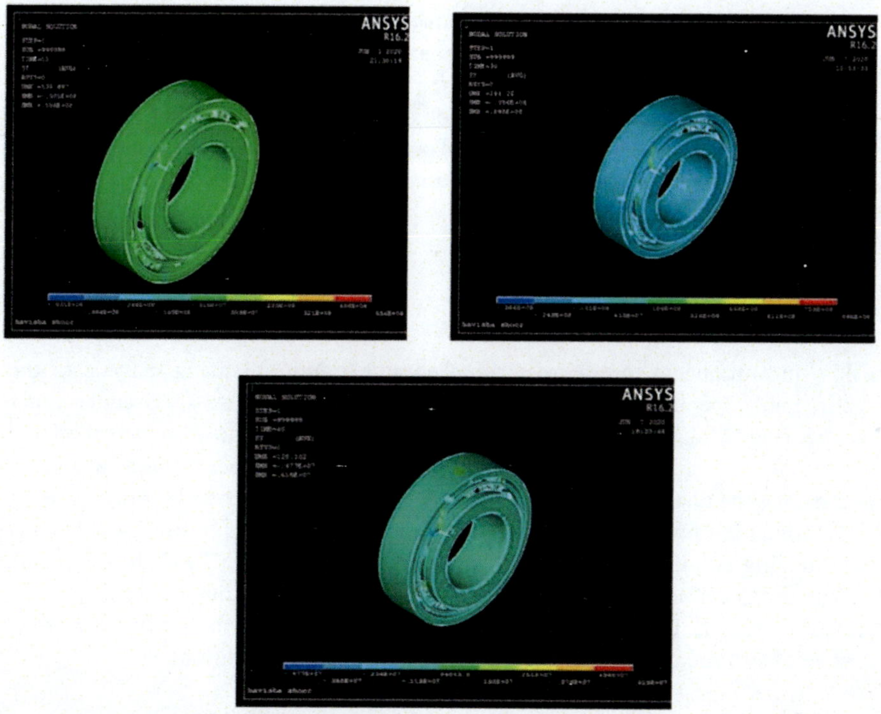

Fig. 7 Transient analysis of contact stresses

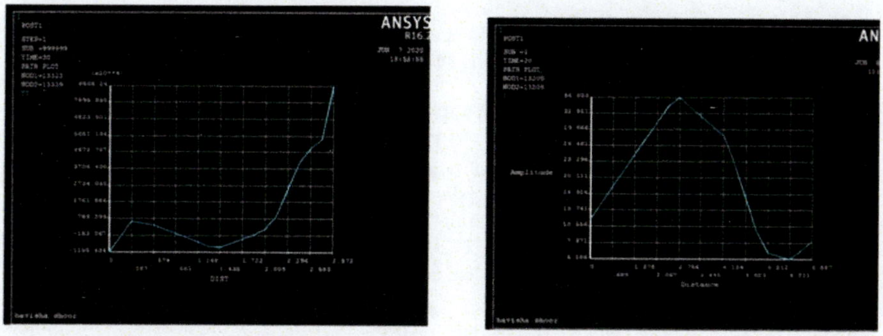

Fig. 8 Graph showing maximum amplitude and stresses

transient behavior of taper roller bearing. Displacement at 10 s was found 1.1 mm which is further increased to 36.5 mm at 20 s may be due to contact between taper roller and defect on outer race region further value of displacement decreases to 27.9 mm at 40 s. Simulated result revealed the maximum stress generated at outer race of taper roller bearing. At 10-s stress noted was 55 MPa and value increased to 85.68 MPa at 40 s, which indicates that as defect propagates value of stress also increasing. Maximum variation of contact stress found from 10 to 40 s 35.80%. Figure 8 shows the graph in the vicinity of maximum stress region nodes are selected in region where maximum stress is generated and graph is plotted. Maximum stress noted was 85.68 MPa in the defective region of outer race and also shows the graph of maximum displacement of 36.5 mm. As per simulated results fluctuating stresses were obtained which vary with respect to time.

5 Conclusion

In this study transient three-dimensional nonlinear finite element analysis of the experimental taper roller bearing was carried out using the ANSYS code. Taper roller bearing (Make: NBC, Model: 30,205) was chosen and drawing was prepared using CAD software as per manufactures catalogue. This paper explains a process to determine the contact stresses in the outer raceway of a taper roller bearing using FEM. Simulation revealed the transient behavior of taper roller bearing subjected to radial loading conditions. Maximum stress noted was 85.68 MPa with percentage in variation 35.80% from 10 to 40 s even maximum stress is found in region where defect was created. Hence FEM can also be used as condition monitoring to predict the state of system at particular time under given loading conditions.

References

1. Al-badour F, Sunar M, Cheded L (2011) Vibration analysis of rotating machinery using time–frequency analysis and wavelet techniques. Mech Syst Signal Process 25:2083–2101. https://doi.org/10.1016/j.ymssp.2011.01.017
2. Singh G, Kumar R, Singh M, Singh J (2017) Detection of crack initiation in ball bearing using FFT analysis. Int J Mech Eng Technol 8(7):1376–1382
3. Kumar R, Jena DP, Singh M (2010) Identification of inner race defect in radial ball bearing using acoustic emission and wavelet analysis. ISMA2010 International Conference on Noise and Vibration Engineering, Leuven, Belgium, 20–22 Sept, pp 2883–2891
4. Cambow R, Singh M, Bagha AK, Singh H (2018). To compare the effect of different level of self- lubrication for bearings using statistical analysis of vibration signal. Mat Today Proceed 5(14): Part 2, 28364–28373
5. Scari S (2013) Finite element analysis of tapered roller bearing : complete model X simplified 1730–1735
6. Xu J, Zhang J, Huang Z, Wang L (2015) Calculation and finite element analysis of the temperature field for high-speed rail bearing based on vibrational characteristics. J Vibroengi- Neering 17:720–732
7. Hatt F (1988) Finite element analysis of elastomeric components. Elastomerics 120:24–27
8. Šulka P, Sapietová A, Dekýš V, Sapieta M (2018) Static structural analysis of rolling ball bearing. MATEC Web Conf 244.https://doi.org/10.1051/matecconf/201824401023
9. Tang Z, Sun J (2011) The contact analysis for deep groove ball bearing based on ANSYS. Procedia Eng 23:423–428. https://doi.org/10.1016/j.proeng.2011.11.2524

10. Choudhury A, Tandon N (2000) Application of acoustic emission technique for the detection of defects in rolling element bearings. Tribol Int 33:39–45. https://doi.org/10.1016/S0301-679X(00)00012-8
11. Minhas N, Nikhil N, Banwait SS (2019) Vibration analysis and fault identifications of rolling element bearings-a review. Int J Mech Prod Eng Res Dev 9:1133–1142. https://doi.org/10.24247/ijmperdaug2019117
12. Xin HR, Zhu L (2014) Contact stress FEM analysis of deep groove ball bearing based on ANSYS workbench. Appl Mech Mater 574:21–26. https://doi.org/10.4028/www.scientific.net/AMM.574.21
13. Purushothaman P, Thankachan P, John' S, Purushothaman MP (2014) Hertz contact stress analysis and validation using finite element analysis. www.ijraset.com
14. Long JM, Wang GF, Feng XQ, Yu SW (2012) Two-dimensional Hertzian contact problem with surface tension. Int J Solids Struct 49:1588–1594. https://doi.org/10.1016/j.ijsolstr.2012.03.017
15. Nguyen-Schäfer H (2019) Computational tapered and cylinder roller bearings
16. da Silva Scari A, Américo P, Magalhães A (2014) Influence of the number of rollers on a tapered roller bearing. J Mech Eng Autom 4:560–564. https://doi.org/10.17265/2159-5275/2014.07.003
17. Ezaki Y, Terasawa H, Wada T (2012) Vibration analysis for tapered roller bearing fatigue prevention. J Syst Des Dyn 6:665–675.https://doi.org/10.1299/jsdd.6.665

Cooling System Design of PPE with Simulation

Sagarjit Das, Md. Wasim, Lipa Manna, and Soumyajit Roy

1 Introduction

The biggest name of terror in the world is COVID-19 at present time, and people are well aware of its tremendous effect. In this situation, the biggest warriors of COVID-19 are the doctors, police, and other health workers, especially in the hospital. To secure from COVID-19, they have to wear a PPE kit, on duty. This PPE kit protects those from any kind of virus contamination but wearing PPE during duty is terrifying to them. Because wearing a PPE kit for a long time causes excess sweat and at a time it will lead to headaches and breathing problems. The body temperature will also increase that may cause illness. For these problems, a system device is introduced which will give a cooling effect with safety precautions keep in mind. So, in this case, it will be easy to do something for a long time using the PPE kit. It is of very low cost, which makes it affordable to all. A series of simulation experiments using the device were carried out and the result was satisfactory.

2 Background

Farhang et al. [1] have proved that, in the experiment, a large percentage of workmen found the PPE uncomfortable (other than face mask). It has also come to a verdict to sort out the agonizing factors in the kit.

John et al. [2] have caught sight of the PPE and found that the manufacturers give the majority of their concern to the safety of the clothing, not in coziness. It shows the solacement of the kit plays a role and should be emphasized. John et al. [3] have done field test, from which the uneasiness from the PPE is highlighted that there

S. Das (✉) · Md. Wasim · L. Manna · S. Roy
MCKV Institute of Engineering, Howrah, West Bengal 711204, India

A. Parey et al. (eds.), *Recent Trends in Engineering Design*, Lecture Notes in Mechanical Engineering, https://doi.org/10.1007/978-981-16-1079-0_17

is extreme discomfort in the clothing. The factors affecting over here are mainly hotness.

Farhang [4] made the survey and as a result, we can see only 25% of the employee was comfortable in PPE and rest were not. With the study, it came to know the main factors which are resisting the employees to use it are discomfort, irritated skin, sweat, etc. It has also suggested the optimum goal, i.e., restful condition, permanence, etc.

Eun-Kee et al. [5] have sketched to output that the use of protective clothing is lost in farers clone to the messiness of it and lack of knowledge. Ikuharn et al. [6] have investigated the health condition of workers at the construction plot. Several health-hazardous can be seen in the process. Concerning this, we can assume the health status of health workers, fire workers, and farmers using PPE in adverse conditions.

Lausent et al. [7] have found that using PPE will increase body temperature. Making extension of the theory can say that labors working in the field under scorching sun and using PPE will lead to health hazards.

Ingvar [8] has made an approach where the report has been given that the evaporative sweat is responsible for hampering the heat transfer in between protective equipment which may cause the workers to get tired. Here comes the fact about the efficiency ratio of pervasion to determine the heat exchange during evaporation. The result shows that if the transfer of heat exchange due to precipitation and climatic contrast needs to be explained, more experimental works are required.

A viewpoint has been given by Parsons [9], where he is talking about anatomical objectives about the human body temperature which should not much differ from the normal body temperature and the amount of sweat is needed to control under a level to avoid irritation. These objectives involve the maintenance equilibrium of heat to keep the normal body temperature 37 °C but the heat equilibrium characteristics of protective equipment are complicated for which some important components are not perfectly estimated.

Patrick et al. [10] have stated atmospheric volume of protective equipment plays an important role in feeling the alleviation of operators while wearing it. This volume can be evaluated as both clothed and unclothed cases. This report also says that there should be an airing index which will maintain the cooling. It is also signified that this atmospheric volume has an impact on an insulator that is why a critical point in cooling effect is necessary to get comfort.

Candas et al. [11] have presented how the temperature of the human body changes with and without the clothing. In clothed system, more sweat is made than in the non-clothed system, in an airless condition. For a non-clothed system, the increased pressure of vapor decreases evaporation and as a result the body is getting wet that decreases the sweat dissipation which is not found in clothed system. This is under the same steam pressure. As a result, workers will feel cool while wearing protective equipment in airless (less-airflow) conditions.

Here comes the most important aim that is given by Raimundo et al. [12] is to examine the properties of protective equipment in case of an extreme condition like firefighting. For that, they have proposed an improved version of the multi-branch temperature control model, which has the capability to reserve a high-temperature heat that will help the users to survive during extreme situations. The outcomes have shown that the protection of employees is necessary for that high temperature in the way to improve the quality of protective equipment.

Goldman [13] has stated that the values of insulation and vaporization impedance are the most important keys to get the most perfectly thermal shielding in clothing. Protection and responsibilities should be taken when air pass ability changes greatly or when unwanted air transfer occurs. Newer studies have described the required improvement to measure the flow of air, where thermal reaction of human body is explained as the parameters of equipment, amount of heat and surroundings.

3 Modeling

The modeling of the "PPE Cooler" along with the mechanical design and electronics circuit has been described below. The components were tabulated in Table 1 with Specifications. The schematic diagram of the PPE cooling system setup was given in Fig. 1 along with the fixture on PPE. The pictorial view of the device was shown in Fig. 2 along with circuit diagram.

4 Experimental Procedure

The experimental setup along with specification was tabulated in Table 2 and discussed in Fig. 1.

To simulate the experiment with "PPE Cooler" the device was attached to the shoulders on both sides with a strap from the top of the box so that the system can be carried easily. Another most effective part in that box was open where the fans were kept and there PPE kit was cut. N95 Covers were placed at that gap. So, while providing cool air, if any nano-sized virus or harmful particles were with that air, they will get stuck by that layer and only fresh air will enter.

The same person was fitted with PPE kit at the same temperature of 25 °C and humidity of 47%, but this time the ventilation system was set at the back, which will give cooling effect. After every 45 min, both temperature and humidity were measured again in the same way to observe variation. The experiment has been carried out on multiple times and on multiple people.

5 Results and Discussion

The result of the simulation experimentation has been tabulated and thereafter discussed briefly.

Table 1 Specification and use of the components for "PPE cooler"

	Components	Use	Specification	Number
1	CPU fan	Produce cooling effect	9 Blades, 12 V, 0.30A, Material-Plastic,	2
2	Lithium-ion battery	To Run Fans For 5 Hours (Continuous)	3.7 V, 2900 Mah	3
3	Storage box	Hold the whole system	Dimension-90 mm × 240 mm × 60 mm, material-plastic	1
4	IC CD4017 (CMOS Decade Counter IC)	Used for low range counting applications	High speed 16 pin CMOS Decade counter, 10 decoded outputs, Wide supply voltage range from 4 to 14 V, typically + 5 V, Maximum Clock Frequency: 5.5 Mhz, 16-pin PDIP, PDSO	1
5	1N4007 Rectifier Diode	Used to convert alternating current to direct current	Maximum Recurrent Peak Reverse Voltage 1000 V, Maximum RMS Voltage 700 V, Peak Forward Surge Current: 30A, Maximum DC Blocking Voltage 1000 V, Average Forward Current: 1.0 A, Maximum Instantaneous Forward Voltage: 1.0 V	1
6	5-PIN 5 V Relay	1. Used for Switching Circuits 2. To Control On/Off for Heavy Loads At A Determined Time/Condition	Voltage across coil: 5 V DC, Maximum AC load current: 10A @ 250/125 V AC, Compact 5-pin configuration with plastic molding, Operating time: 10 ms Release time: 5 ms, Maximum DC load current: 10A @ 30/28 V DC	1

(continued)

Table 1 (continued)

	Components	Use	Specification	Number
7	LED	Here LEDS ware used as battery charge level indicators	Weather resistance, UV Resistant, 5 mm Round Standard Directivity, Forward Voltage: 1.8–2.4 V, Forward Current: 30 Ma, Reverse Voltage: 5 V, Operating Temperature: −25 to +80 °C, Luminous Intensity: 20 mcd	Green-1 Red-1 White-1
8	IR (Infra Red) Sensor Module	Used for Sense a Remote Control	5VDC Operating voltage, I/O pins are 5 and 3.3 V compliant, Range: Up to 20 cm, Adjustable Sensing range, Built-in Ambient Light Sensor, 20 mA supply current, mounting hole	1
9	4.7 μF/50 V Electrolytic Capacitor	Used For Flashing Circuits	Value: 4.7 μF, Temperature: −40 to +85 °C, Type: Radial Tolerance: 20%, Rated Voltage: 50 V	1
10	Resistor	Circuit element (Reduce current flow, adjust signal levels)	Resistance (ohm): 330, 1000 Resistor Type: Carbon Film, Size: Standard, Power (Watts): 0.25	$330\ \Omega^{-1}$ $1000\ \Omega^{-1}$

5.1 Results

The humidity and temperature inside the PPE was measured with "PPE Cooler" and without using the "PPE Cooler". The Graph was plotted in Fig. 3 and the Problems Experienced during Experiment was tabulated in Table 3.

5.2 Discussion

Above mentioned, two types of tests were performed; one with "PPE Cooler" and another was without "PPE Cooler". The humidity graph shows that the humidity

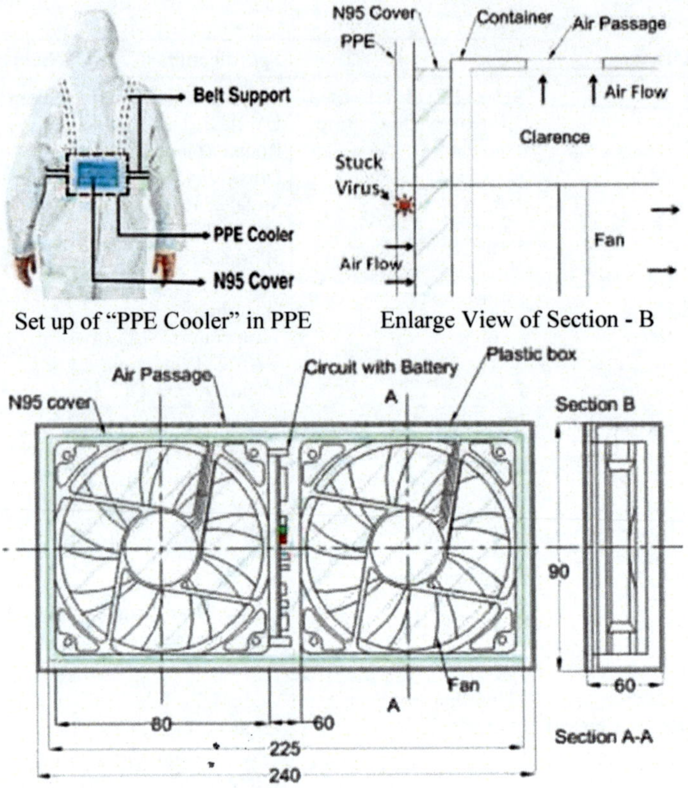

Fig. 1 Schematic diagram of the "PPE cooler"

increases for both the case, but the rate of increment were much higher for the experiment without using the "PPE Cooler", whereas using "PPE Cooler", the increase in humidity was tolerable for more than 5 h.

The same result can be shown for the graph of temperature varying with time.

The slope of the graph was greater in the process of not using ventilation than the slope of the graph in the process of using ventilation. So when the device was not used in the PPE kit, body temperature increases rapidly but when the "PPE Cooler" was used, temperature increases at a very low rate, which makes the PPE kit less painful to everyone.

6 Conclusion

The experiment shows that, while using the PPE kit without "PPE Cooler", sweating starts after a few moments. This leads to breathing problem. Dry throat and sweating in the head later make the user feel very uncomfortable.

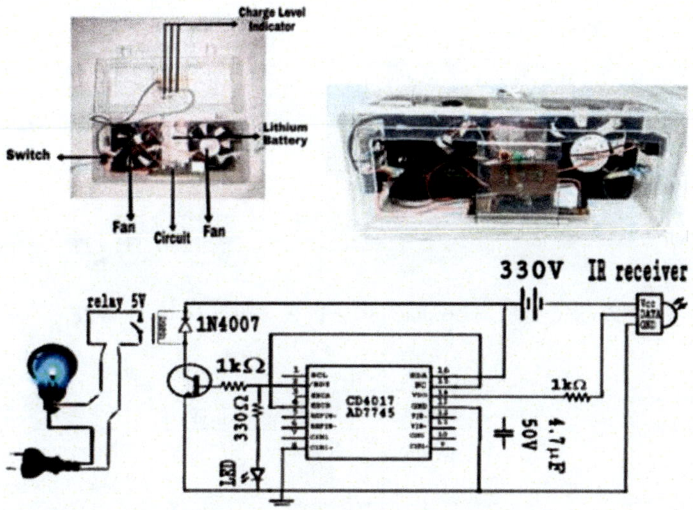

Fig. 2 Pictorial View of PPE cooling system setup (Top Left), front view (Top Right), and circuit diagram for "PPE cooler" (Bottom)

Table 2 Specification and use of the components used for simulation

	Components	Use	Specification
1	PPE Kit	Designed to guard the health of by minimizing the exposure to a virus, bacteria etc.	Approved by Positex, Medical Disposable Protective Coverall Suits, For High-Risk User, PP Non-Woven with Shoe Cover and Face Shield
2	N95 Cover	Use to protect from airborne particles	Brand: Gleam, Respirator Type: Air-purifying Respirator, Anti-Virus With Melt Blown Fabric Layer ISO CE Certified
3	Temperature- Humidity Clock	Use to measure humidity and temperature	Brand- HTC-1, Temperature unit C/F, LCD display
4	PPE Cooler	Use to maintain Temperature and Humidity inside PPE	Discussed in Table 1

Using the "PPE Cooler", sweat does not start in that huge rate within 2 h. After a long time sweat increases, this may cause a little bit of dry throat. The ""PPE Cooler"" lowers the humidity inside the PPE kit, as a result, the sweat evaporation rate will increase faster and body temperature will be in a stable condition.

The designed "PPE Cooler" cools for 5 h by running continuously. But if required, it can be run for more hours by adding an extra Lithium-Ion battery or by running with the break with the same setup. Even if more air discharge is required then a large diameter fan can be installed in the same setup.

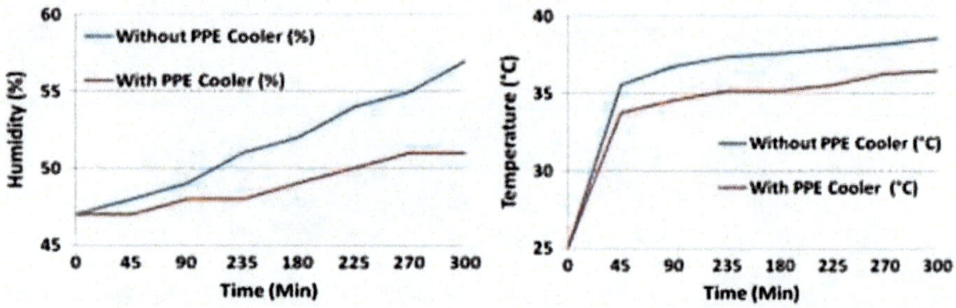

Fig. 3 Graphs for humidity and temperature inside PPE, with and without "PPE cooler"

Table 3 Problems experienced during experiment

	Time span	Without "PPE cooler"	With "PPE cooler"
1	45 min	Sweat	No sweat
2	1 h 30 min	Sweat	Very little sweat
3	2 h 15 min	Sweat and dry throat	Sweat
4	3 h	Very high sweat	Increase in sweat
5	3 h 45 min	Breath rate high	Increase in sweat
6	4 h 30 min	Wet head and face	Increase in sweat
7	5 h	Wet clothed PPE	Increase in sweat and dry throat

The '"PPE Cooler"' is a combination of small parts, if one element gets damaged that can be repaired or replaced separately which is a great way to avoid huge costs of replacing the whole device system. So, it can be concluded, this device system is effective and affordable for this type of cooling activities.

References

1. Akbar-Khanzadeh F, Bisesi MS, Rivas RD (1995) Comfort of personal protective equipment. Butterworth Heinemann 26(3):195–198. https://doi.org/10.1016/0003-6870(95)00017-7
2. Anthony AJD, Houshang S (1988) Ergonomics aspects of personal protective equipment: its use in industrially developing countries. J Human Ergol 17:67–79. https://doi.org/10.11183/jhe1972.17.67
3. Anthony AJD, Houshang S (1990) Adaptation to discomfort in personal protective devices: an example with safety helmets. Ergonomics 33(2):137–145. https://doi.org/10.1080/00140139008927105
4. Akbar-Khanzadeh F (1998) Factors contributing to discomfort or dissatisfaction as a result of wearing personal protective equipment. J Human Ergol 27(1, 2):70–75. https://doi.org/10.11183/jhe1972.27.70
5. Park E-K, Hannaford-Turner K, Lee HJ (2009) Use of personal protective equipment in agricultural workers under hot and humid conditions, industrial health 47:200–201. https://doi.org/10.2486/indhealth.47.200

6. Morioka I, Miyai N, Miyashita K (2006) Hot environment and health problems of outdoor workers at a construction site industrial health 44(3):474–480.https://doi.org/10.2486/indhealth.44.474
7. Grélot L, Koulibaly F, Maugey N, Janvier F, Foissaud V, Aletti M, Sagui E (2015) Moderate thermal strain in healthcare workers wearing personal protective equipment during treatment and care activities in the context of the 2014 ebola virus disease outbreak. JID 2016:213. https://doi.org/10.1093/infdis/jiv585
8. Holmér I (1995) Protective clothing and heat stress. Ergonomics 38(1):166–182. https://doi.org/10.1080/00140139508925093
9. Parsons KC (1988) Protective clothing: heat exchange and physiological objectives. Ergonomics 31(7):991–1007. https://doi.org/10.1080/00140138808966738
10. Sullivan PJ, Mekjavic IB, Kakitsuba N (1987) Determination of clothing microenvironment volume. Ergonomics 30(7):1043–1052. https://doi.org/10.1080/00140138708965994
11. Candas V, Candas V, Hoeft A (1995) Clothing, assessment and effects on thermophysiological responses of man working in humid heat. Ergonomics 38(1):115–127. https://doi.org/10.1080/00140139508925089
12. Raimundo AM, Figueiredo AR (2009) Personal protective clothing and safety of firefighters near a high intensity fire front. Fire Saf J 44(4):514–521. https://doi.org/10.1016/j.firesaf.2008.10.007
13. Goldman RF (1974) Clothing design for comfort and work performance in extreme thermal environments. Trans New York Acad Sci 36(6Series II):531–544. https://doi.org/10.1111/j.2164-0947.1974.tb01599.X

Parametric Optimization of Hole Quality in Laser Drilling Kevlar/Basalt Hybrid FRP Composite

Kaushal Pratap Singh, Gavendra Norkey, and Girish Dutt Gautam

1 Introduction

With the rise of advance materials in the various engineering sectors, the rigid design, complex shape, and unusual size of the work piece confined the use of traditional machining methods. This situation arises the demand for non-traditional machining techniques for such materials [1]. In industries, Electron Beam Machining (EBM), Electro-Chemical Machining (ECM), Electrical Discharge Machining (EDM), Laser Beam Machining (LBM) Abrasive Water Jet Machining (AWJM), etc. are major types of non-traditional machining processes which are using in industries for machining advance materials. These processes are widely replacing conventional machining techniques due to their ability to produce complex design and intricate shape on work piece [2].

LBM is one of the most applicable machining processes which is suitable for all materials like metals, non-metals, and composite materials. In LBM, a high energy laser beam is strike on the work surface and removes materials through melting and vaporization.

In LBM, various factors like beam, material, and machining parameters significantly affect the quality characteristics of machined work surface. In laser drilling these quality characteristics can be classified as hole circularity, heat-affected zone (HAZ), recast layer thickness, etc. there were a lots of investigations have been conducted during past to analyze the effect of vital drilling parameters on the quality charactertics of drilled hole. It has been observed that in these studies experiments were performed by using the design of experiments (DOE) techniques. These

K. P. Singh (✉) · G. Norkey
Department of Mechanical Engineering, Lovely Professional University, Jalandhar, Punjab 144411, India

G. D. Gautam
Department of Mechanical Engineering, Mangalmay Institute of Engineering and Technology, Greater Noida, Uttar Pradesh 201310, India

A. Parey et al. (eds.), *Recent Trends in Engineering Design*, Lecture Notes in Mechanical Engineering, https://doi.org/10.1007/978-981-16-1079-0_18

approaches are superior to other techniques due to their systematic and scientific way to define the number of experiments with limited resources [3, 4]. Nd:YAG and CO_2 are two most popular types of laser systems. Nd:YAG is highly preferable on CO_2 in industries due to its ability to operate in pulsed mode and develop higher peak power. In recent years, hybrid fiber reinforced polymer composites receives the attention of researcher due to their low cost and improved mechanical properties. In the present research article, authors are used a Kevlar-29 and basalt fibers-based hybrid composite sheet as work piece during laser drilling. Kevlar-29 has extraordinary thermal, mechanical, and antiballistic properties and is widely replacing Glass and Carbon fiber reinforced polymer (FRP) composites in automobile industries. On the other side, Basalt fibers are extracted from natural volcanic rock. Basalt fibers are harmless and eco-friendly materials. From sometimes, the application of basalt fibers is utilized in the automobile, marine, construction, and aerospace sectors [5–7].

Fiore et al. [7] calculated the tensile strength of hybrid basalt and glass fiber sheet and revealed that the tensile strength has been increased by the mixing of glass fiber. Abrate et al. [8] studied that machining of basalt fiber composite materials is difficult because of the non-uniformity and sensitivity toward heating of the matrix whereas high abrasiveness of the reinforcing agents. Mathew et al. [9] investigated that Nd:YAG laser can satisfactorily cut the FRP composite. They also developed the predictive models for heat-affected zone and taper of the cut surface using response surface methodology (RSM). Hocheng et al. [10] observed that in conventional drilling of FRP, delamination is a major issue which affects the hole quality. Tsao et al. [11] identified a correlation between the process parameters and induced delamination during drilling of a CFRP laminate. Davim et al. [12] investigated the equivalent delamination factor (Fed) during core-saw drilling-induced delamination. Gaugel et al. [13] studied about the machinability of FRP composite and observed that requirement of advanced geometry cutting tools. Petrucci et al. [14] identified the ability to machine basalt-glass hybrid composite laminates. In literature, it has been observed that conventional methods have various limitations in machining of FRP composites and resulted good quality characteristics will not be found.

Therefore, the attentions of the researchers are shifted toward the application of LBM for machining of FRP composites. Solati et al. [15] observed that the hole quality has been most affected significantly from heat-affected zone of GFRP composite during laser drilling. Gautam and Pandey [16] identified the optimum levels of process parameter to reduce the kerf deviations at both sides during laser cutting of KFRP composite. Gautam and Mishra [17] acquired the optimum levels of laser parameters to improve the kerf quality of the performed laser cut by using Firefly Algorithm. Gautam and Mishra [18, 19] also identified the optimum settings of laser parameters for cutting of hybrid FRP composite laminate. In all studies, they considered lamp current, pulse width, pulse frequency, compressed air pressure, and cutting speed as input laser parameters and kerf quality characteristics as output responses. Mishra et al. [20] revealed that standoff distance (SOD) significantly affects the geometry of top and bottom kerf deviation during laser cutting of basalt-glass-kevlar 29 hybrid FRP composite laminates. In this article, the hole circularity will be maximized for the fabrication of Kevlar 29-Basalt hybrid composite through the laser beam drilling. For doing so, the hybrid fiber composite has been fabricated with the help of hand lay method and examined its hole circularity. After that drilling operation has performed by using pulsed Nd:YAG laser. Moreover, Taguchi optimization technique has been utilized to find the effects of drilling parameters on hole circularity.

2 Methodology, Material, and Experimentation

The methodology of this study has been divided into four phases. The first phase focuses on the fabrication of KFRP and BFRP hybrid fiber composite laminates with the help of hand layup method. Composite sheet has been fabricated in the laboratory environment using epoxy resin-520 with hardner-509 manufactured and supplied by Electro coating and Insulation Technical Pvt. Ltd., Pune, India. The orientations of Basalt-Kevlar-29 hybrid fiber mats in fabricated laminates have been kept [B-0°/K-0°/B-45°/K-45°/B-0°] for the five layers. The orientation of fibers affects the physical properties of fabricated composite. Therefore, to ensure the smooth flow of epoxy resin, the orientation angle of (0°/45°) has been taken. A mild steel mould has been used to fabricate the composite laminate. The dimensions of the mould were 150 mm × 150 mm × 20 mm. Silicon spray has been used to work as releasing agent. Resin glue has been prepared with the help of using a magnetic stirrer to ensure homogeneous mixture of epoxy and hardener (5:1 ratio proportion). Thereafter, mats of fiber fabrics have been placed according to the defined the stacking sequence with the intermediate layers of resin glue. At last, mould has been tighten by the bolts and rest for curing at room temperature. After 24 h, mould opened and fabricated laminate has been released and cut into the required size. The thickness of the laminate is 1.2 mm (Fig. 1).

Fig. 1 KFRP-BFRP hybrid fiber composite specimen

Table 1 Details of process parameters for KFRP-BFRP hybrid fiber composite

Symbol	Factor	Unit	Level 1	Level 2	Level 3
I	Lamp current	Amp	160	180	200
f	Pulse frequency	Hz	20	25	30
p	Air pressure	kg/cm^2	8	9	10

In the second phase, A 250 W pulsed Nd: YAG laser system has been used to perform drilling experiments on the fabricated composite laminate. This system is developed at Raja Ramanna Center of Advanced Technology (RRCAT), Indore, India. An assist gas has been compressed throughout the experimentation. Whereas standoff distance and impact angle of incident beam kept constant at 1 mm and 90°, respectively, for whole range of experiments. Three laser drilling parameters such as lamp current (I), pulse frequency (f), and air pressure (p) have been selected as an variable input parameters and listed in Table 1, with their respective levels.

Taguchi method has been selected to design experiments. Therefore, total 9 experiments have been conducted. After performing experiments, hole circularity has been measured for 9 different combinations of input laser drilling parameters, viz., lamp current, pulse frequency, and air pressure. Further, in the next phase, the Taguchi approach has been applied for optimization of hole circularity. Moreover, in this phase, a confirmation of experiment has been also done to validate the optimal result by using Taguchi method. In the fourth phase, identify the analysis of parametric effects for all process parameters.

KFRP and BFRP hybrid fiber composite having the nature of fiber pull out from the sides during drilling because of this sample preparation process for measuring the quality characteristics has been done very carefully.

For this research work, the test for determining the hole circularity has been performed on the peak of the sample variable. After performing the experiments, measurements of four diameters $d1$, $d2$, $d3$, and $d4$ have been taken on one side of each drilled hole along the circumference at an interval of 45° angles. One of the sample measurements has been shown in Fig. 2. Measurements have been taken with the help of anoptical microscope (OM-JEOL), Japan, placed in the Indian Institute of Technology (IIT), Kanpur, India. The circularity of the hole has been calculated by using the Eq. 1.

$$\text{Hole}_{\text{circularity}} = \frac{d\,\text{min}}{d\,\text{max}} \tag{1}$$

where Hole$_{\text{circularity}}$ represents the hole circularity, d min represents the minimum diameter and d max represents the maximum diameter of the individual holes.

The images obtained from OM have been adopted for the measurement of the Hole circularity for every KFRP and BFRP hybrid fiber composite sample. The measured values of hole circularity are tabulated in Table 2.

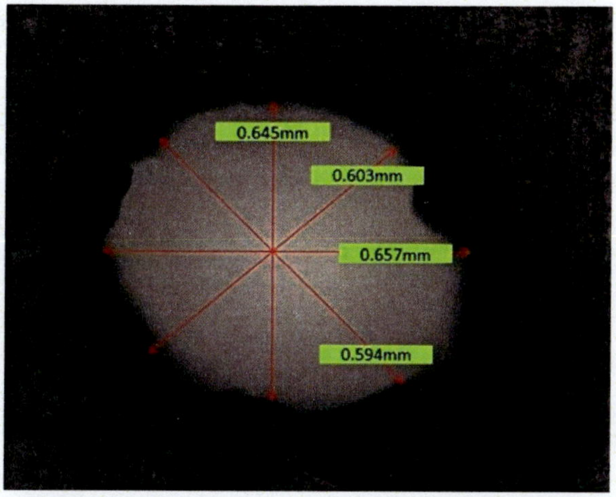

Fig. 2 Hole diameter measurements

Table 2 The response of hole circularity under different process parameters

Exp. No.	Process parameters factors and their level			Response
	I	*f*	*p*	Hole circularity
1	160	20	8	0.9169
2	160	25	9	0.6650
3	160	30	10	0.9006
4	180	20	9	0.7895
5	180	25	10	0.7706
6	180	30	8	0.8584
7	200	20	10	0.8908
8	200	25	8	0.8007
9	200	30	9	0.8081

3 Taguchi Methodology

A Japanese engineer G. Taguchi proposed many approaches for parametric design which are sometimes called "Taguchi Method". These methods have been utilized various types factorial designs. Taguchi introduced experimental design for:

- Designing products as per robust to environmental conditions;
- Designing and developing products/processes for robust to component variations;
- Minimizing variation around a target value.

The techniques of Taguchi are applicable widely. Basically Taguchi method has been employed an orthogonal array by considering the whole parameter with a few number of experiments. The loss function is defined with the deviation between the experimental value and the desired value. For finding the loss function from desired value corresponds to performance characteristics, Taguchi method can be implemented. After that it is further determined the signal-to-noise (*S/N*) ratio η. Usually there are three types of the performance characteristic in the *S/N* ratio, the first one is, the smaller-the-better, second one is,the larger-the-better, and the last one is, nominal-the-best. The *S/N* ratio for each level of process parameters is computed based on the *S/N* analysis. On the basis of the performance characteristic, the larger *S/N* ratio indicates the better performance characteristic Therefore, highest *S/N* ratio η indicates the optimal level of the process parameters. Furthermore, a statistical analysis of variance (ANOVA) is performed for finding statistically significant process parameters. In this paper, Taguchi method is used to obtain optimal laser beam machining performance in drilling operation by using drilling process parameters.

Nominal is the best:

$$^{S}/_{N_N} = 10\text{Log}\left(\frac{\overline{y}}{s_y^2}\right) \tag{2}$$

Higher is the better:

$$^{S}/_{N_H} = -10\text{Log}\left(\frac{1}{n}\sum_{i=1}^{n}\frac{1}{y_i^2}\right) \tag{3}$$

Smaller-is-the better:

$$^{S}/_{N_S} = -10\text{Log}\left(\frac{1}{n}\sum_{i=1}^{n}y_i^2\right) \tag{4}$$

where y is the mean of observed value, S_y^2 is the variance of y, n is the number of observations. The goal of this research has to produce maximum hole circularity in laser drilling operation. Larger hole circularity values represent better or improved hole quality. Therefore, a larger is the better quality characteristic which has been implemented and introduced in this study.

4 Results and Discussion

An analyzing the experimental data on the basis of signal/noise ratio is the main essential criterion in the Taguchi method. According to Taguchi, *S/N* ratio should be higher for getting the optimum drilling condition. Each factor corresponds to their level is equally distributed in the L_9 experiments of orthogonal array, due to which, it is easy to find the effect of each factor at each level. Hole circularity is calculated from mean *S/N* ratio for all process parameters at each level and the main effect

Table 3 The result of *S*/*N* ratio under different process parameter

Exp. No.	*S*/*N* ratio	Exp. No.	*S*/*N* ratio	Exp. No.	*S*/*N* Ratio
1	−0.75356	4	−2.05296	7	−1.00440
2	−3.54357	5	−2.26342	8	−1.93060
3	−0.90936	6	−1.32621	9	−1.85070

represented the value of hole circularity for each experiment. Average *S*/*N* ratio for each parameter at each level is given in Table 3.

According to the Taguchi design, the level values are given in Table 1 which is obtained from MINITAB 16 Program. Accordingly, Fig. 3 and Tables 4 and 5 showed that the higher value for the factor lamp current (*I*) at level third, factor pulse frequency (*f*) at level first and the factor air pressure (*p*) at level first. Therefore experiments will be conducted (311) as 200 amp for lamp current, 20 Hz for pulse

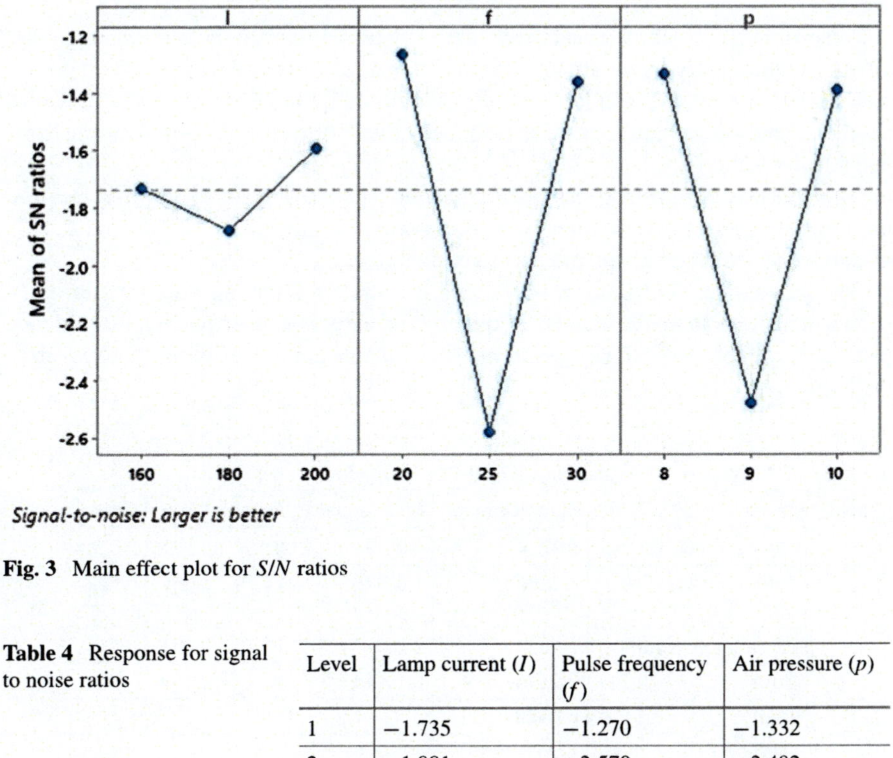

Fig. 3 Main effect plot for *S*/*N* ratios

Table 4 Response for signal to noise ratios

Level	Lamp current (*I*)	Pulse frequency (*f*)	Air pressure (*p*)
1	−1.735	−1.270	−1.332
2	−1.881	−2.579	−2.482
3	−1.595	−1.362	−1.392
Delta	0.286	1.309	1.146
Rank	3	1	2

Table 5 Response for means

Level	Lamp current (*I*)	Pulse frequency (*f*)	Air pressure (*p*)
1	0.8275	0.8657	0.8587
2	0.8062	0.7454	0.7542
3	0.8332	0.8557	0.8540
Delta	0.0270	0.1203	0.1045
Rank	3	1	2

frequency, and 8 kg/cm^2 for air pressure to obtain drilling condition. Effectiveness was observed at three depth levels.

The value of hole circularity for each factor level shows the main effect plots for the average 9th experimentation. The specimen subjected to higher pulse frequency or higher air pressure pulse frequency showed larger hole circularity. The flame current has low effect on hole circularity compared to other two parameters according to Fig. 3.

According to previous researchers, they suggested that the hole circularity was affected significantly by the flame current. The air pressure had a lower effect on the hole circularity but according to the observation in Tables 3 and 4. It signifies that the major role participation of pulse frequency then pulse frequency and air pressure had a lower effect.

The analysis of variance has been also calculated and resulted the p values for the flame current, pulse frequency, and air pressure are 0.020, 0.529, and 0.413, respectively, which confirms the previous statements.

The percentage contribution of various parameters affecting hole circularity is shown in Table 6. In this case pulse frequency and air pressure are the most significant parameters as compared to lamp current affecting hole circularity of the material after drilling.

Table 6 ANOVA for hole circularity of KFRP-BFRP Hybrid fiber composite

S. No.	Factor	DF	Sum of square	Mean square	*F*	*P*	Contribution
1	Lamp current (*I*)	2	0.1224	0.06119	0.56	0.642	0.020
2	Pulse frequency (*f*)	2	3.2030	1.60149	14.58	0.064	0.529
3	Air pressure (*p*)	2	2.5037	1.25183	11.40	0.081	0.413
4	Error	2	0.2196	0.10981		0.036	
5	Total	8	6.0486				

5 Conclusions

The Nd:YAG laser drilling of KFRP/BFRP hybrid fiber composite has been deliberated in this paper for obtaining the precise result on hole quality based on Taguchi methodology. The result has been analyzed using ANOVA and it has been concluded that the proposed method is applicable for predicting more accuracy in result. Other conclusions are summarized below:

1. The condition for Optimum drilling which related to maximum −1.595 S/N ratio value of the larger hole circularity value for drilling operation (311) was found to be 200 A for the lamp current, 20 Hz for the pulse frequency, and 8 kg/cm^2 for the air pressure.
2. By using Taguchi experimental design, the number of experiments was reduced in the same or similar area in drilling operations to determine optimum drilling conditions. More accurate results will be applicable in academic and industrial uses.
3. The most significant parameters are pulse frequency and air pressure as compared to lamp current affecting hole circularity of the material after drilling.

References

1. Lau WS, Lee WB (1995) Unconventional machining of composite materials. J Mater Process Technol 48:199–205
2. Riveiro QF, Lusquinos F, Comesana R, Pou J (2010) Parametric investigation of CO2 laser cutting of 2024-T3 alloy. J Mater Process Technol 210:1138–1152
3. Lau WS, Lee WB (1992) Pulsed Nd:YAG laser cutting of carbon fiber composite materials. Annals CIRP 39(1):179–182
4. Sivarao ATJS, Ammar S (2010) DOE based statistical approach in modeling of laser processing—review and suggestion. Int J Eng Technol 10:4
5. Montgomery DC (1997) Design and analysis of experiments, 4th edn. Wiley, New York
6. Choudhury IA, Shirley S (2010) Laser cutting of polymeric materials: an experimental investigation. Opt Laser Technol 42:503–508
7. Abrate S (1997) Machining of composite materials. In: Composites engineering handbook (A 9811526 01-24), vol 11. Marcel Dekker, New York, pp 777–810
8. Mathew J, Goswami G, Ramakrishnan N, Naik N (1999) Parametric studies on pulsed Nd: YAG laser cutting of carbon fibre reinforced plastic composites. J Mater Process Technol 198–203
9. Hocheng H, Tsao CC (2003) Comprehensive analysis of delamination in drilling of composite materials with various drill bits. J Mater Process Technol 140(1):335–339
10. Tsao CC, Hocheng H (2004) Taguchi analysis of delamination associated with various drill bits in drilling of composite material. Int J Mach Tool Manuf 44(10):1085–1090
11. Davim JP, Rubio JC, Abrao AM (2007) A novel approach based on digital image analysis to evaluate the delamination factor after drilling composite laminates. Compos Sci Technol 67(9):1939–1945
12. Gaugel S, Sripathy P, Haeger A, Meinhard D, Bernthaler T, Lissek F et al (2016) A comparative study on tool wears and laminate damage in drilling of carbon-fiber reinforced polymers (CFRP). Compos Struct 155:173–183

13. Solati A, Hamedi M, Safarabadi M (2019) Combined GA-ANN approach for prediction of HAZ and bearing strength in laser drilling of GFRP composite. Opt Laser Technol 113:104–115
14. Gautam GD, Pandey AK (2018) Teaching learning algorithm based optimization of kerf deviations in pulsed Nd:YAG laser cutting of Kevlar-29 composite laminates. Infrared Phys Technol 89:203–217
15. Gautam GD, Mishra DR (2019) Firefly algorithm based optimization of kerf quality characteristics in pulsed Nd:YAG laser cutting of basalt fiber reinforced composite. Compos Part B 176(107340):1–75
16. Gautam GD, Mishra DR (2019) Teaching learning algorithm based optimization of kerf deviations in pulsed Nd:YAG laser cutting of Kevlar-29 composite laminates. FME Trans 47:560–575
17. Gautam GD, Mishra DR (2019) Dimensional accuracy improvement by parametric optimization in pulsed Nd:YAG laser cutting of Kevlar-29/basalt fiber-reinforced hybrid composites. J Brazil Soc Mech Sci Eng 41(284):1–22
18. Mishra DR, Gautam GD, Prakash D, Bajaj A, Sharma A, Bisht R, Gupta S (2020) Optimization of kerf deviations in pulsed Nd:YAG laser cutting of hybrid composite laminate using GRA. FME Trans 48:109–116
19. Gautam GD, Pandey AK (2018) Pulsed Nd:YAG laser beam drilling: a review. Opt Laser Technol 100:183–215
20. Ross PJ (1988) Taguchi techniques for quality engineering. McGraw-Hill, New York

Flywheel Energy Application in Commercial and Agricultural Field: A Typical Review

Sagar D. Shelare, Ravinder Kumar, and Pravin B. Khope

1 Introduction

A system for storing energy assumes a significant role in today's world. An upcoming world without systems for storing energy is just like a computer device with no hard disc. There are several energy storage devices utilized in manufacturing like battery energy storage [1], thermal energy storage [2], air energy storage (compressed) [3], energy storage system with flywheel technology and flywheel.

Flywheel which is additionally called a mechanical battery [4], might be used as a primary system of energy storage and it is the oldest of all [5]. The flywheel is utilized in machines fills in like vault, which stores when energy supply is extra than required and releases it when the necessary energy is more prominent than a requirement. Flywheels restrict adjustment in rotational speed, which helps predictable turn of a shaft during intermittent torque is applied by the capacity source, or when the discontinuous force, like a piston pump, is set on it. The function of the engine flywheel is to make a smooth transmission to wheels. In order to ensure the constant output flywheel is attached with the same link of the shaft. Whenever a shaft receives power higher than the average value then it is received and stored by flywheel for a moment. Figure 1 shows the typical casting of different shapes of a flywheel.

Energy stored is corresponding to the mass and rotational velocity square of a flywheel. This can be comprehended by equations of energy (E) and moment of inertia (I) as in Eqs. (1) and (2), respectively, [6],

$$E = \frac{1}{2}\left(I \times \omega^2\right) \quad (1)$$

S. D. Shelare · R. Kumar (✉)
School of Mechanical Engineering, Lovely Professional University, Phagwara, Punjab, India

P. B. Khope
Department of Mechanical Engineering, Priyadarshini College of Engineering, Nagpur, Maharashtra, India

A. Parey et al. (eds.), *Recent Trends in Engineering Design*, Lecture Notes in Mechanical Engineering, https://doi.org/10.1007/978-981-16-1079-0_19

Fig. 1 The typical casting of different shapes of a flywheel

$$I = \left(r^2 \times m \times h\right)/2 \tag{2}$$

where r radius; m mass and h height of the flywheel.

The flywheel is generally attached at one end of the shaft in order to have uniform torque. They store energy when it is in excess and releases it when desired. Hence a flywheel can be considered as a reservoir of energy, which gives energy at the desired time. A flywheel is needed in two types of machines, one in which the operation is intermittent. In this type of machine, the flywheel absorbs energy from a power source during the ideal period and delivers a large amount of stored energy in a very short useful portion of the cycle. The machines of this type include press, riveting machines, etc. So it is obvious that in these machines, a large amount of energy is required during the working period for performing the operation. Without Flywheel, one should be compiled to install a motor at very high power, but during the rest of the cycle, when no energy is required, the motor will be running at no load. So by using a flywheel, a smaller power unit can be used and large power can be obtained for a shorter period. In the second type of machine, the necessity of flywheel is realized where torque generated is in the form of peaks and troughs (non-uniform torque), as in the case of I.C engines, steam engines, etc. Here flywheel smoothens out the speed fluctuations caused by the non-uniform flow of power from piston during the working stroke.

When the shaft receives the varying output the whole mechanical system becomes unstable and shaft will unable to work. To ensure the constant output flywheel is attached with the same link of shaft. Whenever a shaft receives power higher than the average value then it is received and stored by flywheel for a moment. This power is disbursed when shaft receives powerless than the average power. Usually, the power, energy, and velocity all are taken analogously with each other. Flywheel absorbs kinetic energy through its inertial mean. The storing power of kinetic energy depends upon density, volume, and distribution of mass about the axis. Many papers are reviewed in this with conventional and software methods. Due to a number of models of engine and their performance in between cycles the flywheel designed

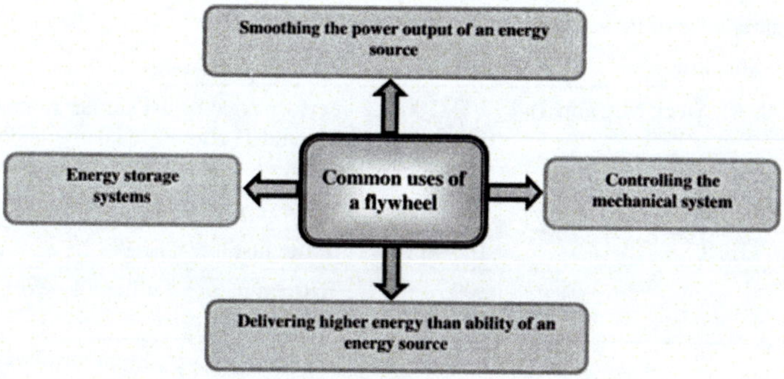

Fig. 2 Common uses of a flywheel

today in such a way so that the total storage capacity of flywheel can be increased for given constraints. In modern days the compactness is also one of the basic needs of life that enhances the system appearance.

Flywheel has its uses in most of the areas. Figure 2 shows the common uses of a flywheel.

2 Flywheel Materials and Shapes

Figure 3 shows the typical cross section of the flywheel. In this figure Do, D, and d are the outside, mean, and internal diameter of the rim respectively. Also, a and b are sections of the arm and width of the flywheel.

Flywheels are made using a wide variety of materials; based on the applications. C.I. flywheels are utilized into older steam engines. Flywheels utilized in automobiles are produced by C.I., st., or al. Flywheels made from composite or high-quality steel were designed in anticipation of their use in systems for motor energy storage

Fig. 3 Typical cross section of the flywheel [6]

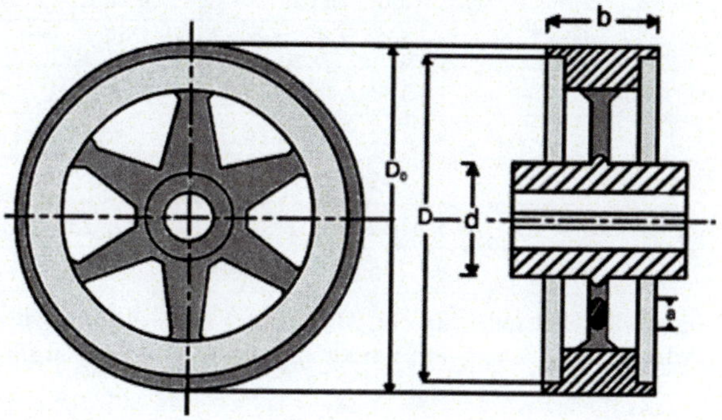

Table 1 Significance of flywheel materials

Material specification	M in kJ/kg	Remarks
High strength Al, Steel, Mg, and Ti Alloy	100–200	Performance is just similar to one another. Mg and Ti alloys are costly than steel and Al alloys
Composites: CFRP	200–500	Work very well
Composites: GFRP	100–400	Better just like CFRP
Beryllium	300	Toxic in nature. Difficult to work. Expensive in price
Cast Iron	8–10	Higher density makes it's suitable for limited velocity and strength
Lead Alloy	3	

and stoppage. high-quality steel or composite has been planned in favor of its utilization into motor energy stockpiling and stopping mechanisms. The proficiency of the flywheel is controlled by the most extreme measure of energy store over unit weight. Stored energy increases with flywheel's rotational speed; notwithstanding, the stresses likewise increment.

Various materials used to make the flywheel along with their significance/drawbacks are shown in Table 1

The alloy steel, carbon steel 1065, maraging steel, and stainless steel are mostly used to make a flywheel. Within the upcoming year, increasingly, appropriate characteristics and materials will be picked to investigate that will give progressively exact outcomes and a similar strategy of materials for different uses.

Table 2 shows calculated values for various physical values of the flywheel required for storing 250 J.

The various shapes of the available materials for the flywheel are as shown in Table 3. Disc of lavals and solid disc is made up of metals while composites are preferred to make a thick ring and thin ring. Value of shape constant (K) for disc of lavals, solid disc, thick ring, and thin ring are 1, 0.606, 0.305, and 0.50, respectively.

Table 2 Various physical values of the flywheel for storing 250 J

Material	R	M	ω	ϱ	η
Al. Alloy	153	3.3	2406	0.0421	151,515
C. I	104	16.6	1465	0.0084	30,121
Composite: CFRP	195	1.0	3382	0.1389	500,000
Composite: GFRP	150	3.8	2323	0.0365	131,579
Maraging steel	145	4.4	2218	0.0316	113,636

Note R flywheel radius in cm, M flywheel mass in gm, ω flywheel angular velocity in rpm, ϱ flywheel energy density in kWh/kg, η flywheel efficiency in J/kg

Table 3 Various shapes of flywheel

Material	Shape type	Shape constant value	Shape
Metal	Disc of lavals	1.00	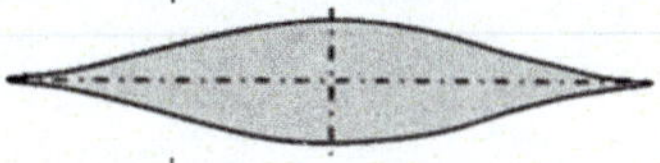
Metal	Solid disc	0.606	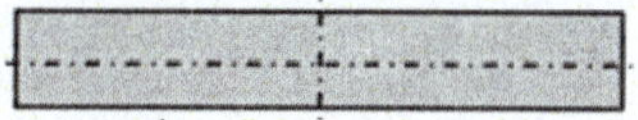
Composite	Thick ring	0.305	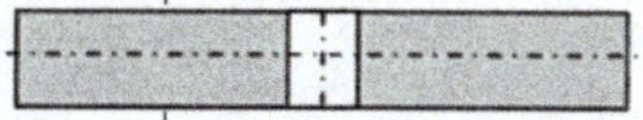
Composite	Thin ring	0.50	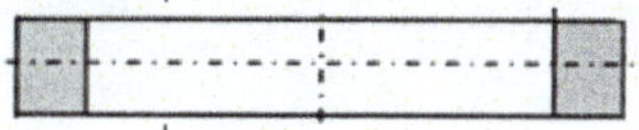

3 Recent Advances in a Flywheel

Ha et al. considered flywheel plans comprising numerous composite shells combined with obstruction fit so as to pre-stress a composite into pressure. A group found ideal measurements for such a structure [7] and found that the subsequent plans could be scaled [8, 9]. They additionally created and displayed plans for split-type hubs [10] and composite hubs [11], the last of which has additionally been the subject of several patents [12]. Authors additionally created and displayed plans for split-type center points [8] and composite center points [11], the last of which has additionally been the subject of a few licenses [12]. These investigations have been joined by the assembling and testing of the subsequent plans [13], approving the discoveries.

Flywheel supportive networks have likewise observed critical advances. Flywheels are generally enclosed inside the vacuum chamber to diminish aerodynamic energy misfortunes [14]. Magnetic bearings have been used on fixed flywheels and propose altogether diminished friction on customary rolling bearings [15, 16].

Zakiuddin et al. developed chaff cutter which works on HPFM. Author also established generalized models for the chaff cutter operated by HPFM for responses [17]. Wagh et al. have developed a novel Flour Kneading Machine with HPFM. This machine found its suitability in pantry, educational institutes, hall kitchen, and several industries [18]. Shelare et al. have developed Charoli nut desheller which works on HPFM. Author also formulated the mathematical model for the responses [20].

4 Agricultural Application of Flywheel

Several researchers used flywheel for the storage of human energy. Dr. J. P. Modak and his team have developed several machines for agricultural mechanization. Machines which was also called a processing unit works on the pedaling unit which mainly comprises of bicycle unit and flywheel. In this pedaling unit, flywheel stores the energy which has generated by the peddler and utilizes the same for the working of various agricultural machines. The various authors like Bhatkulkar and team, Dhale and team, Ganorkar and team, Ghuge and team, Khope and team, Khope and team, Khope and team, Moghe and team, Nimbarte and team, Sonawane and team, Sonde and team, Undirwade and team, and Waghmare and team have been developed the process units like Nursery Fertilizer Mixer, Oil press, Water Pump, Novel Gearbox, Chaff cutting machine, Forge Cutter, Ice cream Making Machine, Turmeric Polishing Machine, Garlic Peeling Machine, Sugar Cane Sprout Cutter Machine, Wood Chipper Cutter, Sliver cutting machine, and Stirrup Making Machine, respectively.

5 Results and Discussion

The various flywheel-assisted machines for agricultural purpose are given below in Table 4.

The above machine found cost-effective, can be used for human-operated process machines having intermitted working. The efficiency of the machine finds higher at low input.

6 Conclusions

The energy storing system has the prospective to employ a key role in facilitating the addition of renewable resources. Numerous advantages can be gained by utilizing efficient energy storage. This paper provides a review of storage devices especially mechanical energy with the help of flywheel. The total storage capacity of flywheel gives uniformity in speed at changing loads. This factor has been considered by many authors. Energy storage is the main scope for a flywheel. Various factors affected the performance of the flywheel like material and shapes are studied. Many machines have been developed by many researchers for the agricultural purpose with a flywheel and found efficient and reliable for work.

This paper gave an up-to-date technical review of the flywheel in different commercial and agricultural processes. Till date no thresher machine is developed with flywheel, so there is a scope of development as most of the machines with flywheel working safely.

Table 4 Various agricultural applications of flywheel

Agricultural machine	Specification/significance/limitations	References
Chaff cutting machine	Developed chaff cutting machine to overcome uncertainly of electricity. The developed machine is worked with the help of HPFM. The machine is proper for marginal farmers having 2–3 cattle	[19]
Chaff cutter	Developed chaff cutter which works on HPFM. Established generalized models for the chaff cutter operated by HPFM for responses	[17]
Charoli nut desheller	Developed mathematical model of Charoli nut desheller	[20]
Flour Kneading Machine	Developed a novel Flour Kneading Machine with HPFM. This machine found its suitability in pantry, educational institutes, hall kitchen, and several industries	[18]
Forge Cutter	A newer cutter is made up which different than other fodder cutters that will work on the non-conventional energy source. The machine runs with the help of flywheel energy	[21]
Garlic Peeling Machine	Developed Garlic Peeling Machine assisted by a flywheel. The machine is cost-effective and useful for human-operated machines having intermitted working	[22]
Ice cream Making Machine	Developed human-powered Ice cream making machine assisted by a flywheel. In machine wooden having a mixture of ice and salt, where ice cream ingredients are stirred and blended with flywheel energy	[23]
Novel Gearbox	Developed a novel gearbox for the HPFM. Gearbox runs smoothly with the help of flywheel	[24]
Nursery Fertilizer Mixer	Developed human-powered nursery fertilizer mixer assisted by flywheel (HPFM). The machine is economically feasible, can be used by a normal worker and it minimizes the time	[25]
Oil press	Developed an approximate generalized data-based model by varying independent variables for an oil press. The recognized database will be helpful for lower to medium output oil press	[26]

(continued)

Table 4 (continued)

Agricultural machine	Specification/significance/limitations	References
Oil expeller	Developed Oil expeller which works on HPFM. Established generalized models for the Oil expeller phenomena operated by HPFM for responses	[27]
Sliver cutting machine	Designed and developed a sliver cutting machine assisted by a flywheel. The operation of the machine is successful and highly reliable	[28–30]
Stirrup Making Machine	Developed a human-powered flywheel motor for stirrup making machine. The machine is cost-effective and useful for human-operated machines having intermitted working. The efficiency of the machine finds higher at low input	[31, 32]
Sugar Cane Sprout Cutter Machine	The machine is useful for cutting a sugarcane bud that is helpful for sugarcane sown. The machine reduces economic loss and human effort	[33]
Turmeric Polishing mc	Designed and developed a turmeric polishing Machine. Machine work well with flywheel energy	[34]
Water Pump	Developed human-powered water pump assisted by a flywheel. The machine can lift a moderate amount of water at a considerable height	[35]
Wood Chipper Cutter	Developed wood chipper cutter assisted by a flywheel. The machine is cost-effective and useful for human-operated machines having intermitted working. Also experimental plan was established to measure human energy	[36]

References

1. Mousavi GSM, Nikdel M (2014) Various battery models for various simulation studies and applications. Renew Sustain Energy Rev 32:477–485. https://doi.org/10.1016/j.rser.2014.01.048
2. Li G (2015) Energy and exergy performance assessments for latent heat thermal energy storage systems. Renew Sustain Energy Rev 51:926–954. https://doi.org/10.1016/j.rser.2015.06.052
3. Satkin M, Noorollahi Y, Abbaspour M, Yousefi H (2014) Multi criteria site selection model for wind-compressed air energy storage power plants in Iran. Renew Sustain Energy Rev 32:579–590. https://doi.org/10.1016/j.rser.2014.01.054
4. Bitterly JG (1998) Flywheel technology: past, present and 21th century projections. IEEE AES Syst Manag 13(8):13–16
5. Strzelecki R, Benysek G (2008) Power electronics in smart electrical energy networks. Springer Press, London

6. https://mechanical-design-handbook.blogspot.com/2007/12/flywheels.html. Accessed: 30 July 2020
7. Kyu Ha S, Kim DJ, Sung TH (2001) Optimum design of multi-ring composite flywheel rotor using a modified generalized plane strain assumption. Int J Mech Sci 43(4):993–1007. https://doi.org/10.1016/S0020-7403(00)00047-3
8. Ha SK, Han HH, Han YH (2008) Design and manufacture of a composite flywheel press-fit multi-rim rotor. J Reinf Plast Compos 27(9):953–965. https://doi.org/10.1177/0731684407086625
9. Van Rensburg J, Petrus J, Groenwold AA, Wood DW (2013) Optimization of cylindrical composite flywheel rotors for energy storage. Struct Multidiscip Optim 47(1):135–147
10. Ha SK, Kim MH, Han SC, Sung TH (2006) Design and spin test of a hybrid composite flywheel rotor with a split type hub. J Compos Mater 40(23):2113–2130. https://doi.org/10.1177/0021998306061324
11. Kim SJ, Hayat K, Nasir SU, Sung KH (2014) Design and fabrication of hybrid composite hubs for a multi-rim flywheel energy storage system. Compos Struct 107(1):19–29. https://doi.org/10.1016/j.compstruct.2013.07.032
12. Bakholdin D, Bosley RW, Rosen HA (1996) Flywheel rotor with conical hub and methods of manufacture therefore. U.S. Patent 5,566,588, issue 1, pp 1–5
13. Hayat K, Kim SJ, Lee YH, Kwon JD, Kim KT, Hockney D, Arseneaux J, Jung SY, Ha SK (2011) Design, fabrication and testing of a hybrid composite flywheel rotor and hub. In: ICCM international conferences on composite materials, pp 1–6
14. Jian L, Chau KT, Yu Gong JZ, Jiang CY, Li W (2009) Comparison of coaxial magnetic gears with different topologies. IEEE Trans Magn 45(10):4526–4529. https://doi.org/10.1109/TMAG.2009.2021662
15. Bleuler H (2009) Magnetic bearings: theory, design, and application to rotating machinery. Springer Science and Business Media
16. Strasik M, Hull JR, Mittleider JA, Gonder JF, Johnson PE, McCrary KE, McIver CR (2010) An overview of boeing flywheel energy storage systems with high-temperature superconducting bearings. Superconductor Sci Technol 23(3). https://doi.org/10.1088/0953-2048/23/3/034021
17. Zakiuddin KS, Singh MP, Modak JP (2019) Mathematical modeling and simulation of chaff cutter energized by human powered flywheel motar. In: Uhl T (ed) Advances in mechanism and machine science. In: IFToMM WC 2019. Mechanisms and machine science, vol 73. Springer, Cham. https://doi.org/10.1007/978-3-030-20131-9_350
18. Sanket W, Bawiskar HS, Singh R, Dhumal S, Kardile C (2019) Design of flour kneading machine by using human powered flywheel motor (HPFM). Trends Mach Des 6(1):19–24. https://doi.org/10.37591/tmd.v6i1.2635
19. Khope PB (2013) Establishing empirical relationship to predict the chaff cutting phenomenon energized by human powered flywheel motor (Hpfm). IOSR J Agric Veterin Sci 3(2):35–39. https://doi.org/10.9790/2380-0323539
20. Shelare SD, Kumar R, Khope PB (2021) Formulation of a mathematical model for quantity of deshelled nut in charoli nut deshelling machine. In: Prakash C, Krolczyk G, Singh S, Pramanik A (eds) Advances in metrology and measurement of engineering surfaces. Lecture notes in mechanical engineering. Springer, Singapore
21. Khope PB, Modak JP (2013) Development and performance evaluation of a human powered flywheel motor operated forge cutter. Int J Sci Technol Res 2(3):146–149
22. Dhande HK, Shelare SD, Khope PB (2020) Developing a mixed solar drier for improved postharvest handling of food grains. Agric Eng Int CIGR J 22(4):17–27
23. Ramawat RB, Khope PB, Choudhary PS (2014) Design and performance evaluation of pedal operated ice-cream making machine. Int J Eng Res Technol 3(2278–0181):1780–1783
24. Ghuge VD, Modak JP (2014) Experimental data based model for time to exhaust flywheel energy in a human powered flywheel motor driven system having a novel gearbox. Int J Res Eng Adv Technol 1(6):1–5
25. Bhatkulkar HS, Modak JP (2014) Design and development of nursery fertilizer mixer energized by human powered flywheel motor. Int J Res Emerg Sci Technol 1(5):69–73

26. Dhale AD, Modak JP (2012) Formulation of experimental data based model for oil press using human powered flywheel motor as energy source. Agric Eng Int CIGR J 14(3):218–229
27. Sheikh SM, Zakiuddin KS (2019) History of human powered oil expeller: a literature review. In: Zhang B, Ceccarelli M (eds) Explorations in the history and heritage of machines and mechanisms. History of mechanism and machine science, vol 37. Springer, Cham. https://doi.org/10.1007/978-3-030-03538-9_7
28. Undirwade SK, Singh MP, Sakhale CN (2015) Experimental investigation of processing time, number of slivers and resistive torque required for human powered bamboo sliver cutting operation. J Bamboo Rattan 14(1–4):33–51
29. Jawalekar SB, Shelare SD (2020) Development and performance analysis of low cost combined harvester for rabi crops. Agric Eng Int CIGR J 22(1):197–201
30. Undirwade SK (2018) Development and optimization of experimental data based models for bamboo sliver cutting by using human powered flywheel motor. Int J Mech Prod Eng Res Dev (IJMPERD) 8(1):1007–1020. https://doi.org/10.24247/ijmperdfeb2018121
31. Waghmare SN, Sakhale CN, Tembhurkar CK, Shelare SD (2020) Assessment of average resistive torque for human-powered stirrup making process. In: Iyer B, Deshpande P, Sharma S, Shiurkar U (eds) Computing in engineering and technology. Advances in intelligent systems and computing, vol 1025. Springer, Singapore
32. Waghmare SN, Shelare SD, Tembhurkar CK, Jawalekar SB (2021) Development of a model for the number of bends during stirrup making process. In: Prakash C, Krolczyk G, Singh S, Pramanik A (eds) Advances in metrology and measurement of engineering surfaces. Lecture notes in mechanical engineering. Springer, Singapore
33. Sonawane RA, Waybhase JR, Yengupatla RB, Shumail S (2017) Design and development of sugar cane sprout cutter machine by human powered flywheel motor concept. Int Res J Eng Technol (IRJET) 4(1):119–124
34. Moghe SM, Zakiuddin KS (2013) Design and development of turmeric polishing machine energized by human power flywheel motor—a past review. In: 1st international and 16th national conference on machines and mechanisms, INaCoMM, pp 920–923
35. Ganorkar AP, Zakiuddin KS, Hussain HA (2012) Development of pedal operated water pump. J Jpn Soc Pediatric Surg 5(2):359. https://doi.org/10.11164/jjsps.5.2_359_3
36. Sonde VM, Shirpurkar PP, Giripunje MS, Ashtankar PP (2021) Experimental and dimensional analysis approach for human energy required in wood chipping process. In: Hassanien A, Bhatnagar R, Darwish A (eds) Advanced machine learning technologies and applications. AMLTA 2020. Advances in intelligent systems and computing, vol 1141. Springer, Singapore. https://doi.org/10.1007/978-981-15-3383-9_61

Cyclostationary Analysis-Based Gearbox Fault Diagnosis Under Varying Speed

Vikas Sharma, Anand Parey, and Ram Bihari Sharma

1 Introduction

Gearbox of a wind turbine operates at uncertain wind speed due to uncontrollable and harsh environment conditions. Signal-processing techniques and condition indicators (CIs) are the backbone of any fault detection algorithms developed for a gearbox. When a gearbox is operating, its health can be monitored continuously by frequency content explored using appropriate signal processing technique [1]. Ignorance of any fault in the gears of a gearbox could possibly lead to a catastrophe. Various signal processing techniques consisting of time domain [2], frequency domain [3–5], and time–frequency domain [6] along with CIs have been used widely to state the health of gearbox [4, 6, 7]. These techniques process the gearbox signals (noise, structural vibrations, and gear vibrations) acquired by the transducers. They measure the energy of the signal statistically and quantify the level of vibrations generated due to fault.

Time synchronous averaging (TSA), a widely explored technique, provides an accurate estimate of the gear tooth meshing vibration signal under constant speed conditions [8]. To diagnose gear fault at the incipient stage, kurtosis was applied to TSA signals [9]. TSA can be used for tacho supported applications only. When a gear meshes with another gear, amplitude and phase modulation occurs due to the presence of pitch errors, eccentricity, wear, and non-uniform load. Consequently, these modulations appear as sidebands around the gear mesh frequency (GMF) and its harmonics in the frequency spectrum. For an incipient fault, the sideband activity

V. Sharma (✉)
Department of Mechanical Mechatronics Engineering, The LNM Institute of Information Technology, Jaipur, India

A. Parey
Discipline of Mechanical Engineering, Indian Institute of Technology Indore, Indore, India

R. B. Sharma
Automobile Engineering Department, RJIT, BSF Academy, Tekanpur, Gwalior, Madhya Pradesh, India

A. Parey et al. (eds.), *Recent Trends in Engineering Design*, Lecture Notes in Mechanical Engineering, https://doi.org/10.1007/978-981-16-1079-0_20

is weak and masked by noise components, making it difficult to explore them using Fourier spectrum alone. Spectral kurtosis (SK) is a measure of energy in frequency domain, used for detection of early faults in presence of heavy overlapping noise [10].

Further, if there is periodic variation of nth order moment or the statistical moments of a non-stationary signal varies with time, and then such non-stationary vibration signals of gearbox are treated as cyclostationary signals [11, 12]. On evaluating cyclic cumulants and cyclic spectrum, cyclostationary analysis detects modulation caused due to faults. CIs based on cyclostationarity signal can detect fault effectively [13].

Most of the fault diagnosis methods were developed under either constant speed or speed specific conditions, rely on the assumption of hypothetical data, computational complexity, and signal stationarity. Conversely, gearboxes often operate under varying conditions, and their vibration signals are multi-component, modulated, and non-stationary. To the best of our knowledge, the research on a gearbox under fully varying speed is very limited. The main contributions of this work are

1. A novel analysis exhibiting a effectiveness of fourth-order of cyclostationary analysis, thereby detecting fault transients caused by meshing of faulty gears;
2. The speed variation performed to examine the performance of the cyclostationary analysis resembles the real-time speed variation (i.e., run-up, random fluctuation, and coast-down), not yet studied.

The manuscript reviews the theoretical background of all the signal processing methods followed by experimental investigation. The experimental findings, discussion about the diagnostics capability of the each method are also reported.

2 Theoretical Background

2.1 *Time-Domain technique- TSA*

To examine the occurrence of faults, TSA extracts the gear tooth meshing vibrations thereby attenuating the non-synchronous components present in the gearbox vibration signal. It enables the average signal to exhibit the meshing of teeth over an entire revolution. [8]. TSA for a raw vibration signal $y(t)$, is expressed as [8]

$$\overline{y}(t) = \left(\frac{1}{N}\right)\sum_{n=0}^{N-1} y(t - nT) \tag{1}$$

where T is shaft rotation period and N is number of averages. TSA segments the vibration signal synchronous with shaft rotation, gear rotation, etc. to perform the ensemble averaging to extract the signal of interest. After extracting the ensemble averaged signal, kurtosis can directly be applied to comment about the gear health.

2.2 Kurtosis

It shows the peakedness of probability density function (PDF) of a signal [14]. For a healthy gear vibration signal, kurtosis is approximately 3 [15]. It is defined as fourth-order moment normalized by the square of variance of a signal. It is given by

$$K = \frac{N \sum_{i=1}^{N} \left(Y_i - \overline{Y}\right)^4}{\left(\sum_{i=1}^{N} \left(Y_i - \overline{Y}\right)^2\right)^2} \tag{1}$$

2.3 Frequency Domain technique- Spectral Analysis

Measurement in frequency domain states the amplitude of different frequencies present in time-domain signal. Using discrete Fourier transform (DFT) various components, i.e., GMF, harmonics of GMF, sidebands, shaft rotation frequency, etc. can be observed. For healthy gearbox running at constant speed, the GMF dominates the frequency spectrum. Condition of gear health can be evaluated by comparing the healthy spectrum to the faulty gear spectrum. With the increasing concern of early gear fault diagnosis under heavy noise, SK has been used, as energy normalized fourth-order moment of the real part of the DFT. It detects non-Gaussian components generated due to gear fault. Antoni et al. [10, 16, 17] introduced the SK technique into mechanical fault diagnosis. The SK of a signal $y(t)$ can be evaluated as [17]

$$K_y(f) = \frac{C_{4Y}(f)}{S_{2Y}^2(f)} = \frac{S_{4Y}(f)}{S_{2Y}^2(f)} - 2, \qquad f \neq 0 \tag{2}$$

where $C_{4Y}(f) = S_{4Y}(f) - 2S_{2Y}^2(f)$, $f \neq 0$ is fourth-order spectral cumulant, $S_{2Y}^2(f)$ is the power spectral density of $y(t)$. The energy measurement of the complex envelop $H(t, f)$ is given by second-order instantaneous moment: $S_{2Y}(f) = \langle H|(t, f)|^2 \rangle$. Based on SK, Kurtogram a 2-D map, is used for fault diagnosis, exhibiting the values of SK for frequency and bandwidth [17].

2.4 Cyclostationarity-Based Kurtosis

Cyclostationary signals are defined by the periodicity of moments. When the first-order moment of the signal $y(t)$, i.e., mean $m_Y(t)$ is periodic with period T, then it is said to be a first-order cyclostationary (CS1) signal, expressed as

$$m_Y(t) = E\{y(t)\} = m_Y(t+T) \tag{3}$$

where $E\{\cdot\}$ implies ensemble average. The signals which exhibit periodicity in auto-correlation function (ACF) are termed as a signal with second-order cyclostationarity (CS2) [12, 13]. Either amplitude modulation or frequency modulation or both appearing in stochastic process are examples of CS2 signals. ACF of a signal $y(t)$ is a periodic function with period T and is expressed as

$$R_{YY}(t,\tau) = E\left\{y(t+\overline{\beta}\tau)y(t-\beta\tau)*\right\} = R_{YY}(t+T,\tau) \tag{4}$$

where τ is the time lag, $\beta + \overline{\beta} = 1$ and * refer to the conjugate. ACF is independent of t for 2nd order stationary signal such that $R_{YY}(t,\tau) = R_{YY}(\tau)$. Cyclic behavior is observed for a gearbox under operation which is due to the inherent parameters such as excitation by fault, speed variations, and load. It exhibits that such signals are fundamentally cyclostationary in angular domain [12]. Therefore, time-domain vibration signal can be resampled by an angular variable ϑ and a cyclic period Θ in the angular domain with angle lag as ϕ. Thus, a signal $y(t)$ is defined in angular domain as $y(\vartheta)$. Consequently both Eq. (4) and (5) can be rewritten as

$$m_Y(\vartheta) = E\{y(\vartheta)\} = m_Y(\vartheta + \Theta) \tag{5}$$

$$R_{YY}(\vartheta,\varphi) = E\left\{y\left(\vartheta + \overline{\beta}\varphi\right)y\left(\vartheta - \overline{\beta}\varphi\right)^*\right\} = R_{YY}(\vartheta + \Theta,\varphi) \tag{6}$$

Fundamentally, a gearbox vibration signal is a combination of various orders of cyclostationarity. Researchers have cited that vibrations of first-order periodicity are caused by the imbalanced and misalignment present in the rotating machines. Whereas modulated vibrations generated due to wear, friction force or impacts between teeth belongs to second-order periodicity [13, 18]. Thus, to detect and to state about the fault, these different orders have to be explored. Therefore, cyclic cumulants can be used to express the indicators of cyclostationarity (ICS_{IY} and ICS_{2Y}):

$$\text{ICS}_{1Y} = \frac{\sum_{\alpha\in\overline{A},\alpha\neq 0}\left|C_{1Y}^{\alpha}\right|^2}{\left|C_{2Y}(0)\right|^1} \tag{7}$$

$$\text{ICS}_{2Y} = \frac{\sum_{\alpha\in\overline{A},\alpha\neq 0}\left|C_{2Y}^{\alpha}(0)\right|^2}{\left|C_{2Y}^{0}(0)\right|^2} \tag{8}$$

where $\overline{A}$ is the set of interested cyclic frequencies which is a subset of A. These indicators are normalized by the energy of residual signal and therefore they are dimensionless. For a signal $y(t)$, skewness and kurtosis are defined as the energy normalized third and fourth-order cumulants respectively and can be evaluated as

$$\mathrm{ICS}_{nY}^{0} = \frac{C_{nY}^{0}(0)}{\left|C_{2Y}^{0}(0)\right|^{n/2}}, \quad n = 3, 4 \tag{9}$$

For evaluating cyclostationary-based kurtosis, $n = 4$ is considered. To summarize all the available spectral information, the cyclic cumulants at zero lag were used.

3 Experimental Setup

3.1 Test Rig

The experimental investigations were performed on a Drivetrain Diagnostic Simulator (Fig. 1), which utilizes a 2.237 kW, 3phase, 0–3000 rpm AC drive motor simulating a WT. The operational speed and load can be varied at the same time with the help of a computer. The vibration signals were acquired from single-stage gear set of 20° pressure angle, further specifications of gearbox are mentioned [4]. Gear tooth faults, i.e., cracked tooth, chipped tooth, and missing tooth (Fig. 1) have been considered. Experiments were performed for all the test gears at varying speed at no load, 40% load, i.e., 32 Nm and 80% load, i.e., 64 Nm, respectively. The profile of speed as shown in Fig. 2 is considered in accordance with fluctuating wind speed observed in wind turbines [19, 20]. Such varying speed profile was never analyzed. The vibration signals were acquired with a sampling rate of 12.5 kHz.

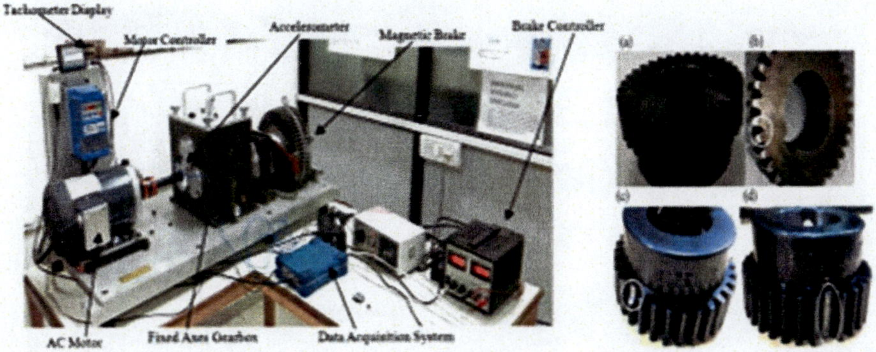

Fig. 1 Experimental setup and different gears **a** healthy gear; **b** cracked tooth gear; **c** chipped tooth; **d** missing tooth gear

Fig. 2 Profile of fluctuating speed

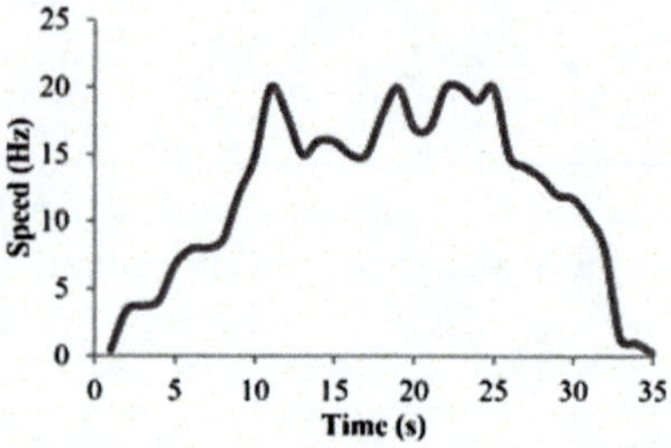

4 Results and Discussion

The vibration signals acquired for the gear tooth faults under for 40% and 80% load are shown in Fig. 3. On observing Fig. 3, it was noticed that the amplitude of time-domain signals for healthy, crack, and missing tooth fault at 40% load appears nearly same (Fig. 3a, b, d), whereas for chipped tooth the amplitude and the density of peaks was found increased comparatively (Fig. 3c). Thus, from Fig. 3, it becomes difficult to state the fault from raw vibration signals. Conversely, at 80% load, an increase in the amplitude of vibration signal was observed (Fig. 3). It implies that the impulses which were observed weak for missing tooth gear at 40% load were observed strong and visible at 80% load. However, the impressions of faults to distinguish from raw vibration signals and kurtosis (Table 1) were not clear. Both varying speed and

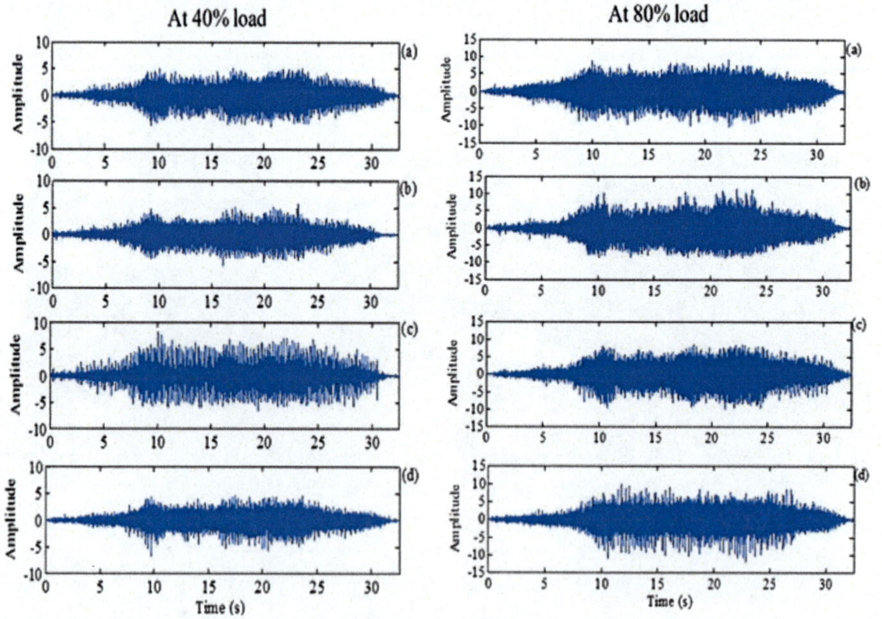

Fig. 3 Gear vibration signals, **a–d** corresponds to different gear health at different load

Table 1 Value of kurtosis for raw time-domain vibrations signals and TSA signals

Load (%)	Gear							
	Raw signal				TSA signal			
	Healthy gear	Crack tooth gear	Chipped tooth gear	Missing tooth gear	Healthy gear	Crack tooth gear	Chipped tooth gear	Missing tooth gear
0	6.0294	6.3093	9.1206	5.9104	4.9584	6.8745	5.9120	11.9496
40	6.2245	5.8465	8.7754	6.9715	5.1045	4.9876	5.9951	8.4513
80	6.8515	6.1514	9.5718	9.8898	4.7845	5.5978	5.4879	5.7978

localized tooth fault govern the raw vibration signal, as a result the variance of the signal is very high. Therefore, kurtosis of healthy gear vibration signal at different loads was found nearly 6. Further, at 40 and 80% load the value of kurtosis for crack tooth gear was found less than the value of healthy gear. Thus, gear faults cannot be detected from raw vibrations signals.

4.1 Analysis Using Time Synchronous Averaging and Kurtosis

TSA of raw vibration signal was performed, a well-proven in wind turbine cases [15]. For constant speed conditions, resampling approach followed by a simple linear interpolation, for equal number of samples, was adopted. The procedure of TSA has been adopted, i.e., resampling and ensemble averaging are performed. Figure 4 shows the TSA signals and it was observed that resampled TSA have a limited capability to detect fault at the aforementioned varying speed. The occurrence of transients was irregular as shown in Fig. 4 for chipped tooth and missing tooth; conversely, no transient was visible in case of cracked tooth. A possible reason could

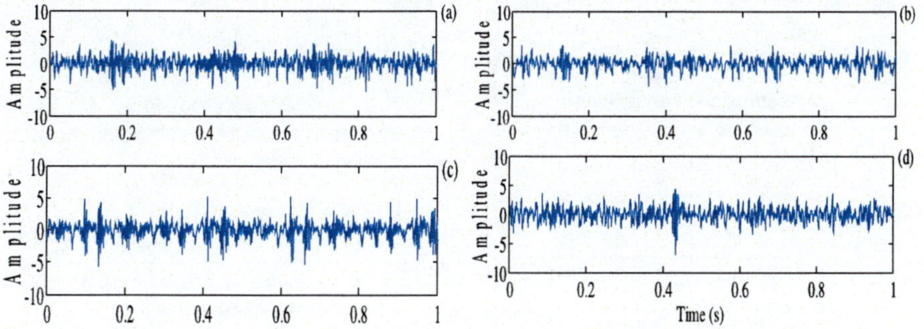

Fig. 4 TSA signals obtained under fluctuating speed for different gear health conditions at 40% load

be inseparability of transients towards the end of the rotation due to different phases, because of modulations. The kurtosis value calculated from the TSA signals and proved to be unreliable (Table 1). This proves that the TSA fails to diagnose fault at aforementioned varying speed. Kurtosis of both raw signals and TSA signals when calculated doesn't indicate the fault present in the signals for different gear health.

4.2 Analysis Using Spectral Kurtosis

The spectral smearing in frequency domain encourages the use of SK, which shows the kurtosis as a function of frequency. Figure 5 shows the power spectrum and the spectral kurtosis with respect to normalized frequency plotted for different gear health considered. The y-axis of power spectrum is amplitude, whereas x-axis has no unit. It exhibited the peaks observed in power spectrum of different gear health have a variation in their amplitude. The amplitude of peaks for both healthy and the

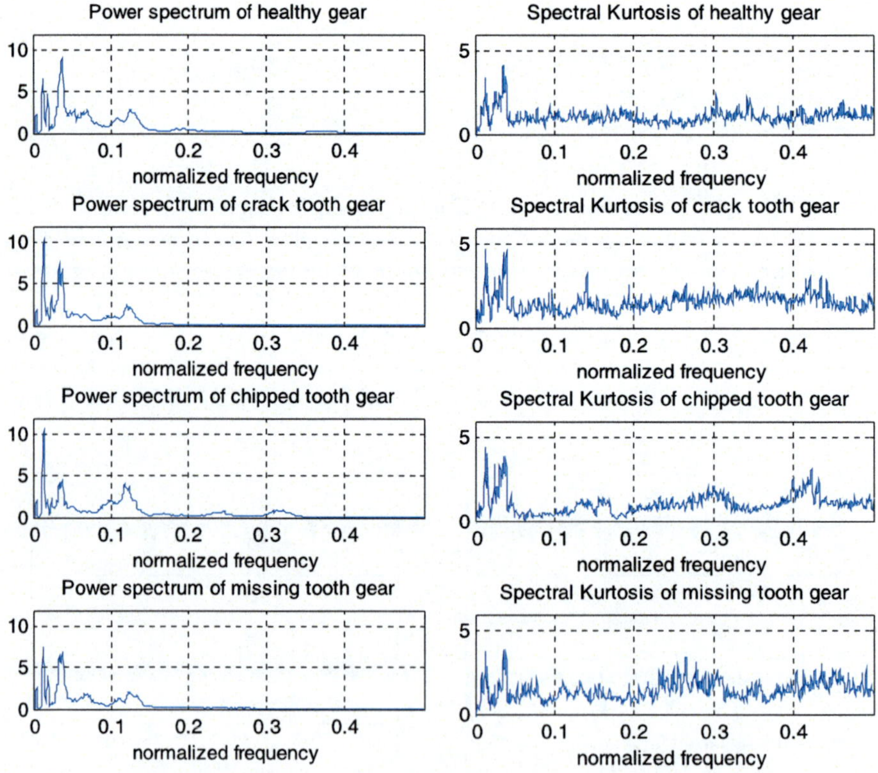

Fig. 5 Power spectrum and spectral kurtosis for different gear health at 40% load

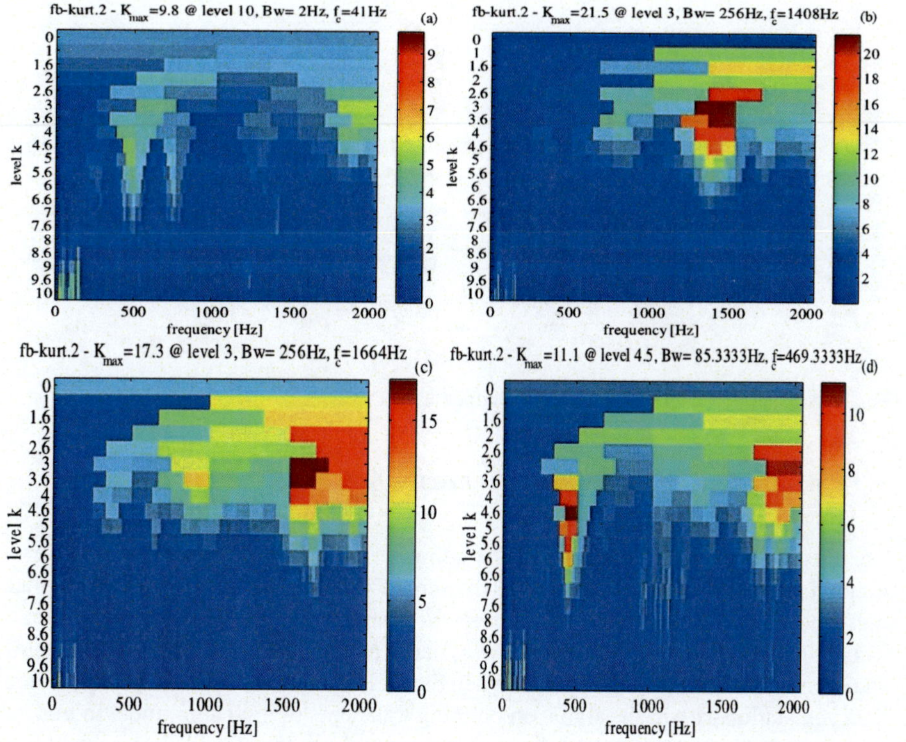

Fig. 6 Kurtogram is plot obtained from the waveform in Fig. 3 for different gear health

missing tooth gear is less than the amplitude of peaks for crack and chipped tooth gear.

At 32 Nm load the value of kurtosis for healthy gear was observed as 9.8, whereas for crack tooth, chipped tooth, and missing tooth is found as 21.5, 17.3, and 11.1, respectively. Further, it can be noticed from Fig. 6 that the level of faults observed was in a range of 3–5, but healthy gear reflects highest kurtosis at level 9.8 for 40% load. Thus, it can be said that both SK and power spectrum of the different gear health have reflected ambiguous results. A similar behavior of SK is observed at 80% load. This deviation of results reflects the insensitivity of SK towards the fault diagnosing mechanism of a gearbox.

Thus, at aforementioned varying speed profile both conventional spectral analysis and SK failed to reveal fault symptoms when the vibration signal is non-stationary.

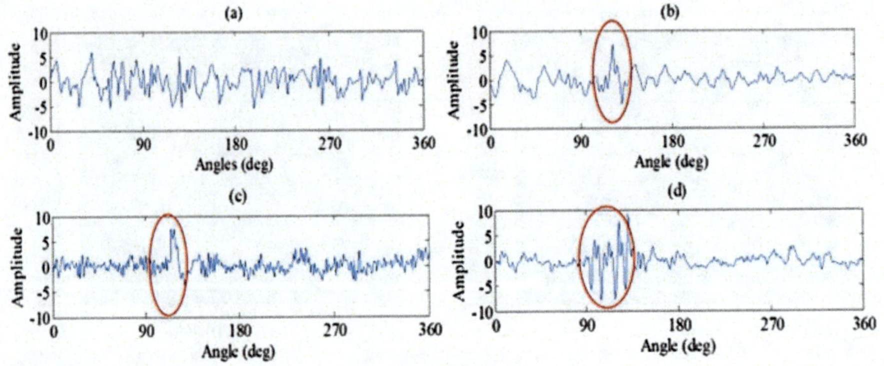

Fig. 7 Angular domain average vibration signals at 40% load

4.3 Analysis Using Cyclostationarity Analysis

The theory of cyclostationarity will be competent for fault detection when a time-domain signal is resampled in angular domain. Therefore, resampling in angular domain was performed in pursuit of uniform samples/revolutions using tachosignal. Thereafter, angular domain averaging was performed to identify the first-order cyclostationary present in the signal as shown in Fig. 7. It is the first step to find the presence the cyclostationarity in the signal on noticing transients of high amplitude. In Fig. 7a no transient impulse was noticed, correlating it to a healthy gear. Whereas the transients peaks were observed in Fig. 7b–d, when compared to each other. The transients observed in Fig. 7d have a larger duration in terms of angle as compared to Fig. 7a, which is possible indication of vibrations generated due to the impact of the missing tooth. Further, the residual signal derived by cyclic average vibration signals was used to calculate the fourth-order cumulants, i.e., cyclostationarity based kurtosis. In order to evaluate the cyclic cumulants, cyclic frequency and cyclic orders were calculated first. For the present study, the normalized cyclic frequency was evaluated as 0.03125 and cyclic orders are set up to a range of 0–70. From healthy gear to missing tooth gear, the amplitude of cyclic cumulants was found increasing as highlighted by encircling them for both 40% load and 80% load (Fig. 8). The peaks for cyclic cumulants appear at same orders, but their amplitude has gained twice at 80% load. The peaks for cracked tooth and chipping tooth were observed in between the cyclic order of 30–40. But for missing tooth the peaks were observed from the cyclic order 25–40.

Cyclostationarity-based kurtosis (ICS_{4Y}) has been calculated for each fault and its response was worth noting (Table 2), as the fault can be directly identified using a simple threshold of 3.5. ICS_{4Y} associated with the healthy gear was observed low, suggesting that no fault development on the gear. Whereas ICS_{4Y} exhibits a sudden increase of CS4 cyclic cumulant for gear crack. Although the speed of gearbox was varied, ICS_{4Y} shows significant impact on fault diagnosis.

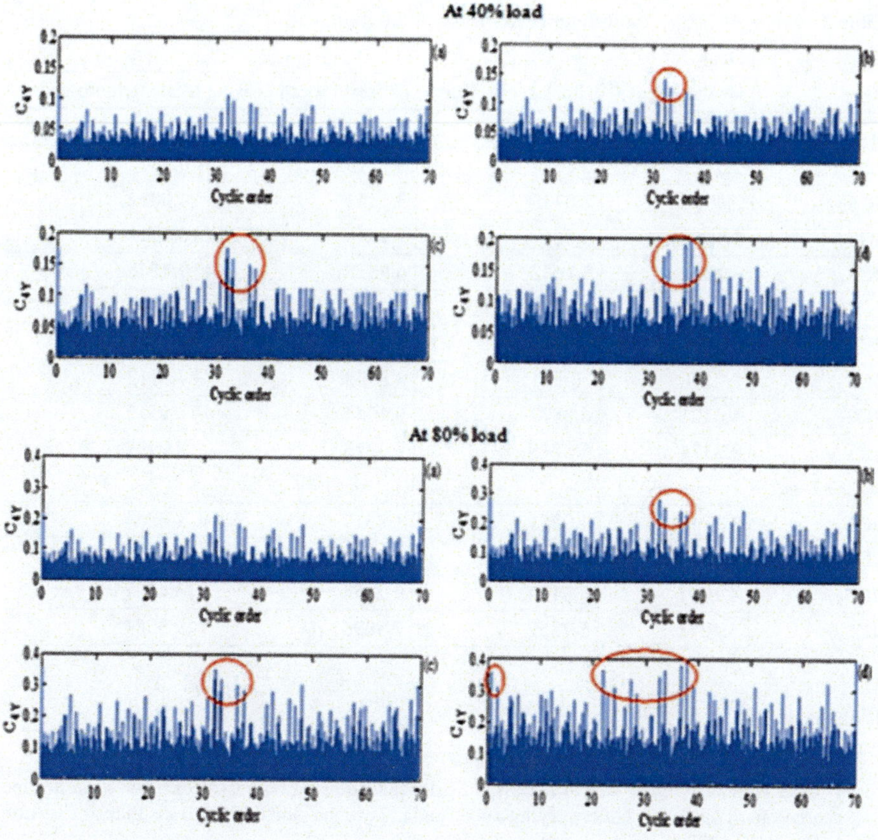

Fig. 8 Estimated fourth-order cyclic cumulant for different gear health angular domain vibration signals

5 Conclusion

Experimental investigation was conducted for gearbox fault diagnosis imitating wind speed variation. The resampled TSA was not capable extract fault transients under varying speed. Similarly, the SK was not found effective in noticing the fault symptoms excited by the gear faults. Conversely, cyclostationary analysis attenuated the non-synchronous components indicating the presence of fault symptoms by CS1 components. Further, CS4 cylic cumulant was found increased at the cyclic orders of gear tooth for both 40% load and 80% load. ICS_{4Y} supports the presence of localized faults too. Cyclostationary analysis completely outperforms the results of resampled TSA signal, spectral kurtosis—kurtogram and it could prove to be a robust approach.

Table 2 Values of $ICSn_Y$ for different gear health at different load

CIs	Gear			
	Healthy gear	Crack tooth gear	Chipped tooth gear	Missing tooth gear
a *No load*				
ICS_{1Y}	0.5121	0.6678	0.7815	0.8312
ICS_{2Y}	2.4751	2.8795	3.1354	4.0012
ICS_{3Y}	3.9557	5.6647	7.9945	10.4715
ICS_{4Y}	**3.2491**	**5.1642**	**6.9815**	**9.1548**
b *40% load*				
ICS_{1Y}	0.8874	0.9574	1.2157	1.5541
ICS_{2Y}	1.2198	3.0541	3.8745	4.5213
ICS_{3Y}	4.0121	6.8877	9.1044	12.9515
ICS_{4Y}	**3.3371**	**5.8745**	**7.0015**	**8.9995**
c *80% load*				
ICS_{1Y}	0.9748	1.0251	1.2517	1.6564
ICS_{2Y}	2.1145	3.1917	3.6615	4.5553
ICS_{3Y}	4.1001	7.0114	9.2255	13.4507
ICS_{4Y}	**3.4071**	**5.7857**	**7.8095**	**9.5501**

References

1. Wang J, Cheng F, Qiao W, Qu L (2018) Multiscale filtering reconstruction for wind turbine gearbox fault diagnosis under varying-speed and noisy conditions. IEEE Trans Industr Electron 65(5):4268–4278
2. Bechhoefer E, Qu Y, Zhu J, He D (2013) Signal processing techniques to improve an acoustic emission sensor. In: Annual conference of the prognostics and health management society, vol 4, 1–8
3. Barszcz T, Randall RB (2009) Application of spectral kurtosis for detection of a tooth crack in a planetary gear of a wind turbine. Mech Syst Sig Process 23:1352–1365
4. Sharma V, Parey A (2017) Frequency domain averaging based experimental evaluation of gear fault without tachometer for fluctuating speed conditions. Mech Syst Sig Process 85:278–295
5. Saidi L, Ali JB, Benbouzid M, Bechhoefer E (2016) The use of SESK as a trend parameter for localized bearing fault diagnosis in induction machines. ISA Trans 63:436–447
6. Parey A, Tandon N (2007) Impact velocity modeling and signal processing of spur gear vibration for the estimation of defect size. Mech Syst Sig Process 21:234–243
7. Sharma V, Parey A (2016) Gear crack detection using modified TSA and proposed fault indicators for fluctuating speed conditions. Measurement 90:560–575
8. Braun S (2011) The synchronous (time domain) average revisited. Mech Syst Sig Process 25:1087–1102
9. Ahamed N, Pandya Y, Parey A (2014) Spur gear tooth root crack detection using time synchronous averaging under fluctuating speed. Measurement 52:1–11
10. Antoni J, Randall RB (2006) The spectral kurtosis: application to the vibratory surveillance and diagnostics of rotating machines. Mech Syst Sig Process 20:308–331
11. Sharma V, Parey A (2017) Use of cyclostationarity based condition indicators for gear fault diagnosis under fluctuating speed condition. In: Chaari F, Leskow J, Napolitano A, Zimroz R,

Wylomanska A (eds) Cyclostationarity: theory and methods III. Applied condition monitoring, vol 6. Springer, Cham
12. Antoni J, Bonnardot F, Raad A, Badaoui M, El. (2004) Cyclostationary modelling of rotating machine vibration signals. Mech Syst Sig Process 18:1285–1314
13. Zhu ZK, Feng ZH, Kong FR (2005) Cyclostationarity analysis for gearbox condition monitoring: approaches and effectiveness. Mech Syst Sig Process 19:467–482
14. Sharma V, Parey A (2016) A review of gear fault diagnosis using various condition indicators. Procedia Eng 144:253–263
15. Samuel PD, Pines DJ (2005) A review of vibration-based techniques for helicopter transmission diagnostics. J Sound Vib 282:475–508
16. Antoni J (2006) The spectral kurtosis: a useful tool for characterizing non-stationary signals. Mech Syst Sig Process 20:282–307
17. Antoni J (2007) Fast computation of the Kurtogram for the detection of transient faults. Mech Syst Sig Process 21:108–124
18. Kidar T, Thomas M, El Badaoui M, Guilbault R (2013) Diagnosis of gear faults by cyclostationarity
19. https://drømstørre.dk/wp-content/wind/miller/windpower%20web/en/tour/wres/variab.htm
20. Wagner R, Antoniou I, Pedersen SM, Courtney MS, Jørgensen HE (2009) The influence of the wind speed profile on wind turbine performance measurements. Wind Energy Int J Progr Appl Wind Power Convers Technol 12(4):348–362

Control System for Hybrid Journal Bearing: A Fuzzy Approach

Anamika Yadav and Pooja Pathak

1 Introduction

Hybrid bearing has the function of hydrostatic and hydrodynamic bearing both. Hybrid bearing has an edge over poorly hydrostatic bearing by tolerating substantial load over and above the normal design load. These bearings can take heavy dynamic loading that has wide variation in the direction of rotation. The blank space in the bearing surfaces is avoided in hybrid bearing, so that it can maximize the hydrodynamic effect. The bearing has performed the required support to the shaft only if the proper lubrication is maintained in clearance space. The lubrication system of hybrid bearing works as human vein and oil works as blood. So, the researches have always thought about the advancement towards the lubrication for improving and hassle-free working of a turbine using hybrid bearing. Here author uses the concept of fuzzy logic for the prediction and controlling of oil pressure [1]. Authors suggested a decision-making tool for selection procedure of fluid film journal bearing [2]. In this work authors presented a mechatronic system called active hydrostatic journal bearing that is a need of fast-budding manufacturing environment to improve the dynamics and static characteristics of hydrostatic journal bearing [3]. In this work authors developed a neural-fuzzy system with the help of optimal membership functions to generate the input linguistic variables of a system [4]. The researchers developed fuzzy orthonormal-ARX adaptive fuzzy logic-structure feedback observer a fuzzy-based anomaly-diagnosis algorithm on Rolling-element bearings (REBs) that is used in several activities of non-linear rotating machines [5].

Fuzzy logic suggests the selection of type of bearing. Stability analyses for the hybrid bearing have been done by some rechargers. Different oil pump that is used in this lubrication system ensures the proper lubrication for the bearings. Types of

A. Yadav · P. Pathak (✉)
Department of Mathematics, IAH, GLA University, Mathura, Uttar Pradesh, India
e-mail: pooja.pathak@gla.ac.in

A. Parey et al. (eds.), *Recent Trends in Engineering Design*, Lecture Notes in Mechanical Engineering, https://doi.org/10.1007/978-981-16-1079-0_21

pumps that are mainly used are emergency oil pump, main oil pump, and auxiliary oil pump.

1.1 Main Oil Pump (MOP)

The main oil pump provides oil for all the requirements for all the journal bearings in the turbines. It generates high oil pressure. It is attached directly to the turbine shaft at either turbine or at generator end so, it operates with the speed of shaft of turbine. At the time of starting and slowdown of turbine, the main oil pump does not get proper number of rotation to generate required all pressure. At 90–95% of turbine speed, the main oil pump can generate required pressure.

1.2 Auxiliary Oil Pump (AOP)

It has following two functions:

1. It operates at the time of startup and shutdown [6] when the turbine does not have proper number of RPM for the main oil pump to provide eccentrically pressure.
2. It is a standby lube oil pump for the bearings when the main oil pump fails.

It has full flow and provides full pressure that will full fill the turbine and generators oil requirement. When the main oil pump fails, the auxiliary oil pump supply oil to all the bearings and controlled the full power operation.

It operates automatically through a pressure switch that is fitted in the discharge pipe of main oil pump. The pump powered by a alternative current electric motors of fourth class.

1.3 Emergency Oil Pump (EOP)

When the turbine slows down from operating speed to stopping condition, it takes 20–40 min for this complete process. At this time turbine could not provide sufficient electric power to the AOP that will result in reduce oil pressure. Consequently, bearings will damage Due to overheating. This loss is protected by providing sufficient number of alternate power supply lube oil pumps that will insure the lube oil flow.

Emergency oil pump does not provide full flow and full pressure. It is a pump that works only for some emergency time. It cannot fulfill the requirement for continuous period of time.

2 Lubrication System of Hybrid Bearing

The maximum amount of lubricating oil is held in lube oil sump. It has capacity of 4–10 m^3, or it depends on the unit's size. Lube oil pump sucks oil through this tank and discharging with high pressure (generally 500–1000 kPa gauge pressure). From the discharging pipe most of oil is providing to the bearing ad rest of the oil is provided to the control system. The control unit is discharging oil through a pump to generate pressure, so that it provides sufficient pressure to control valves with a smallest piston possible. As the pressure is high, the required area will be less. Hence, the oil volume is also being less. This provides faster response time as the time required oil flow is very less.

The pressure that is required in control oil system is very high compared to pressure in bearing operation. High pressure in bearings requires high capacity of seals so that, there is no leakage of oil from the bearing. For this purpose, pressure reducing valve is provided midway in the oil discharging pipe that goes to the bearings.

Before reaching oil to bearings, it is pasted through filters, so that any solid contaminants present in oil can be removing. The temperature of oil that is following through the bearing should be controlled carefully. For this purpose, coolers are installed in between the distribution pipe. That cools the oil to the required extent. The viscosity of oil affects the flow of oil hence temperature should not go down below the required temperature. Control of the following oil temperature must be taken proper care. For keeping the temperature of bearing oil within limit coolers are installed if the temperature of oil is very less then it results in increased viscosity of oil by within the proper flow of oil in bearings does not happen. Hence, coolers are set to maintain the oil pressure between 45 and 48 bar.

Figure 1 is line diagram of lubricating oil system for a hybrid bearing. Figure 2 explains the unwrapped internal surface structure of a hybrid bearing having oil inlet arrangement.

Figure 3 represents the different oil pump arrangements for 500 MW turbine having six hybrid journals bearing to support the turbine shaft efficiently.

3 Fuzzy Logic Approach

Fuzzy logic was introduced by a Prof. Zadeh [7–9] to introduce the linguistic variable in controlling the engineering devices. It deals with uncertainty and vagueness. Fuzzy logic is used by many researcher to tackle the uncertainty in journal bearing performance [10–13]. The performance and stability of journal bearing is dependent on the groove dimensions and groove locations [14–16]. The membership function of a fuzzy set is represented by, $\tilde{\mu}_A(X)$, defined as follows:

$$\tilde{\mu}_A(X) :\to [0, 1] \tag{1}$$

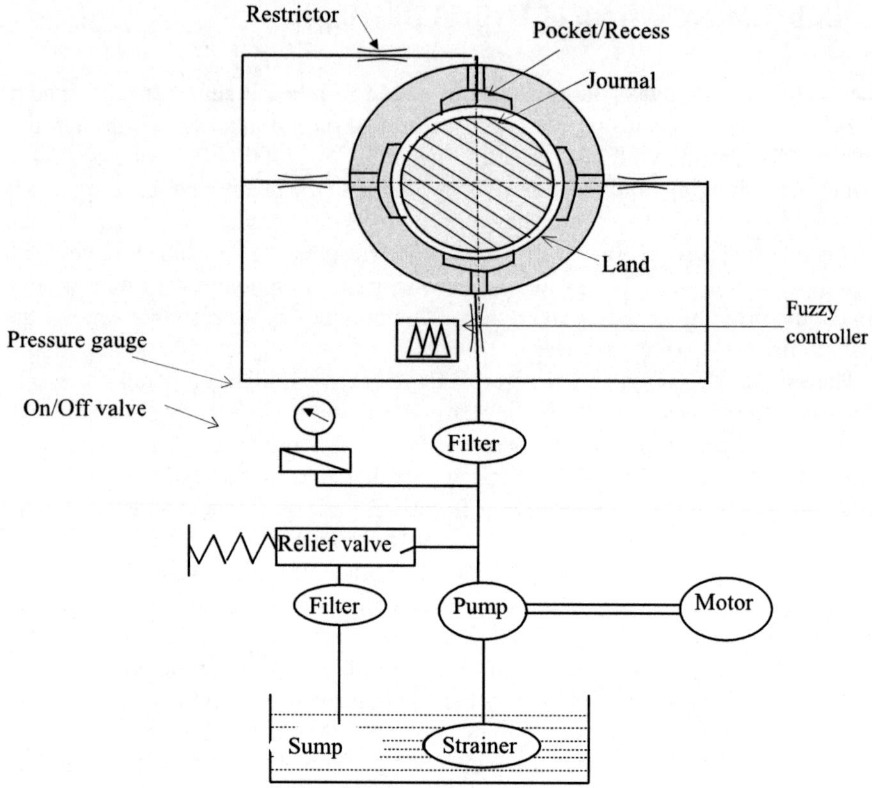

Fig. 1 Lubrication arrangement of a hybrid journal bearing

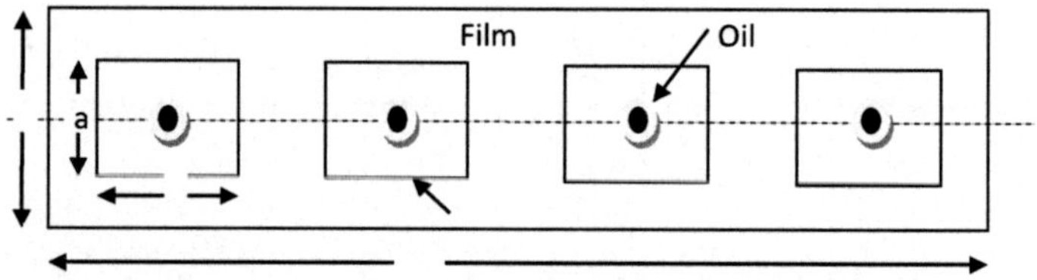

Fig. 2 Unwrapped hybrid journal bearing surface

$\tilde{\mu}_A(X)$ gives the degree of belongingness of x in fuzzy set $\tilde{A}$. Where the degree of membership value "0" corresponds to the absolute non-membership and the "1" corresponds to the absolute membership. Fuzzy Inference is the key to map fuzzy input to crisp output.

Fuzzy Inference system consists of four important sections:

1. The "Fuzzification unit" that converts the information in the form of linguistic variable of minimum film thickness/oil pressure to fuzzy values.

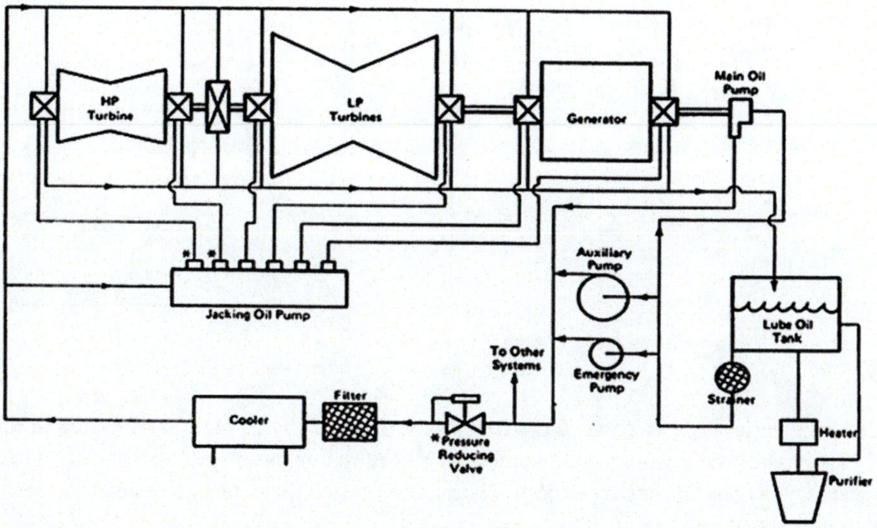

Fig. 3. Lubricating oil system of turbine unit of capacity of 500 MW

2. The "Inference Mechanism" is used to evaluate and to process control rules.
3. The "Rule Base unit" converts the input parameter into decision output variables.
4. The "Defuzzification unit" changes the fuzzy output into its equivalent crisp values.

A fuzzy membership function $\widetilde{\mu}_A(X)$ gives the degree of belongingness of element x of the universe of discourse. Some membership functions are predefined. The simplest form of membership function is straight line. Apart from this other membership functions are triangular, trapezoidal, π-shaped, S-shaped, etc. In this work, triangular membership functions are taken as shown in Fig. 4. Triangular membership function can be calculated as follows:

$$\tilde{\mu}(X; a, c, b) = \begin{cases} 0 & \text{if } X < a \\ \frac{X-a}{c-a} & \text{if } a \le X \le c \\ \frac{b-X}{b-c} & \text{if } c \le X \le b \\ 0 & \text{if } X > b \end{cases} \tag{2}$$

where *a*, *b*, and *c* are defined by experts (Fig. 4).

Fuzzy Inference is used for building fuzzy rules that mimics the human thinking and works over the linguistic variables. Linguistic variables may also be termed as fuzzy variables. The fuzzy variables that are used in this paper are LOW (L), MEDIUM (M), and HIGH (H) for all input parameters. MOP (Main Oil Pump), AP (Auxiliary Pump), and EOP (Emergency Oil Pump) as output parameter in the fuzzy model. Using these fuzzy variables, fuzzy if–then rules are to be defined for fuzzy inference. Rules are generally represented as "if X is $\widetilde{A}$ then Y is $\widetilde{B}$" where X and Y

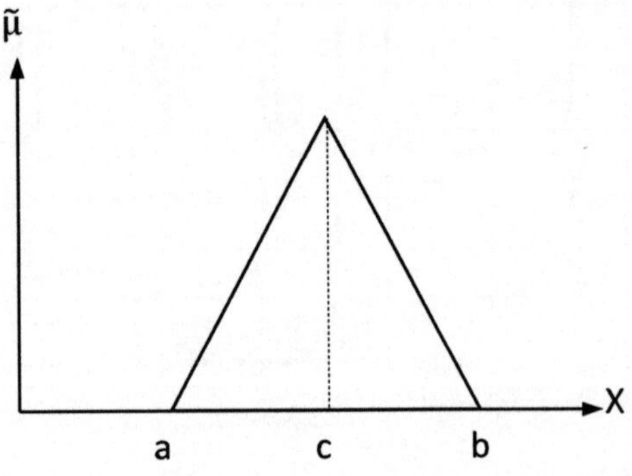

Fig. 4 Triangular membership function, $\tilde{\mu}\ (X; a, c, b)$

are fuzzy variables and $\tilde{A}$ and $\tilde{B}$ are fuzzy sets, obtained by its membership function. If part of the rule is called the antecedent and may consist of multiple inputs with operator AND or OR between them. Then part is called the consequent and may also include several outputs.

4 Fuzzy Expert System

Here in this work fuzzy logic-based smart controller is used to select pump to maintain the minimum film thickness in the bearing. The fuzzy expert system for selection of correct pump is designed around the "Thickness of fluid film (X_1), Journal speed (X_2), Pressure in the bearing (X_3)". Selection of the membership function for an inference system is developed with the opinion of the domain experts. The domain experts identified are from Hindalco Industries, Sonbhadra, U.P, and Jindal Power, Raipur, Chattishgarh.

4.1 Algorithm for Fuzzy Expert System-Based Selection of Fluid Film Bearing

An algorithm for the fuzzy-based selection of pump in fluid film bearing is given below.

Step 1: Define membership function for "thickness of fluid film" (h/c), It ranges from 0.005 to 0.2, "Pressure" (0 to 6.9 MPa) "Speed of journal" is calculated by speed of journal/maximum rated speed of journal (0 to 1).

Step 2: Normalization of the inputs by dividing maximum value of corresponding input.

Step 3: Mapping crisp inputs into their respective membership function to determine their degree of membership i.e. fuzzy inputs.

Step 4: Determination of the rules that fire from the rule base corresponding to these fuzzy inputs.

Step 5: Calculation the degree of truth (R), of each rule by evaluating the parameters in antecedent values with AND operator and also its consequents.

Step 6: The inference engine aggregates consequences of all relevant fired rules into a single fuzzy set for same output (MOP, AP, IOP) by Root Sum Square (RSS) method.

Step 7: Defuzzification method is applied to the fuzzy output to convert it into crisp output with the help of Center of Gravity (CoG) method.

Step 8: Output of the fuzzy expert system about type of bearing.

4.2 *Fuzzification*

Fuzzification is the process that converts linguistic valueinto its equivalent fuzzy value. This is accomplished with the help of the membership functions as discussed earlier in this paper. In this work triangular membership function is used.

Fuzzification is done by selecting input values of parameters at the x-axis and its corresponding value at y-axis vertically to the upper boundaries of membership function to determine the degree of membership. Construction of fuzzy subset and determination of membership function is the very first step in the advances or development of fuzzy logic-based expert system. As shown in Eqs. (3)–(4). Figure 5 shows the membership function for inputs index of Fluid film thickness (X_1).

$$\tilde{\mu}_{\text{LOW}}(X_1) = \begin{Bmatrix} 0 & \text{if } X_1 < 0 \\ \frac{X_1}{0.005} & \text{if } 0 \le X_1 \le 0.005 \\ \frac{0.01 - X_1}{0.005} & \text{if } 0.005 \le X_1 \le 0.01 \\ 0 & \text{if } X_1 > 0.1 \end{Bmatrix} \quad (3)$$

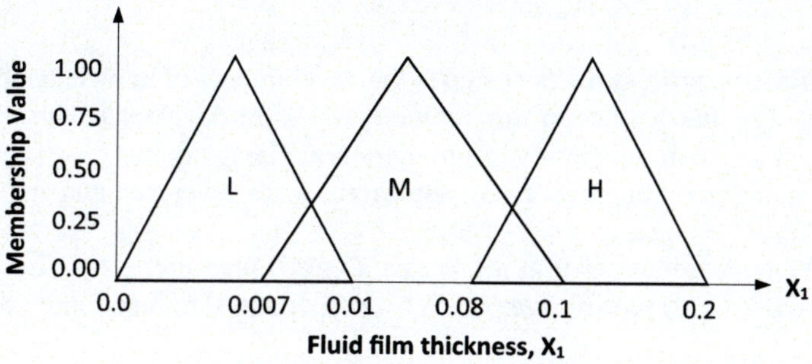

Fig. 5 Membership function of input, fluid film thickness (X_1)

$$\tilde{\mu}_{\text{MEDIUM}}(X_1) = \begin{Bmatrix} 0 & \text{if } X_1 < 0.007 \\ \frac{X_1-0.007}{0.043} & \text{if } 0.007 \le X_1 \le 0.05 \\ \frac{0.1-X_1}{0.05} & \text{if } 0.05 \le X_1 \le 0.1 \\ 0 & \text{if } X_1 > 0.1 \end{Bmatrix} \quad (4)$$

$$\tilde{\mu}_{HIGH}(X_1) = \begin{Bmatrix} 0 & \text{if } X_1 < 0.08 \\ \frac{X_1-0.08}{0.07} & \text{if } 0.08 \le X_1 \le 0.15 \\ \frac{1-X_1}{0.05} & \text{if } 0.15 \le X_1 \le 0.2 \\ 0 & \text{if } X_1 > 0.2 \end{Bmatrix} \quad (5)$$

Similarly, membership functions were derived for input parameters pressure (normalized) (X_2) and speed of journal (normalized) (X_3). Their fuzzy sets are given by Eqs. 6–8 and 9–11.

$$\tilde{\mu}_{\text{LOW}}(X_2) = \text{tringle } (X_2; 0.0, 0.225, 0.45) \quad (6)$$

$$\tilde{\mu}_{\text{MEDIUM}}(X_2) = \text{tringle } (X_2; 0.35, 0.55, 0.75) \quad (7)$$

$$\tilde{\mu}_{\text{HIGH}}(X_2) = \text{tringle } (X_2; 0.55, 0.775, 1.0) \quad (8)$$

$$\tilde{\mu}_{\text{LOW}}(X_3) = \text{tringle } (X_3; 0.0, 0.215, 0.40) \quad (9)$$

$$\tilde{\mu}_{\text{MEDIUM}}(X_3) = \text{tringle } (X_3; 0.30, 0.50, 0.75) \quad (10)$$

$$\tilde{\mu}_{\text{HIGH}}(X_3) = \text{tringle } (X_3; 0.60, 0.775, 0.1) \quad (11)$$

5 Rule Base

After fuzzification process the next step is the development of knowledge base, i.e., fuzzy rules. The fuzzy rules in this research work are developed with the aid of domain experts from industries and from literature. The fuzzy rule base is generated for selection of correct bearing. Twenty-seven rules are developed and are shown in Table 1.

where X_1 (I/P) minimum film thickness, X_2 (I/P) pressure in bearing, X_3 (I/P) journal speed, Y (O/P) type of pump used for proper and safe functioning of journal bearing.

Table 1 Sample Rule base for the fuzzy logic-based expert system

Rule No..	X_1 (I/P)	X_2 (I/P)	X_3 (I/P)	Y (O/P)
1	Low	Low	Low	AP
2	Low	Low	Medium	AP
3	Low	Low	High	EOP
4	Low	Medium	Low	AP
5	Low	Medium	Medium	AP
6	Low	Medium	High	HS
7	Low	High	Low	AP
8	Low	High	Medium	AP
9	Low	High	High	MOP
10	Medium	Low	Low	EOP
11	Medium	Low	Medium	EOP
12	Medium	Low	High	EOP
13	Medium	Medium	Low	AP
14	Medium	Medium	Medium	AP
15	Medium	Medium	High	MOP
16	Medium	High	Low	AP
17	Medium	High	Medium	AP
18	Medium	High	High	MOP
19	High	Low	Low	EOP
20	High	Low	Medium	EOP
21	High	Low	High	EOP
22	High	Medium	Low	EOP
23	High	Medium	Medium	AP
24	High	Medium	High	MOP
25	High	High	Low	AP
26	High	High	Medium	AP
27	High	High	High	MOP

6 Conclusion

The fuzzy logic had been used on and off the different pumps to maintain the minimum film thickness between the bearing and the journal for proper functioning of rotary equipment. The developed fuzzy expert system for the operating of different lube oil pumps in hybrid bearing provides the decision support platform to proper operating of bearing. Using this approach of fuzzy logic with hybrid bearing, life of rotary equipment is increase.

References

1. Rehman WUR, Luo Y, Wang Y, Jiang G, Iqbal N, Rehman SUR, Bibi S (2019) Fuzzy logic based intelligent control for hydrostatic journal bearing. Measur Control 52(2–3):229–243
2. Dwivedi VK, Chand S, Pandey KN (2012) Selection of fluid film journal bearing: a fuzzy approach, advances in computer science, engineering & applications, advances in intelligent and soft computing, vol 166. Springer, Berlin, Heidelberg
3. Rehman WU, Luo Y, Wang Y, Jiang G, Iqbal N, Rehman SU, Bibi S (2019) Fuzzy logic–based intelligent control for hydrostatic journal bearing. Measur Control 52(3–4):229–243
4. Sedova NA, Sedov VA, Bazhenov RI (2018) The neural-fuzzy approach as a way of preventing a maritime vessel accident in a heavy traffic zone. Adv Fuzzy Syst, Article ID 2367096, 8 pages
5. Piltan F, Kim JM (2020) Hybrid fault diagnosis of bearings: adaptive fuzzy orthonormal-arx robust feedback observer. Appl Sci 10(10):3587
6. Dwivedi VK, Chand S, Pandey KN (2014) Effects of turbulence on dynamic performance of accelerated/decelerated hydrodynamic journal bearing system. Int J Des Eng 5(3):256–288
7. Zadeh LA (1973) Outline of a new approch to the analysis of complex system and decision processes
8. Zadeh LA (1968) Fuzzy algorithms. Inf Control 12:94–102
9. Zadeh LA (1965) Fuzzy sets, information and control
10. Ahmed DI, Ahmed Kasolang S, Khidhir BA, Mat Taib CF (2013) Application of fuzzy logic approaches to predict maximum oil-film pressure in Journal Bearing. Appl Mech Mater 393:881–887
11. Baskar S, Sriram S, Arumugam S (2018) Fuzzy logic model to predict oil film pressure in a hydrodynamic journal bearing lubricated under the influence of Nano-based bio-lubricants. Energy Sourc Part A Recov Utilization Environ Effects, 1–8. https://doi.org/https://doi.org/10.1080/15567036.2018.1486479
12. Ahmed DI, Ahmed, Kasolang S, Khidhir BA, Yousif BF (2013) Fuzzy logic based model to predict Maximum oil-Film Pressure in Journal Bearing. Res J Appl Sci Eng Technol 6(20):3871–3878
13. Dwivedi VK, Chand S, Pandey KN (2012) Selection of fluid film journal bearing: a fuzzy approach. In: Advances in intelligence and soft computing, vol 166 AISC, Issue VOL 1. Springer, pp 311–320
14. Dwivedi VK, Chand S, Pandey KN (2015) Effect of the groove dimensions and orientation on the static and dynamic performance of non-recessed hybrid journal bearing. Adv Intell Syst Comput 1089 pp. 555–56
15. Dwivedi VK, Chand S, Pandey KN (2016) Stability analysis of twin axial groove hybrid journal bearing. J Appl Fluid Mech 9(6):2763–2768. ISSN- 1735-3572
16. Dwivedi VK, Pathak P, Effect of axial groove location, length & width ratio on the bearing properties and stability. J Computat Appl Res Mech Eng (JCARME). https://doi.org/https://doi.org/10.22061/jcarme.2019.4411.1541

Design and CFD Analysis on Flow Through a Reactive Muffler of Four-Cylinder Diesel Engine

Ujjal Kalita and Manpreet Singh

1 Introduction

Automobiles industry is the most advanced and most developed industry in today's world, without which people could not move in this fast-moving life. Internal combustion engines are the main source of power in automobile vehicles. These engines are operated by taking in the mixture of air and fuel inside the engine or by compressing the air that is taken inside it. It is divided mainly into two types petrol engine (mixture of air and fuel) and diesel engine (only air is taken which is followed by combustion).

The exhaust comes out from these engines at high pressure and it produces an ample amount of noise. According to different researches, it is found that the noise coming out from exhaust is almost ten times higher than any other structural noise that is produced by vehicle. It is our prime concern to reduce this noise as this noise may lead to harmful disease in human body. According to different health organizations, noise from exhaust may cause heart diseases. To reduce this noise the exhaust pressure should be reduced. Muffler is one such instrument that can be used in the exhaust system for this reduction of noise [1]. A muffler is a device used for reducing the noise coming out of the exhaust system of vehicle. Its main components are tubular metal covering, perforated tubes, absorption materials, and expansion chambers. The arrangement of these components inside muffler would allow the flow from the exhaust system to pass through the inlet pipe of muffler and will pass on to the expansion chamber. The expansion chamber and other components would reduce the pressure inside it and ultimately the amount of noise would be reduced. The pressure drops produced in the muffler from inlet to the outlet of muffler would reduce the noise at the exit. Pressure drop or back pressure is one of the main factors which affect the performance of a muffler. While designing a muffler this factor is utmost taken into consideration. In the present scenario there

U. Kalita · M. Singh (✉)
School of Mechanical Engineering, Lovely Professional University, Phagwara, Punjab, India
e-mail: manpreet.20360@lpu.co.in

A. Parey et al. (eds.), *Recent Trends in Engineering Design*, Lecture Notes in Mechanical Engineering, https://doi.org/10.1007/978-981-16-1079-0_22

are two major types of muffler which are incorporated in automobiles, i.e., Reactive muffler and dissipative muffler. Reactive mufflers generally composed of a single chamber or multiple chambers with which a perforated tube could be incorporated in it. The exhaust from the engine would enter to this chamber through the inlet pipe. The gases would expand in the chamber and this will cause a significant amount of pressure drop inside it. Implementation of perforated tube and baffles would help in reducing the pressure to a higher amount. Thus, it can be said that the back pressure produced inside the chamber would ultimately reduce the noise coming out from the outlet of muffler. Dissipative muffler composed of absorptive material inside it. This material will absorb or dissipate the amount of heat produce by the exhaust system and would ultimately reduce the noise. A third type of muffler could also be incorporated which is combination of both the reactive and dissipative muffler, i.e., a hybrid muffler [2, 3].

2 Literature Review

The design of mufflers has become a great topic of research from many years. The new rules set by the European commission would soon be implemented in vehicles, so it has become utmost important to create new design of muffler. According to this new considerations or regulation, lower limit values for passenger cars, buses, and light trucks will be lowered in two steps of 2-decibel A-weighting (dB (A)) each. For heavy-duty vehicles, the amount of reduction will be 1 dB (A) for the first step and it would be 2 dB (A) for the second step [4]. Different researchers have developed different types of muffler in recent years by taking into considered the performance parameters of muffler. Back pressure, flow velocity, transmission loss, insertion loss are the different performance parameters that are taken into consideration. A good design of muffler should give best noise reduction and optimum back pressure for the engine. Usually mufflers are designed by well-established numerical techniques such as boundary element method or finite element method. Different researchers had worked on this design of muffler from last few decades. Davis et.al in 1954 had done the first comprehensive experiment on the design and analysis of muffler. Davis et al. used the acoustic transfer matrix method and studied the noise reduction principle of muffler. Experimental verification is also performed and with respect to it muffler research theory is established [5]. In their report they had done experiment on 77 different single chamber and multiple chamber mufflers. From the experiment they have plotted the attenuation to frequency curve and this result was compared to the theoretical results. Later Fukuda et al. had developed the transfer matrix method (TMM) which is ideally suited for acoustical modeling of cascaded element in automotive mufflers [6]. This strategy made the standing wave factors to move starting with one component then onto the next in course. The investigation of this work is performed through transfer matrix method. Prof. Amar Pandhare et al. in his work had done the CFD Analysis of Flow through Muffler to Select an Optimum model of Muffler for use in a CI Engine [7]. Later Dragos Tutunea et al. had done the

CFD analysis of resistive muffler [8]. Om Ariara Guhan C P et al. had worked on Three Cylinder LCV Exhaust System. They also performed the CFD analysis and determined the Pressure Drop and Uniformity Index [9, 10]. R. Ramganesh et al. in their work studied the flow pattern and prediction the back pressure of inside the silencer by using the CFD tool [11]. Further Ahmed Elsayed et al. worked on baffle and studied the effect of baffle on the performance of exhaust mufflers [12]. Xu et al. had also worked on analysis of flow field by using computational fluid dynamics [13, 14]. Trivedi et al. [15] in their work optimized the muffler by using a fast light-off exhaust system for motorcycle applications. Dere et al. had also performed CFD analysis in their work [16, 17]. Further Lee et al. had worked on both reactive and dissipative mufflers [18].

The main objective of this study is to design an exhaust muffler which could be used in a four-cylinder diesel engine. This muffler could be designed for optimum flow of exhaust gases through it. Back pressure which is important performance parameter for a muffler is taken for study. High back pressure can reduce engine efficiency, so while designing a muffler this parameter is one of the most important parameters. This study mainly focuses on designing an optimized muffler by determining the pressure drop or the back pressure produce inside muffler. For this study CFD analysis is performed in ANSY's software by taking the different boundary conditions applicable for it. The results obtained from it are compared with the already existing muffler of diesel engine.

3 Design Methodology

A properly designed muffler must satisfy five basic criteria for application in any type of vehicle [19].

1. Acoustic criterion: It specifies the minimum noise reduced by a muffler with respect to frequency.
2. Aerodynamic criterion: It specifies the maximum acceptable average pressure drop in a muffler at a given temperature and mass flow rate.
3. Geometrical criterion: It specifies the space constraint in vehicle. The maximum allowable volume and space provided for installing the muffler is taken into consideration.
4. Mechanical criterion: It specifies the materials used for muffler which would be durable and requires less maintenance.
5. Economical criterion: It is important when we look into our design with the marketplace.

For good design of a muffler, seven broad steps are considered which are followed by practical considerations [20].

4 Detail Study

Consider the case of an LCV diesel engine vehicle, i.e., a Maruti Ciaz vehicle.

Step 1: While setting the benchmark, the back pressure, transmission loss, and pressure drop targets are set for the model of engine considered for study.

Engine data for LCV diesel engine vehicle (Ciaz)

$$\text{Bore}(D) = 69.6\,\text{mm},\ \text{Stroke}(L) = 82\,\text{mm}$$

Number of cylinders (n) = 4, Engine Power (P) = 88.7 bhp at 4000 RPM.

Step 2: The target frequencies could be calculated by using the following formulas. These are frequencies are mainly used while calculating the transmission loss for the muffler by making use of maximum engine rpm.

Theoretical Formulation:

$$\text{CFR} = \text{Engine rpm}/60(2 - \text{strokeengines}) \tag{1}$$

$$\text{CFR} = \text{Engine rpm}/120(4 - \text{strokeengines}) \tag{2}$$

$$\text{EFR} = n \times \text{CFR} \tag{3}$$

where CFR = Cylinder Firing Ratio, EFR = Engine Firing Ratio, n = No. of cylinders, rpm = Revolutions per minute,

By considering 4-stroke engine as it is used in Ciaz.

$$\text{CFR} = 4000/120 = 33.33\,\text{Hz},\ \text{EFR} = 4 \times 33.33 = 133.33\,\text{Hz}$$

The first four harmonics should be suppressed as higher order has little effect on noise. In case of perforation hole diameter will also suppress this frequency.

Step 3: Muffler volume calculation.

As per theory of acoustics, the volume of muffler is calculated as follows.

Swept volume of each cylinder (V_s) = (π/4). D^2. L = (3.14 × 69.6^2 × 82)/4 = 0.3119L Total swept volume of the engine in liters = 0.3119 × 4 = 1.247 L

Volume to be considered for design of muffler = 0.5 × V_s × n n is the factor considered while designing muffler.

From different researches it is found that the factor should be between 12 and 25 de- pending on the space constraint [1].

For a Ciaz engine, the factor considered is 22. Therefore, Volume considered = 0.5 × 1.247 = 0.6235 L.

Volume of muffler = factor × volume considered = 22 × 0.6235 L = 13.72 L = 0.01372 m^3.

Volume can be varied depending on the space constraint.

Step 4: Internal configuration and concept designs.

For designing a muffler, the dimensions of muffler are the important parameters. Calculation of diameter of muffler chamber:

$$\text{Volume of muffler}\,(V_{\text{m}}) = (\pi/4).\,D_{\text{m}}^{2}.L_{\text{m}} \tag{4}$$

By taking reference with existing muffler in Ciaz, length considered is 0.5 m or 500 mm.

Therefore, $0.01372 = (3.14 \times D^2 \times 0.5)/4$,

i.e., $Dm = 0.187$ m = 187 mm.

Calculation of diameter of inlet and outlet pipe of muffler.

As per the standards of supercritical grade of muffler, the diameter of body should be three times the exhaust pipe diameter.

i.e., $Dm = 3 \times de$,

i.e., $187 = 3 \times de$,

i.e., $de = 62.33$ mm or 0.06233 m.

Material properties considered

Property	Unit	Values
Density	kg/m^3	8030
Specific heat	J/(kg K)	502
Thermal conductivity	W/(m*K)	16.2
Conductivity type	Isotropic	Isotropic

5 Problem Statement and Formulation

In this case study five different models of muffler are considered for study of pressure drop and velocity variation. Out of this the optimized muffler could be incorporated in Maruti Ciaz. While designing this muffler, volume of muffler is the prime property that is taken into consideration [10].

The existing volume of muffler is considered for study and modification is done in different models considering the existing parameters of muffler. Here baffles are introduced into the chamber and its effect on back pressure or pressure drop is taken for study. The amount pressure drop from inlet to outlet of muffler will give us an effective muffler that could be used in muffler. The different models considered for study are as follows.

Model 1: Simple Reactive muffler.

Model 2: Reactive muffler with a single hole baffle. Model 3: Reactive muffler with two baffles.

Model 4: Reactive muffler with a baffle and tube Model 5: Reactive muffler with a three-hole baffle.

Table 1 Initial values of velocity inlet B.C

Parameters	Values
Density (kg/m^3)	0.696
Viscosity density (kg/m s)	2.7e–05
Enthalpy (j/kg)	749,575.3
Velocity (m/s)	80
Temperature (K)	470

Table 2 Initial values for pressure outlet B.C

Parameters	Values
Gauge pressure (Pa)	0
Temperature (K)	470

6 Boundary Conditions (B. C)

Here in this study to determine the pressure drop, CFD analysis is performed in ANSY's software. This analysis is performed by considering two types of boundary condition.

1. Velocity inlet boundary condition
2. Pressure outlet boundary condition

By taking the reference of different researches performed for diesel engine, the initial values for velocity inlet and pressure outlet boundary conditions are shown in Tables 1 and 2.

This boundary conditions are applied to all the models.

7 Results and Discussions

Applying the boundary conditions to all the models the analysis diagram for total pressure is shown in the below figures.

7.1 Pressure and Velocity Variation in Model 1

Model 1 is a simple reactive muffler. Figures 1 and 2 shows us the variation of pressure and velocity inside this reactive muffler. As there are no obstacles inside this muffler so pressure drop is low.

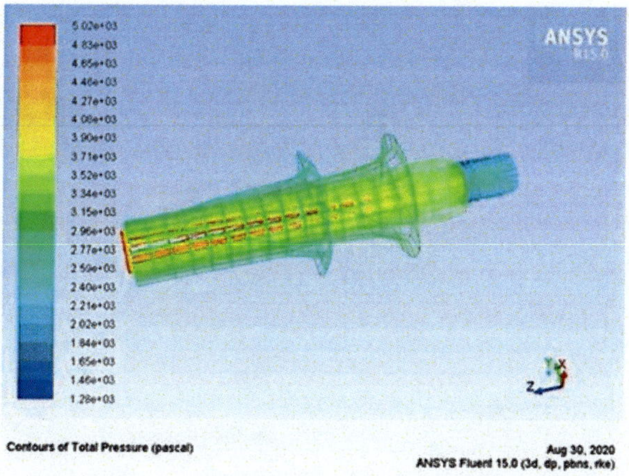

Fig. 1 Pressure variation inside muffler model 1

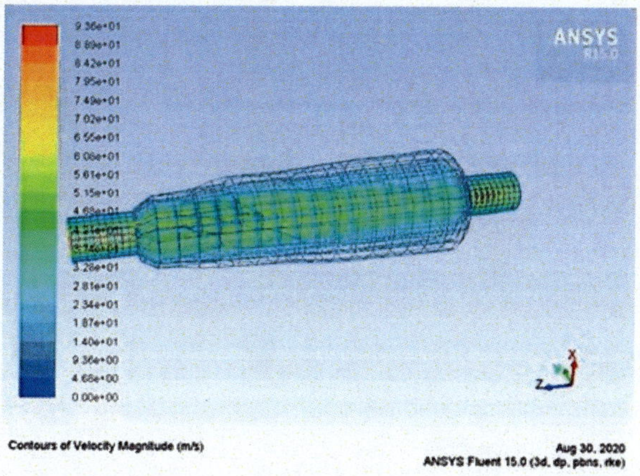

Fig. 2 Velocity variation in model 1

7.2 Pressure and Velocity Variation in Model 2

Figures 3 and 4 shows the pressure and velocity variation inside muffler, when a baffle is placed inside it. It can be seen that pressure increases to a maximum value at point of baffle. As the pressure increases, so the velocity also increases at that point of baffle as shown in Fig. 4.

7.3 Pressure and Velocity Variation in Model 3

Figures 5 and 6 shows the pressure and velocity variation inside muffler, when two baffles are placed inside it. It can be seen that pressure increases to a maximum value at point of first baffle after that pressure drops and after the second baffle the pressure

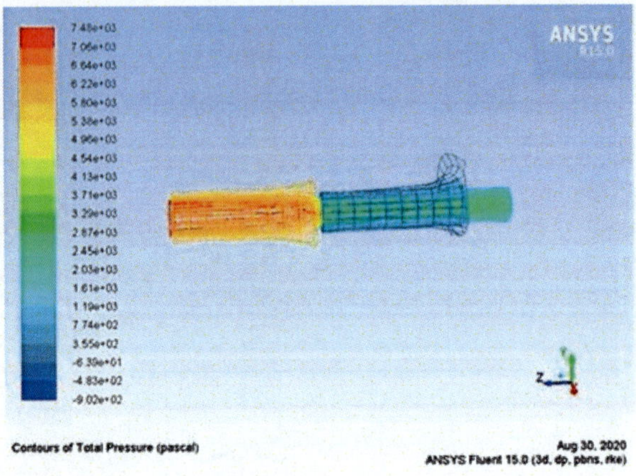

Fig. 3 Pressure variation inside muffler model 1

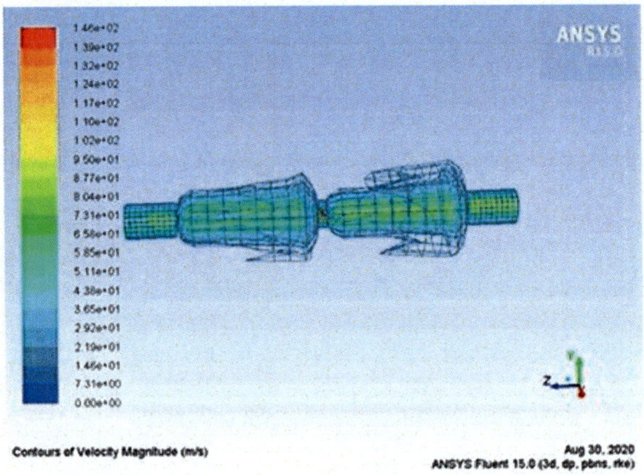

Fig. 4 Velocity variation in model 1

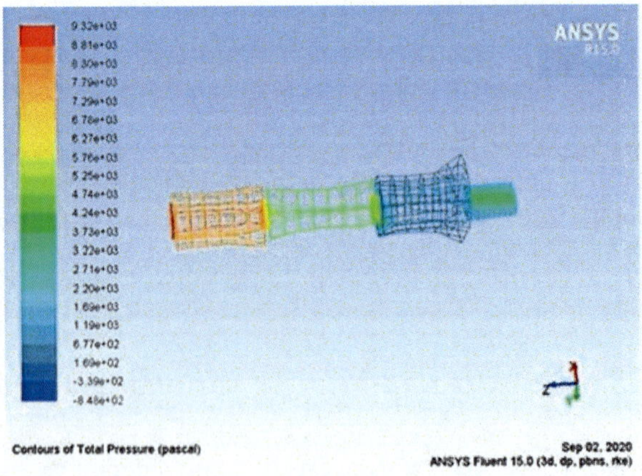

Fig. 5 Pressure variation inside muffler model 3

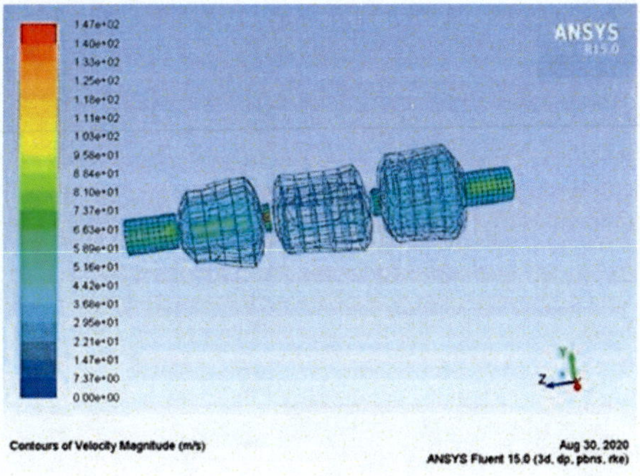

Fig. 6 Velocity variation in model 3

drops by a large amount. The velocity also increases at the junction of both the baffles as shown in Fig. 4.

7.4 *Pressure and Velocity Variation in Model 4*

Here the pressure is high at the inlet but it gradually reduces and as it passes through the tube, the pressure reduces even more as shown in Fig. 7. Maximum velocity attained by this muffler is also quite high but it reduces at the exit of muffler as shown in Figs. 8 and 9.

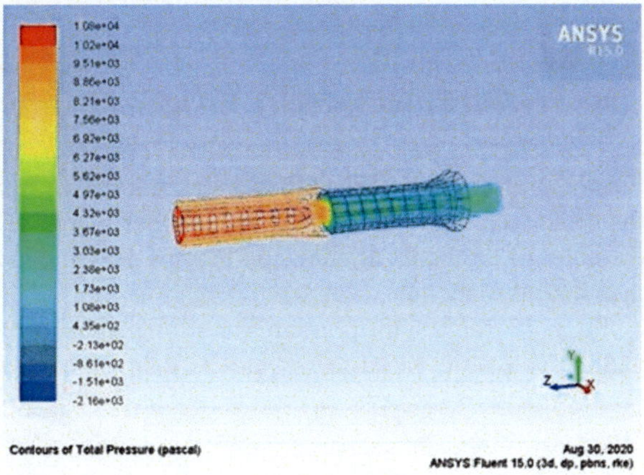

Fig. 7 Pressure variation inside muffler model 3

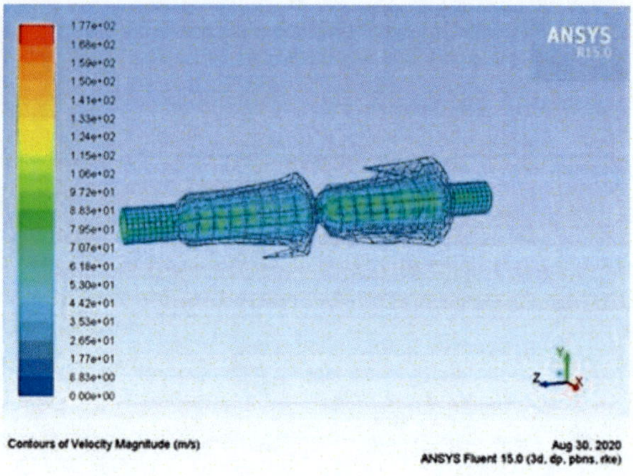

Fig. 8 Velocity variation in model 3

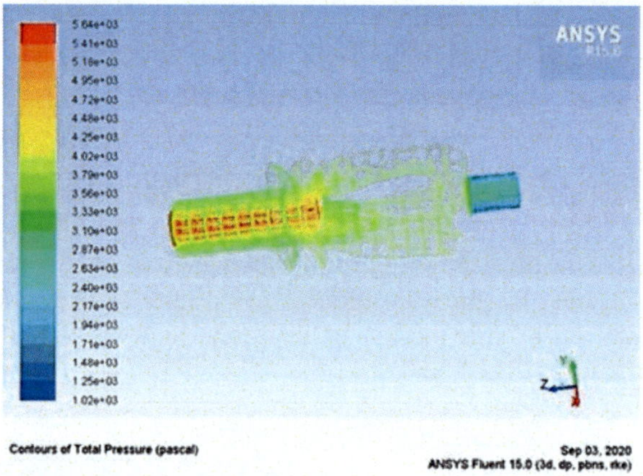

Fig. 9 Pressure variation inside muffler model 5

7.5 Pressure and Velocity Variation in Model 5

From Fig. 9, it can be found that with the increase in number of holes in the baffle, the difference in pressure inside the muffler increases. Velocity also increases near the holes of the baffle as shown in Fig. 10. But this increase in back pressure is less than that of model 4.

From the Table 3, the pressure variation inside the different muffler models is seen and in Table 4 the variation of velocity inside the muffler is observed.

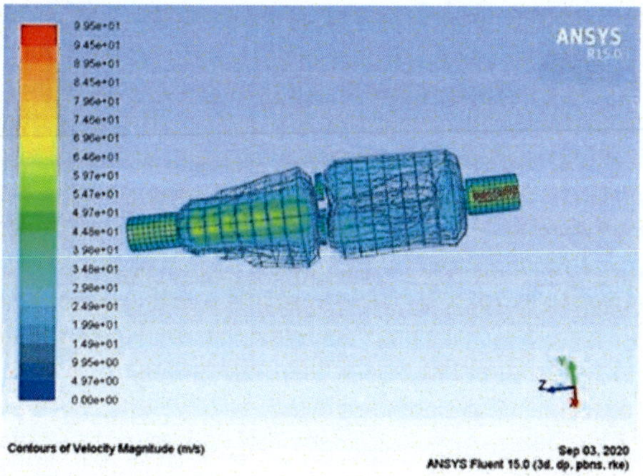

Fig. 10 Velocity variation in model 5

Table 3 Pressure variation inside the mufflers

	Maximum	Minimum	At inlet	At outlet	
1	5020.48	1275.45	2796.35	−1.65	2798
2	7477.41	−901.85	5250	3.38	5246.62
3	9318.28	−847.7	7040.96	2.6	7038.36
4	10,803.53	−2158.47	8574.24	1.59	8572.65
5	5639.84	1016.15	3410.52	4.27	3406.25

Table 4 Velocity variation inside the mufflers

Model	Velocity (m/s)		Velocity (m/s)	
	Maximum	Minimum	At inlet	At outlet
1	93.57	0	80	87
2	146.32	0	80	96.72
3	147.35	0	80	105
4	176.68	0	80	101
5	99.46	0	80	95.12

8 Conclusion

From the work performed by different researchers, it is found that increase of back pressure will reduce the exhaust noise. High back pressure will lead to reduction in more exhaust noise by increasing the pressure drop from inlet to outlet of the muffler. Here CFD analysis is performed for determining the pressure drop inside the muffler. From the analysis performed for different models it is found that when baffle is placed inside the muffler, the back pressure for that muffler is increased. At the position of baffle, the pressure as well as the velocity is increased. Further there is also a substantial increase of pressure when we increasing the number of holes in the baffle. Inserting a tube in the baffle further increases the pressure. Thus, from all the analysis it is found that the pressure drop from the inlet to outlet of muffler is maximum when a baffle as well as a tube is incorporated in the chamber of reactive muffler. There is 67% increase in pressure drop for this case as compared to our simple reactive muffler. For model 2 the increase in pressure drop is 47%, for model 3 and model 5 it is 60.25% and 17.9% respectively. With respect to model 5, back pressure increase by 18% in model 4. Thus, from this analysis of back pressure or pressure drop, it can be concluded that model 4 is the most favorable reactive muffler for use in diesel engine which will eventually reduce maximum noise coming out from exhaust system of vehicle.

References

1. Munjal ML (1987) [M._L._Munjal]_Acoustics_of_Ducts_and_Mufflers_Wit(BookZZ.org).pdf.
2. Le Roy TW (2011) Muffler characterization with implementation of the finite element method and experimental techniques, p 161
3. Liu L, Hao Z, Liu C (2012) CFD analysis of a transfer matrix of exhaust muffler with mean flow and prediction of exhaust noise. J Zhejiang Univ Sci A 13(9):709–716
4. T. H. E. E. Parliament, T. H. E. Council, O. F. The, and E. Union (2014) Regulation (EU) No 540/2014 of the European Parliament and of the Council of 16 April 2014. Off. J. Eur. Union L 158/131, vol. 2013, no. 540
5. Davies POAL, Alfredson RJ (1971) Performance of exhaust silencer. J Sound Vib 15(2):175–196
6. Fukuda M, Muffler. United States Patent 4589517
7. Pandhare PA (2014) CFD analysis of flow through muffler to select optimum muffler model for Ci Engine, vol 4(1), pp 12–19
8. Tutunea D, Calbureanu MX, Lungu M (2013) The computational fluid dynamics (CFD) study of fluid dynamics performances of a resistance muffler. Int J Mech 7(4):401–408
9. Guhan CPOA, Arthanareeswaran G, Varadarajan KN, Krishnan S (2018) Science direct exhaust system muffler volume optimization of light commercial vehicle using CFD simulation. Mater Today Proc 5(2):8471–8479
10. Saripalli P (2015) CFD analysis on flow through a resistance muffler of LCV diesel engine. Int J Sci Technol Soc 3(4):162
11. Ramganesh R, Devaradjane G, Simulation of flow and prediction of back pressure of the silencer using CFD. In: National conference on recent trends and developments in sustainable green technologies. ISSN: 0974-2115
12. Elsayed A, Bastien C, Jones S, Christensen J, Medina H, Kassem H (2017) Investigation of baffle configuration effect on the performance of exhaust mufflers. Case Stud Therm Eng 10:86–94
13. Xu J, Zhou S (2015) Analysis of flow field for automotive exhaust system based on computational fluid dynamics. Open Mech Eng J 8(1):587–593
14. Engineering A, Faurecia CAE, B. Tech, Backpressure study in exhaust muffler of single cylinder diesel engine using CFD analysis shell outlet pipe inlet pipe perforated pipe, pp 1–14
15. Trivedi SM (2017) CFD flow analysis and optimization of exhaust muffler. Int J Res Appl Sci Eng Technol V(VIII): 86–91
16. Dere AA, Singh M, Thakan A, Singh H (2019) Structural optimization of Go-kart chassisby geometrical modifications based on modal analysis. ARPN J Eng Appl Sci 14(18):3234–3240

17. Dere AA, Manpreet Singh, Thakan A, Kumar R, Singh H (2020) Design optimization of Go-Kart chassis frame using modal analysis. Advances in Metrology and Measurement of Engineering Surfaces, pp 171–186, June 2020
18. Selamet A, Lee IJ, Huff NT (2003) Acoustic attenuation of hybrid silencers. J Sound Vib 262(3):509–527
19. Xu MB, Selamet A, Lee IJ, Huff NT (2004) Sound attenuation in dissipative expansion chambers. J Sound Vib 272(3–5):1125–1133
20. Yao Y, Wei S, Zhao J, Chen S, Feng Z, Yue J (2013) Experiment and CFD analysis of reactive muffler. Res J Appl Sci Eng Technol 6(17):3282–3288

Modeling of Chipped Tooth Fault in Straight Bevel Gears

Palash Dewangan, Dada Saheb Ramteke, and Anand Parey

1 Introduction

Dynamic characteristics of a faulty gearbox can be studied using dynamic modeling of gear faults [1]. Tooth breakage is the most common tooth fault in gears, which is a result of tooth surface failure. When sudden enormous stresses develop at the tooth surfaces, partial tooth breakage (chipping) or complete tooth breakage (missing tooth) occurs [2]. In addition, high tooth impact loads may also contribute to tooth breakage [3]. Time-varying gear meshing stiffness is considered as the primary cause of excitation in the gear transmissions, which leads to high vibrations characteristics. In addition, the presence of any fault in gear teeth also alters the dynamics of the gears because fault introduces a decrease in the gear meshing stiffness. Thus, it is imperative to study the dynamics of the gear systems in the presence of faults.

A few works [4–9] may have been found on the modeling of bevel gears without considering faults. These works were focused on the modeling of teeth contact deformation, modal analysis, and effects of assembly and manufacturing error on the dynamic response. However, numerous works can be found on the modeling of faults in spur gears [10]. Chaari et al. [2] proposed a three-dimensional geometry for chipped tooth fault. The geometric parameters are chipped height, chipped width, and chipped thickness. The variation in one of these geometric parameters yields distinct stiffness reduction. Wang et al. [11] used a periodic pulse function multiplied by meshing stiffness function to include the effect of tooth chipping in

P. Dewangan (✉) · D. S. Ramteke · A. Parey
Department of Mechanical Engineering, Indian Institute of Technology Indore, Indore 453552, India
e-mail: phd1701103004@iiti.ac.in

D. S. Ramteke
e-mail: phd1601103002@iiti.ac.in

A. Parey
e-mail: anandp@iiti.ac.in

A. Parey et al. (eds.), *Recent Trends in Engineering Design*, Lecture Notes in Mechanical Engineering, https://doi.org/10.1007/978-981-16-1079-0_23

a planetary gear of a wind turbine gearbox. The results were also validated with experimental results. Li et al. [12] developed a multiple tooth chipping model for fault diagnosis of a compound planetary gear by considering the straight-line shape of the tooth breakage.

On the contrary to that, there is a lack of literature on the modeling of faults in bevel gears. Karray et al. [13] calculated the dynamic response of one-stage spiral bevel gear with a tooth crack defect. Yassine et al. [14] studied the manufacturing and tooth crack defects in a two-stage straight bevel gear using a dynamic model. Later, Karay et al. [15] extended their work under non-stationary operating conditions of a spiral bevel gearbox in the presence of a tooth crack defect.

In the above literature, only the effect of crack fault on the dynamics of bevel gears has been studied. In addition to crack, tooth surface failure such as wear and tooth breakage also occur under high dynamic loads and uncertain loading conditions [16, 17]. Tooth breakage (chipped tooth) may also severely affect the operation of the gearbox and leads to failure. Therefore, it is crucial to study the chipping tooth fault behavior in a bevel gearbox. This motivates the authors to study the dynamics of bevel gearbox with chipped tooth fault. In addition, in the available literature, modeling of chipped tooth fault and its effects on the dynamics of bevel gears have not been addressed.

In this paper, the time-varying meshing stiffness (TVMS) model of a straight bevel gear to account for a chipped tooth fault is proposed. The TVMS model is incorporated in the dynamic model of a one-stage straight bevel gearbox to calculate dynamic responses. Experiments are conducted to validate the simulation results. It is shown that the model can diagnose the missing tooth fault in a one-stage bevel gearbox.

2 Dynamic Modeling of One-Stage Straight Bevel Gearbox

In this study, modeling of a straight bevel gearbox is adopted from Ref. [14], as illustrated in Fig. 1. The model consists of a bevel gear pair with five degrees of freedom for each gear body (pinion/wheel), i.e., three translations and two rotations. The translations are defined by $x_j, y_j, z_j, (j = p, w)$, and rotations are defined as ϕ_j and ψ_j. The rotations of pinion shaft and wheel shaft is defined as θ_j, and the displacements of motor and receiver are defined as θ_m and θ_r, respectively. $k(t)$ is the TVMS, k_{jx} and k_{jy} $(j = p, w)$ are the radial bearing stiffness, k_{jz} is the axial bearing stiffness, $k_{j\phi}$ and $k_{j\psi}$ are the bearing tilt stiffness, and $k_{j\theta}$ is the torsional stiffness of the shaft containing gears j.

The tooth deflection due to meshing of the teeth is given by [14]

$$\delta = \{L\}^T\{q\} \tag{1}$$

where

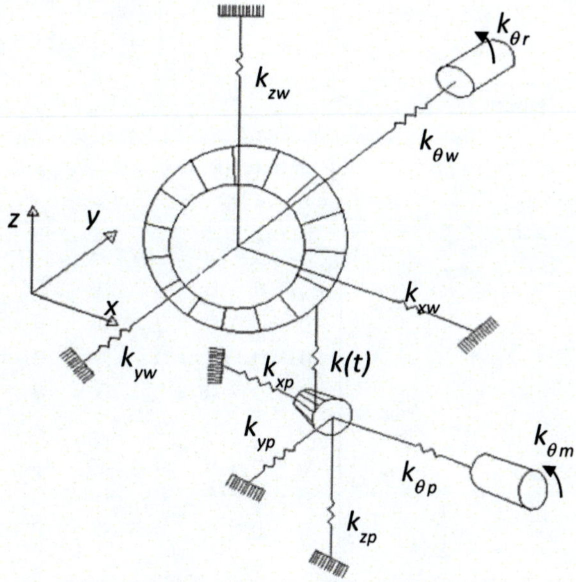

Fig. 1 An illustration of dynamic model of straight bevel gear

$$\{L\} = \{s_1, s_2, s_3, s_4, s_5, s_6, s_7, s_8, s_{10}, s_{11}, 0, s_9, s_{12}, 0\}, \tag{2}$$

and

$$q = \{x_p, y_p, z_p, x_w, y_w, z_w, \phi_p, \psi_p, \phi_w, \psi_w, \theta_m, \theta_p, \theta_w, \theta_r\}^{\mathrm{T}} \tag{3}$$

The coefficients $s_j (j = 1, 2, \ldots 12)$ can be found in Ref. [14]. Lagrange's method is used to obtain the motion equation of the system, which is written as [14]

$$M\ddot{q}(t) + C\dot{q}(t) + [K_s + K(t)]q(t) = F(t) \tag{4}$$

where subscripts m and r indicates motor and receiver, respectively.

In Eq. (2), M is the diagonal mass matrix of the system and can be written as

$$M = diag(m_p, m_p, m_p, m_w, m_w, m_w, I_{px}, I_{py}, I_{wx}, I_{wy}, I_m, I_{p\theta}, I_{w\theta}, I_r) \tag{5}$$

where m_j $(j = p, w)$ is the concentrated mass and I_{ij} $(i = p, w;\ j = x, y)$ is the diametral moment of inertia of the gear bodies. I_m,$I_{p\theta}$, $I_{w\theta}$, and I_r are the diametral moment of inertia of the motor, pinion shaft, wheel shaft, and the load, respectively.

The TVMS matrix is given by [14]

$$K(t) = k(t)\{L\}\{L\}^{\mathrm{T}} \tag{6}$$

The bearing stiffness matrix is given by [14]

$$K_s = \begin{bmatrix} K_T & 0 \\ 0 & K_R \end{bmatrix} \tag{7}$$

where

$$K_T = \text{diag}(\{k_{px}, k_{py}, k_{pz}, k_{wx}, k_{wy}, k_{wz}\}) \tag{8}$$

and

$$K_R = \begin{bmatrix} k_{p\phi} & 0 & 0 & 0 & 0 & 0 & 0 & 0 \\ 0 & k_{p\psi} & 0 & 0 & 0 & 0 & 0 & 0 \\ 0 & 0 & k_{w\phi} & 0 & 0 & 0 & 0 & 0 \\ 0 & 0 & 0 & k_{w\psi} & 0 & 0 & 0 & 0 \\ 0 & 0 & 0 & 0 & k_{p\theta} & -k_{p\theta} & 0 & 0 \\ 0 & 0 & 0 & 0 & -k_{p\theta} & k_{p\theta} & 0 & 0 \\ 0 & 0 & 0 & 0 & 0 & 0 & k_{w\theta} & -k_{w\theta} \\ 0 & 0 & 0 & 0 & 0 & 0 & -k_{w\theta} & k_{w\theta} \end{bmatrix} \tag{9}$$

The proportional Rayleigh damping C can be defined as

$$C = 0.05M + 10^{-4} K_m \tag{10}$$

where K_m is the mean meshing stiffness matrix, i.e., Eq. (6) without $k(t)$.

3 Modeling of Meshing Stiffness

3.1 Meshing Stiffness of Healthy Bevel Gear

In this paper, TVMS for a pair of straight bevel gear is modeled as the square wave approximation. The time evolution of meshing stiffness of a healthy bevel gear pair is illustrated in Fig. 2. The speed of the pinion shaft is taken as $f_p = 7.04\,H_z$. Therefore, the gear meshing frequency (GMF) obtained is $f_m = 126.7\,H_z$, and the meshing period obtained is $T_m = 0.0079\,s$.

3.2 Meshing Stiffness with Chipped Tooth Fault

The meshing stiffness calculation is based on the Tredgold assumption [18]. According to which the straight bevel gears, when projecting on a plane tangent to the back cone, may be considered equivalent to spur gears. The pitch of this

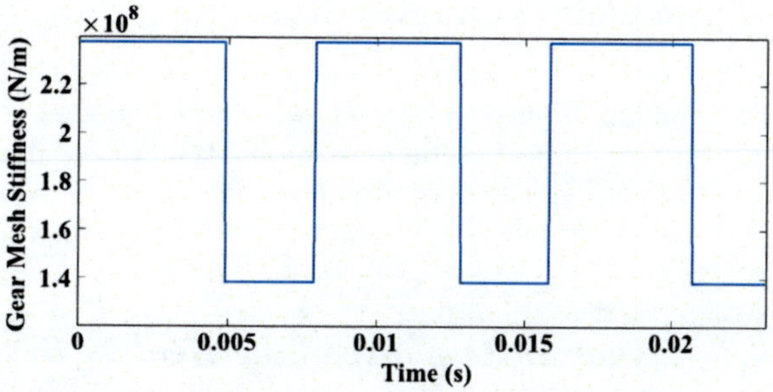

Fig. 2 TVMS of a healthy bevel gear pair

approximated spur gear will be the same as the bevel gear, and the pitch radius will be equivalent to the back cone radius.

For the current study, the chipped tooth fault is modeled on the pinion of a straight bevel gear pair. Chipping of teeth may vary from small to large portions of a tooth along the length, width, and height [2]. In this study, instead of modeling the actual chipping phenomenon, a proportional amount of reduction in stiffness corresponding to chipping is used. It can be observed from Ref. [2] that for 50% of chipped width, around 50% stiffness reduction was obtained. This indicates that chipped width has a considerable influence on tooth stiffness. Based on the observation from Ref. [2], a meshing stiffness with variation in chipped width is proposed. In this model, the meshing stiffness reduction is varied from 0 to 50% for DTPC and is kept the same as the healthy case for STPC because a slight change has been observed for STPC in Ref. [2]. The time evolution of meshing stiffness of a straight bevel gear pair with chipped tooth fault for two meshing periods is illustrated in Fig. 3(a), and for two revolutions of a shaft is illustrated in Fig. 3(b).

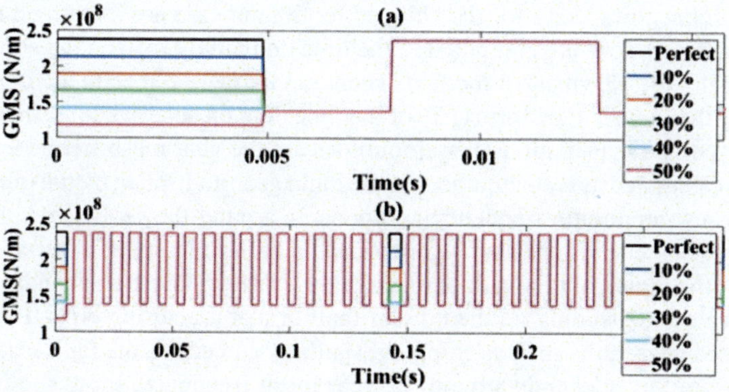

Fig. 3 TVMS of a straight bevel gear pair with chipped tooth fault **a** for two meshing periods, **b** for two rotations of shaft

4 Simulation and Results

For simulation purposes, bevel gear parameters are obtained from an experimental test rig (see Table 1). The mass moment inertia of the motor and receiver are $I_m = 0.0055\ \text{kg m}^2$ and $I_r = 0.1\ \text{kg m}^2$.

4.1 *Dynamic Response of the Healthy Bevel Gear System*

For calculating the dynamic response, as described in Sect. 3.1, the pinion speed is taken as 7.04 Hz, and the corresponding GMF is 126.7 Hz. The equation of motion has been solved using the Newmark β algorithm, which gives all the vibration quantities, i.e., displacement, velocity, and, acceleration. A Fourier transform (FT) is obtained from the time domain acceleration data to generate frequency response. The frequency spectra of the healthy bevel gear at pinion bearing are illustrated in Fig. 4. The GMF and its harmonics can be explicitly observed.

4.2 *Dynamic Response of Bevel Gear with Chipped Tooth Fault*

For calculating the dynamic response of bevel gear for chipped tooth fault, a 30% reduction in the meshing stiffness corresponding to 30% chipped width is considered. The reason for choosing this value is that it is comparable to the experiment (illustrated in Sect. 5). The fault is introduced in the pinion of the straight bevel gear pair. The frequency spectra for the chipped tooth fault case are illustrated in Fig. 5. First, the accelerations are computed in the time domain by solving the equations of motions using the Newmark β method. Then, the frequency spectra are obtained by computing the Fourier transform (FT) of the time domain accelerations. In Fig. 5, the frequency components with higher amplitudes are the gear mesh frequency (GMF) which are calculated as teeth numbers of the input gear multiplied by the input speed. The other low-amplitude frequency components around the GMF are called sidebands. These sidebands are also regarded as the fault feature characteristics which are mainly the indicator of a fault or manufacturing error. These sidebands may have different values depending on the type of fault or manufacturing error. For most of the localized gear faults such as chipping, spalling, and crack, the fault feature characteristics are the sidebands around the gear mesh frequency. From the frequency spectra of Fig. 5, it can be observed that the fault feature characteristics appear as the low-amplitude multiple sidebands $(nf_m \pm nf_p)$ on either side of the gear meshing frequencies and around the harmonics as well, where n is the number of gear mesh harmonics, and f_m and f_p are explained in Sect. 3.1; $2f_m$ and $3\ f_m$ are the second and the third harmonic of the gear mesh frequency, respectively.

Table 1 Design parameters of one-stage bevel gear test rig

Parameters	Module (m)	Mean meshing stiffness (N/mm)	Pressure angle (°)	Number of teeth	Mass (kg)	Inertia (kg m^2)	Bearing stiffness (N/m)	Torsional stiffness (Nm/rad)
Pinion	2	2×10^8	20°	18	0.03	3×10^{-6}	$k_x = 1 \times 10^8 k_y = 1 \times 10^8 k_z = 1 \times 10^8 k_\phi = 1 \times 10^8 k_\psi = 1 \times 10^8$	$k_{\theta p} = 1 \times 10^4 k_{\theta w} = 2 \times 10^4$
Gear	2		20°	27	0.05	1.15×10^{-5}		

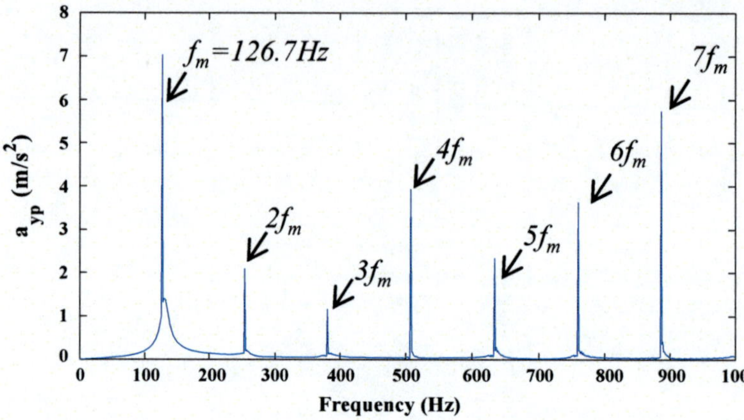

Fig. 4 The frequency spectra of healthy bevel gear at pinion bearing

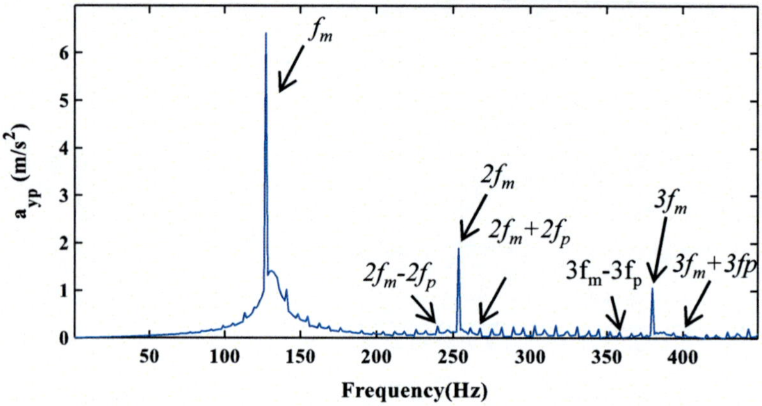

Fig. 5 The frequency spectra of bevel gear with chipped tooth fault at pinion bearing for 30% chipped width

5 Experimental Validation

For the validation of simulated results, an experiment was conducted on a single-stage straight bevel gearbox with a healthy and chipped tooth fault case. A bevel gearbox mounted on a machinery fault simulator was used as a test rig. The healthy bevel gear and bevel gear with a chipped tooth fault used in the experiment are illustrated in Fig. 6a, b. The specification of the gearbox is given in Table 2.

The measured frequency spectra of a healthy bevel gear are illustrated in Fig. 7. The GMF and some of its harmonics are observed from Fig. 7. However, the response also contains some high amplitude noise components. The frequency spectra of the bevel gear test rig with a chipped tooth fault are illustrated in Fig. 8. First,

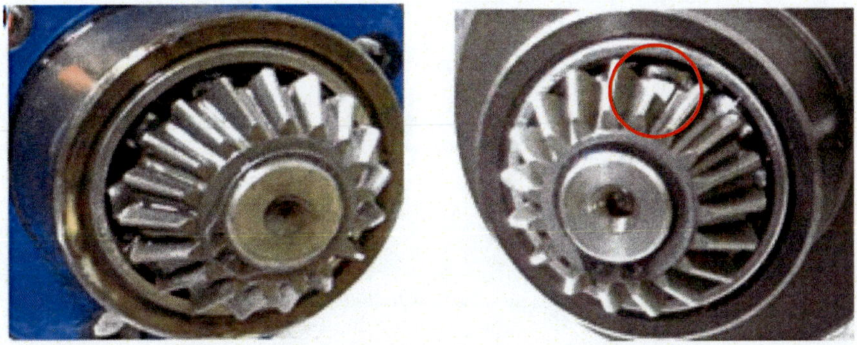

(a) (b)

Fig. 6 Gears with different health condition. **a** Healthy gear, **b** Gear with chipped tooth with 30% loss of width (position highlighted in red)

Table 2 Specifications of the bevel gearbox

Parameters	Value
Gear ratio	1.5:1
Wheel pitch angle	56°19′
Pinion pitch angle	33°41′
Pressure angle for wheel and pinion	20°
Number of teeth in pinion	18
Number of teeth in the wheel	27

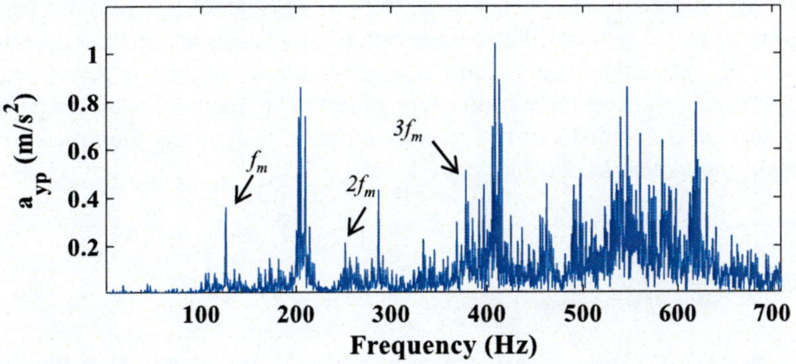

Fig. 7 The frequency spectra of healthy bevel gear experimental signal at pinion bearing

the time domain accelerations are obtained from the accelerometer mounted over the test rig. Then, the frequency domain response is obtained by doing FT of the time domain acceleration signal. The fundamental GMF (f_m) and small amplitude sidebands around it can be observed. Likewise, for the third harmonic (3 f_m) of GMF,

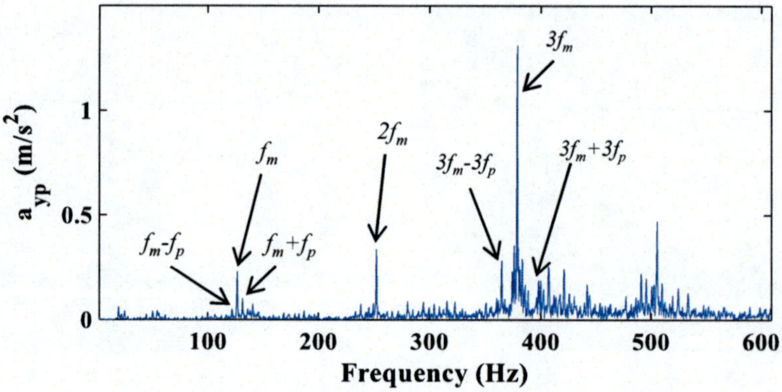

Fig. 8 Frequency spectra (experimental) of bevel gear with chipped tooth fault at pinion bearing

sidebands can be observed around it. However, because of noise in the measurement, sidebands around the second harmonic of GMF are not clear. The comparison of Figs. 5 and 8 shows that results are in good agreement.

6 Conclusion

In the present study, a gear meshing stiffness model of chipped tooth fault is proposed using square wave approximation. In the model, a reduction in the meshing stiffness governed by variation in chipped width is implemented. The meshing stiffness model is then incorporated into the dynamic model of a one-stage straight bevel gearbox. The dynamic response is calculated numerically, and fault feature characteristics are observed. For validation of the simulation result, an experiment is performed. The simulation and experimental results are compared and found to be in good agreement. Therefore, it is concluded that the proposed model successfully identifies the fault in a single-stage straight bevel gearbox.

References

1. Randall RB (1982) A new method of modeling gear faults. Trans Asme J Mech Des 104:259–267. https://doi.org/10.1115/1.3256334
2. Chaari F, Baccar W, Abbes MS, Haddar M (2008) Effect of spalling or tooth breakage on gearmesh stiffness and dynamic response of a one-stage spur gear transmission. Eur J Mech A/Solids 27:691–705. https://doi.org/10.1016/j.euromechsol.2007.11.005
3. Tian X, Zuo MJ, Fyfe KR (2008) Analysis of the vibration response of a gearbox with gear tooth faults. In: Proceedings of IMECE04 2004 ASME international mechanical engineering congress exponential, 13–20 Nov 2004. Anaheim, Calif. USA. 4466, 785–793. https://doi.org/https://doi.org/10.1115/imece2004-59224

4. Bruyère J, Dantan JY, Bigot R, Martin P (2007) Statistical tolerance analysis of bevel gear by tooth contact analysis and Monte Carlo simulation. Mech Mach Theory 42:1326–1351. https://doi.org/10.1016/j.mechmachtheory.2006.11.003
5. Peng T, Lim TC (2018) Influence of gyroscopic effect on hypoid and bevel geared system dynamics. SAE Int J Passeng Cars Mech Syst 2:1377–1386
6. Feng Z, Wang S, Lim TC, Peng T (2011) Enhanced friction model for high-speed right-angle gear dynamics †. J Mech Sci Technol 25:2741–2753. https://doi.org/10.1007/s12206-011-0803-3
7. Chang-Jian CW (2011) Nonlinear dynamic analysis for bevel-gear system under nonlinear suspension-bifurcation and chaos. Appl Math Model 35:3225–3237. https://doi.org/10.1016/j.apm.2011.01.027
8. Peng T, Lim TC, Yang J (2017) Eccentricity effect analysis in right-angle gear dynamics. In: Proceedings of ASME 2011 international design engineering technical conference computing information engineering conference, 1–14
9. Yinong L, Guiyan L, Ling Z (2010) Influence of asymmetric mesh stiffness on dynamics of spiral bevel gear transmission system. Math Probl Eng, 1–13. https://doi.org/https://doi.org/10.1155/2010/124148
10. Liang X, Zuo MJ, Feng Z (2018) Dynamic modeling of gearbox faults: a review. Mech Syst Sig Process 98:852–876. https://doi.org/10.1016/j.ymssp.2017.05.024
11. Wang X, Xu Y, Li T (2016) Dynamic characteristics of wind turbine gearbox with the chipping fault. In: Proceedings of 5th international conference electrical engineering automation control. Lecture Notes Electrical Engineering, 367:567–574. https://doi.org/https://doi.org/10.1007/978-3-662-48768-6
12. Li G, Liang X, Li F (2018) Model-based analysis and fault diagnosis of a compound planetary gear set with damaged sun gear † 32:3081–3096. https://doi.org/https://doi.org/10.1007/s12206-018-0611-0
13. Karray M, Chaari F, Viadero F, Rincon AF, Haddar M (2013) Dynamic response of single stage bevel gear transmission in presence of local damage. New Trends Mech Mach Sci Mech Mach Sci 7:337–345. https://doi.org/10.1007/978-94-007-4902-3
14. Yassine D, Ahmed H, Lassaad W, Mohamed H (2014) Effects of gear mesh fluctuation and defaults on the dynamic behavior of two-stage straight bevel system. Mech Mach Theory 82:71–86. https://doi.org/10.1016/j.mechmachtheory.2014.07.013
15. Karray M, Chaari F, Khabou MT, Haddar M (2018) Dynamic analysis of bevel gear in presence of local damage in nonstationary operating conditions. In: Proceedings of the 7th conference on design and modeling of mechanical systems, CMSM'2017, 27–29 Mar, Hammamet, Tunisia, pp 325–330. https://doi.org/https://doi.org/10.1007/978-3-319-66697-6_32
16. Lafi W, Djemal F, Tounsi D, Akrout A, Walha L, Haddar M (2019) Dynamic modelling of differential bevel gear system in the presence of a defect. Mech Mach Theory 139:81–108. https://doi.org/10.1016/j.mechmachtheory.2019.04.007
17. Park M (2003) Failure analysis of an accessory bevel gear installed on a J69 turbojet engine. Eng Fail Anal 10:371–382. https://doi.org/10.1016/S1350-6307(02)00071-7
18. Elkholy AH, Elsharkawy AA, Yigit AS (1998) Effect of meshing tooth stiffness and manufacturing error on the analysis of straight bevel gears. Mech Struct Mach 26:41–61

Fatigue Crack Growth, Life Estimation, and Comparative Analysis of Carbon SteelsA333 Gr 6 Material Piping Component Using FEA Approach

Dinesh Agarwal, Dinesh Kumar Soni, and S. K. Dhakad

1 Introduction

The satisfactory performance of nuclear reactors (newly established or the old and active reactors) requires the extension of reactor life from 40 to 60 to 100 years. The design of nuclear reactors has all-time big challenges for the design engineer and scientists. The nuclear reactor is an essential part of the nuclear power plant, in which a large quantity of heat energy has been produced as the nuclear reaction takes place. A breakdown in the piping could lead to the loss of fluid (i.e., steam, heavy water, and other nuclear fuels). Accidents may lead to the discharge of radioactive materials. Therefore, the estimation of the remaining age of the piping component under fatigue loading is essential to be determined for the safety of power plants and mankind. In this perspective, small-sized sample testing and sophisticated non-destructive assessment together with modeling and simulation will facilitate us to acquire an in-depth knowledge of fracture mechanisms, microstructure, and structural degradations [1, 2].

Fatigue failure takes place when metal is subjected to a dynamic or variable load and will fail at stress lesser than its ultimate tensile strength [3]. The harm caused all through the fatigue development is cumulative and usually unrecoverable. Detailed understanding of fatigue actions is therefore necessary for components subjected to continuous variable amplitude loading [4].

While experiencing considerable cyclic stress, even high ductile material may undergo fatigue failure over a definite period of time. Constant optimum rhythmic

D. Agarwal (✉)
Government Polytechnic College, Morena, Madhya Pradesh, India

D. K. Soni
Rabindranath Tagore University, Raisen, Madhya Pradesh, India

S. K. Dhakad
Samrat Ashok Technological Institute, Vidisha, Madhya Pradesh, India

A. Parey et al. (eds.), *Recent Trends in Engineering Design*, Lecture Notes in Mechanical Engineering, https://doi.org/10.1007/978-981-16-1079-0_24

stress can produce a part to wear and crack, and henceforth, stress will increase the wear until crack grows and ends into terrible component failure [5].

2 Fatigue Research History

Literature survey of milestones in the area of fatigue fracture analysis in different areas dealing with different materials is as follows.

Wilhelm August Julius Albert issued his first article on metal fatigue in 1837. He formulated a test machine for conveyor chains to be used in the Claustral mines [6].

In 1910, O. H. Basquin suggested a log–log correlation for S–N curves, by means of Wohler's test data. Through the 1920s and 1930s, fatigue had developed as a major field for scientific research [7].

Paul Croce Paris (P. C. Paris) was an American academician, engineering consultant, and researcher in the field of mechanics and fatigue. He firstly introduced Paris' law connecting crack growth rate to the amplitude of the stress intensity factor (SIF) range to sub-critical crack growth under a fatigue stress regime. The basic formula interprets as follows:

$$\frac{\mathrm{d}a}{\mathrm{d}N} = c\Delta k^m$$

where "a" is the crack length, "N" is the number of reversals, and $\mathrm{d}a/\mathrm{d}N$ is the fatigue crack growth rate, which denotes the crack growth for a load cycle. On the right side, C and m are material constants determined experimentally and depend on the environmental factors, stress ratio, and the characteristic specimen size. The alternating stress intensity ΔK is the difference between the maximum and minimum stress intensity factors for each load cycle [8, 9].

M. Bergman (1995) published a paper on "stress intensity factors for circumferential surface cracks in pipes" in a journal and derived a computational method for computation of stress intensity factors using finite element method ABACU. He used pipes with both, cracks located at the in surface and outside surface of pipes with a thickness ratio equal to 5–10. The expansion of pipe taken by him was 15 times the wall thickness of pipe, mesh consists of 20 node brick elements, and the number of elements ranged from 1268 to 2342 depending on the length of crack front. Poisson's ratio was set to 0.3 in calculations. The crack depth ratio ranges from 0.2 to 0.8 and the crack aspect ratio ranges from unity to 16 in the analyses. He analyzed them for ample range of geometry (80 different geometries) and load configuration (6 different load cases) which describes influence functions for the SIF along the whole crack front. SIFs are used to calculate approximately how a crack grows due to the shape of crack or stress on pipes. Stress intensity factor $K_1 = \sqrt{\pi a}\sigma_i f_i$, where f_i is the influence function to the stress component σ_i. Evaluated influence function coefficients for each load case were tabulated for inside and outside cracked specimens [10].

3 Finite Element Approach

By means of FEM (finite element method) and BEM (boundary element method) techniques, it is currently feasible to examine complicated crack configurations in genuine technical structures under complex loads with nonlinear material behavior [11].

Fatigue software that interfaces with FEA programs comprises a database of fatigue properties. The fatigue software computes, where and when fatigue cracks will arise (fatigue hot spots), decides factors of security on operational stresses (for optimizations), and discover the possibility of endurance at different service lives. Results are accessible as contour plots of fatigue lives, stress-safety factors, and possibility of failures, and are plotted by means of standard FEA viewers and graphics software.

In this general idea, four major input factors, namely—geometry, material property, meshing, and loading, are considered to have similar functions. These parameters appear to be the most important input to any software for modeling and simulation. In the case of crack problems, either crack edges (for two-dimensional problems) or both crack surfaces (for three-dimensional problems) have to be allocated separate nodes. Due to the singular stress fields at the crack, in general, a very fine mesh is essential in the crack area. For fatigue crack proliferation, the crack has to be incrementally lengthened. This involves re-meshing in the crack area [12]. After pre-processing, analysis and post-processing will take place in the software based on FEM, and result will be displayed on the screen.

4 Methodology

Piping material selected for the present analysis is ASME SA333Gr6 steel. This is used in the manufacturing of pipes of the Indian Pressurized Heavy Water Reactor (PHWR). Material properties have been taken from literature in which experiments had been conducted on seamless pipes of SA333Gr6 carbon steel material used in Indian PHWRs. All the properties were considered for the analysis and defined in the ANSYS software so that the material for the pipe specimen will be SA 333 Gr 6 carbon steel in the software. The pipes have predefined notch on the outer surface circumferentially. The experiment was simulated to calculate life cycles for the desired boundary condition as mentioned below.

4.1 Pipe Specimen Modeling and FEM Analysis

A set of experiments were conducted on full-scale, four-point bending specimens of seamless pipes of the base material. Pipe was first modeled as shown in Fig. 1 in

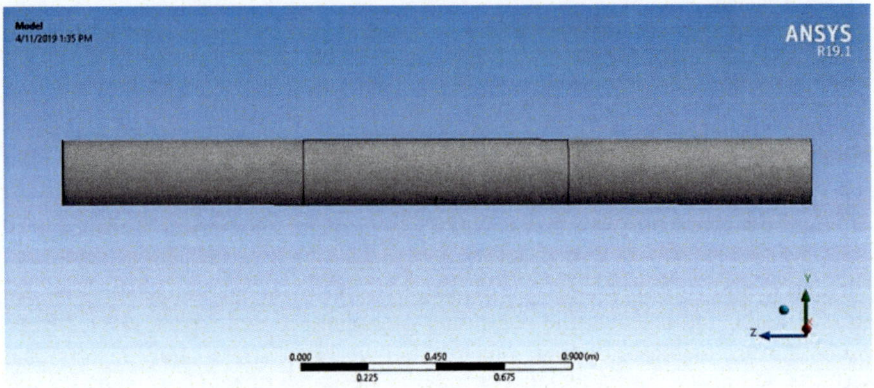

Fig. 1 Model of a pipe specimen

workbench of ANSYS software with distinct dimensions.

PBSC is a pipe specimen without weld seamless pipe of 219 mm outer diameter and 15.38 mm as its total thickness. The total length as outer span of the pipe taken was 2500 mm and inner span was 880 mm on which load has to be applied of 160 kN (P_{max}) and 16 kN (P_{min}). The predefined notch on the pipe is on the outer surface in the center of pipe in the circumferential direction, which was initially 113.0 mm in length and 3.5 mm in depth direction. Specifications considered to model the seamless pipe specimen in the software are given Table 1:

where

PBSC	Pressurized base straight component,
O	Douter diameter of pipe,
t	thickness of pipe,
$2C$	notch length,
P_{max}	maximum load on the pipe,
a	notch depth,
R	load ratio.

Pipe was prepared using the material SA 333 Gr 6 by defining all its properties and simulated as shown in Fig. 2

Pipe was fixed at both the end surfaces. Loading frame was applied on the inner span, i.e., 880 mm in 2500-mm-long pipe specimen (outer span). Loading was applied perpendicular to the inner span edges. Tests have been conducted under cyclic loading at room temperature and air environment.

After modeling the seamless pipe, meshing has been done. Firstly, meshing has been done generally as is defined in the software. Once general meshing with quadratic element order with smooth and fine mesh metrics completed, patch conforming method was chosen in the menu with tetrahedron meshing with definite element size and command was executed. Then around the crack, refinement of meshing was done with sphere of influence (command option in software) with

Table 1 Modeled pipe specimen

Pipe specimen	Notch location metal	OD (mm)	t (mm)	Outer span (mm)	Inner span (mm)	$2C$ (mm)	a (mm)	P_{max} (kN)	R
PBSC	Base	219	15.38	2500	880	113.0	3.50	160	0.5

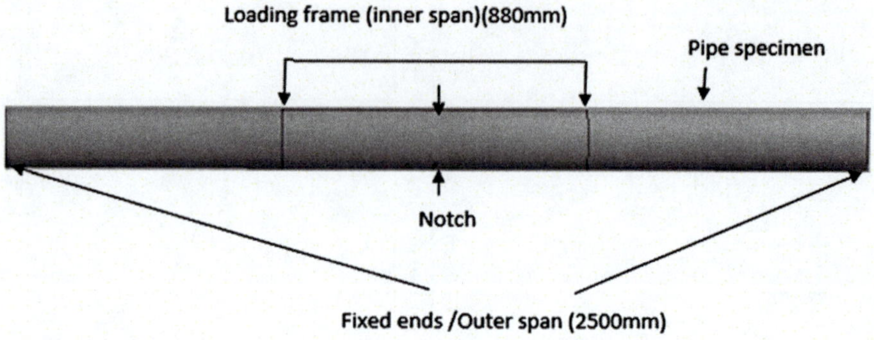

Fig. 2 Test setup for pipe

a generated coordinate system as the center of the crack with fine element size as accepted as per the geometry of the model (but should be done with <1 mm). Creating a finite element mesh with a crack can be a challenging and time-consuming effort. These areas should be re-meshed and further refined to check for solution convergence. This pipe specimen finally generates 39,568 nodes and 28,514 elements (Fig. 3).

The number of cycles on the increment of crack with a certain dimension has been recorded with their respective SIF values on the tips of length as well as depth direction of the crack. Using Paris' law, the crack growth process was repeated up to through-wall crack.

As we started applying load on the specimen, it starts showing deformation. Initially, the deformation was elastic, but repeated loading–unloading cycle on the pipe specimen brings it to the elastic zone so that the permanent deformations increase the crack length. Figures below, showing results of life cycles and variations of SIF values at different crack depths of the pipe specimen when subjected to the experimental conditions, during the crack growth process till through-wall crack, are obtained through ANSYS R19.1 software. Figures 4 and 5 give values for the number of cycles and SIF values at crack initiation, i.e., when crack extended to 0.1 mm depth. Subsequent figures are recorded showing the number of cycles and

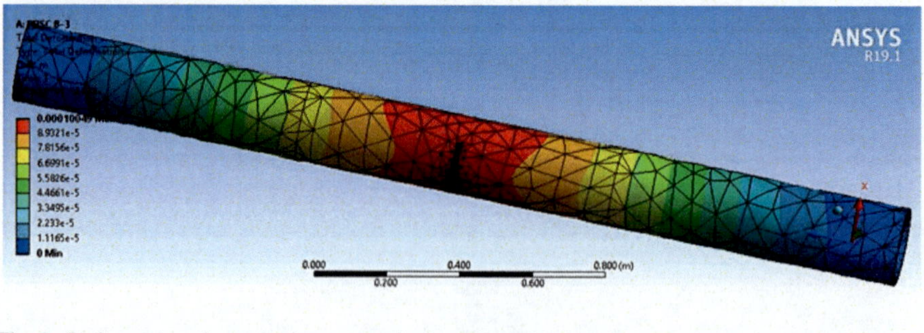

Fig. 3 Deformation shown in the pre-meshed cracked pipe specimen

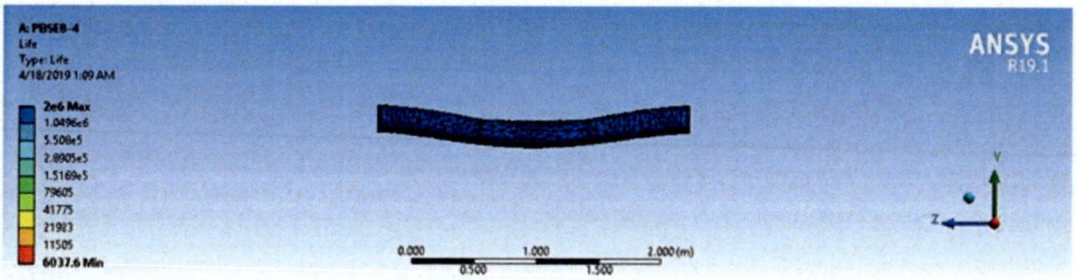

Fig. 4 Fatigue life cycles at 3.6 mm crack depth

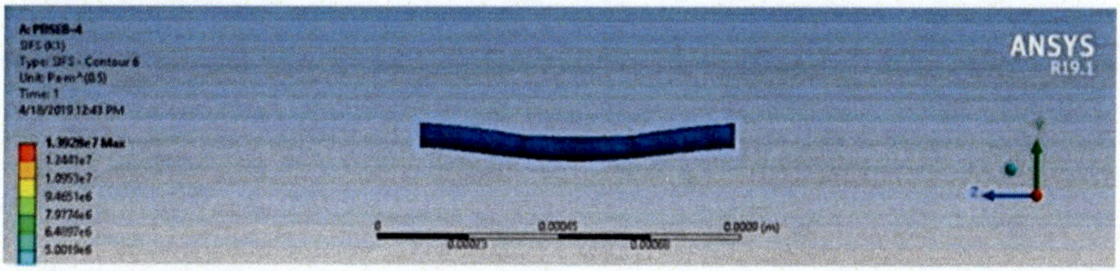

Fig. 5 Variation of SIF at 3.6 mm crack depth

SIF values for the pipe specimen as the crack progresses due to fatigue (Figs. 6, 7, 8, 9, 10, 11, 12, 13, 14 and 15).

Crack initiation depth was considered to be a 0.1 mm increment in the previous crack depth (i.e., crack to grow by 0.1 mm). The software runs to get cycles for fatigue

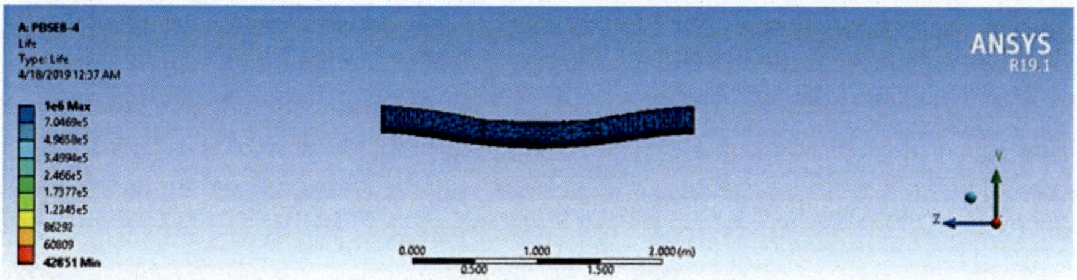

Fig. 6 Fatigue life cycles at 6 mm crack depth

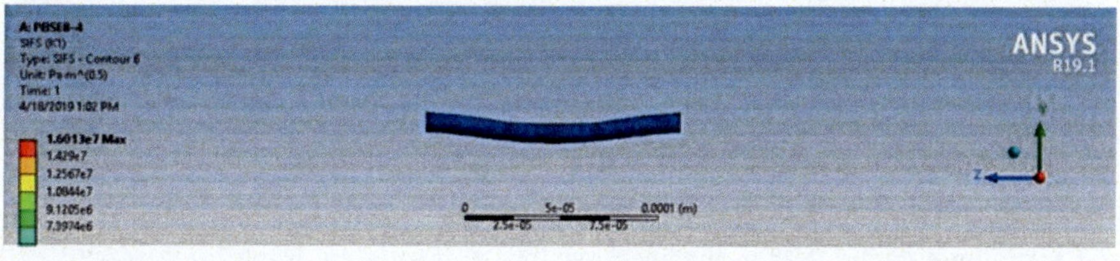

Fig. 7 Variation of SIF at 6 mm crack depth

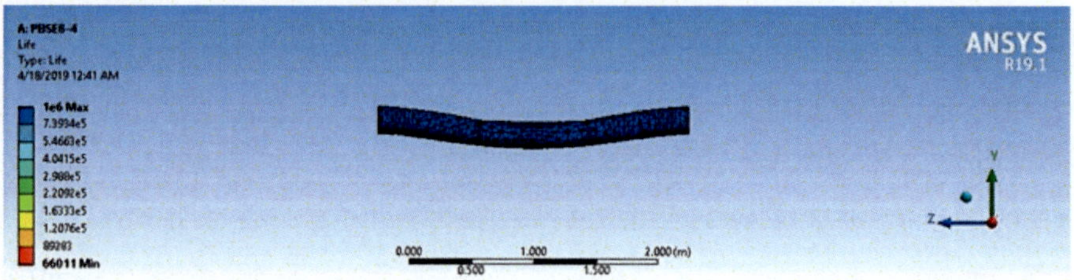

Fig. 8 Fatigue life cycles at 8 mm crack depth

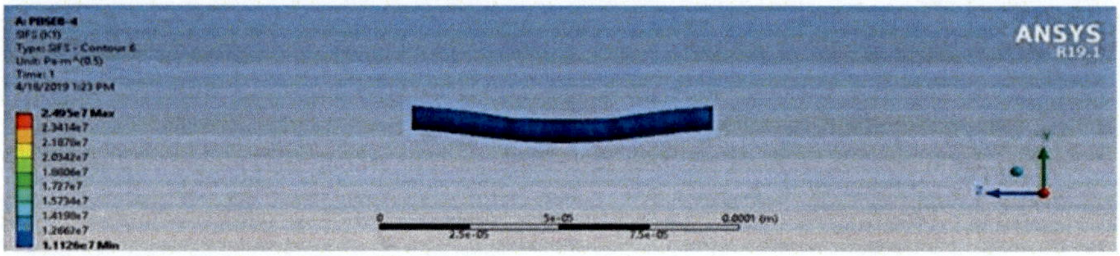

Fig. 9 Variation of SIF at 8 mm crack depth

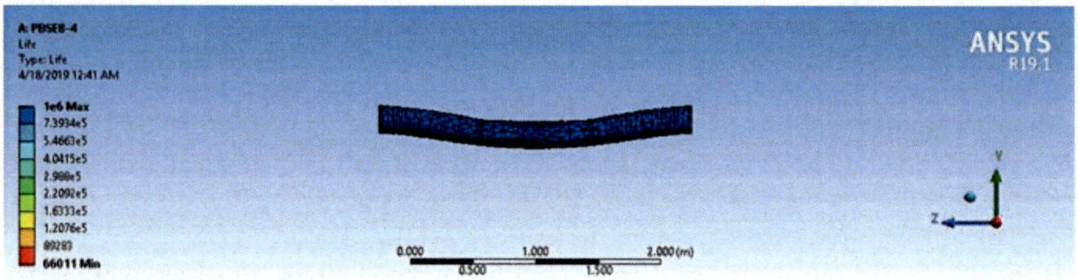

Fig. 10 Fatigue life cycles at 10 mm crack depth

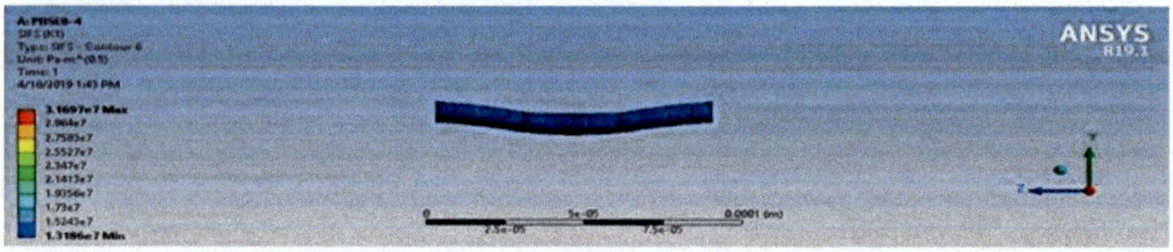

Fig. 11 Variation of SIF at 10 mm crack depth

crack initiation which is found to be 6038. Different values of stress intensity factors and fatigue cycles in the analysis by using ANSYS R19.1 software simulation done for pipe specimen for every crack depth in mm till through-wall crack, as reported after applying a cyclic load to the specimen are presented in graph Figs. 16 and 17, respectively.

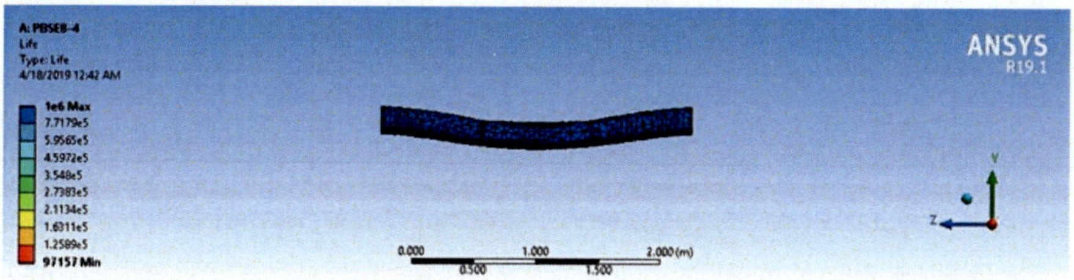

Fig. 12 Fatigue life cycles at 12 mm crack depth

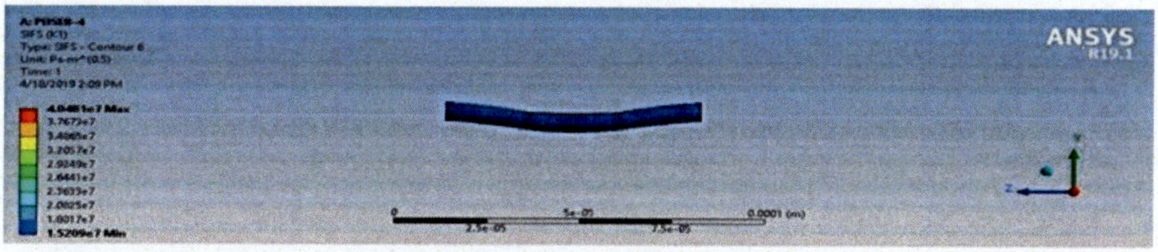

Fig. 13 Variation of SIF at 12 mm crack depth

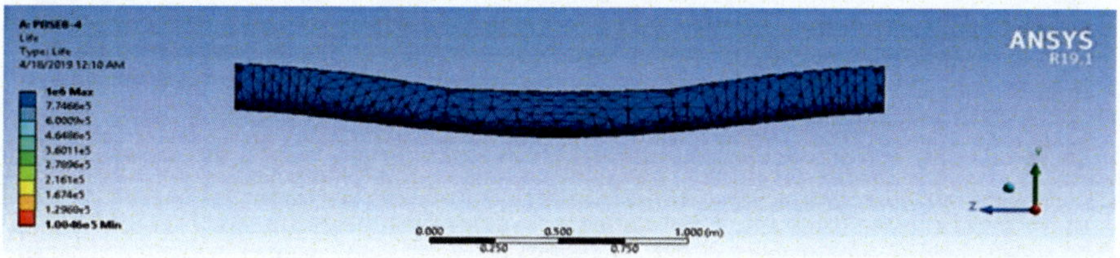

Fig. 14 Fatigue life cycles at 14 mm crack depth

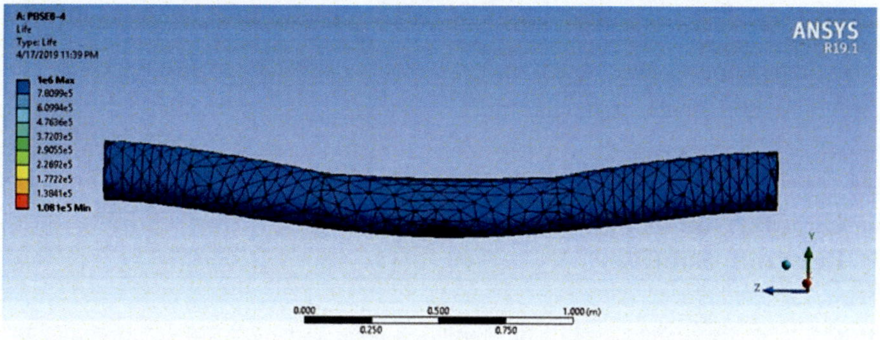

Fig. 15 Fatigue life cycles at 15.38 mm crack depth

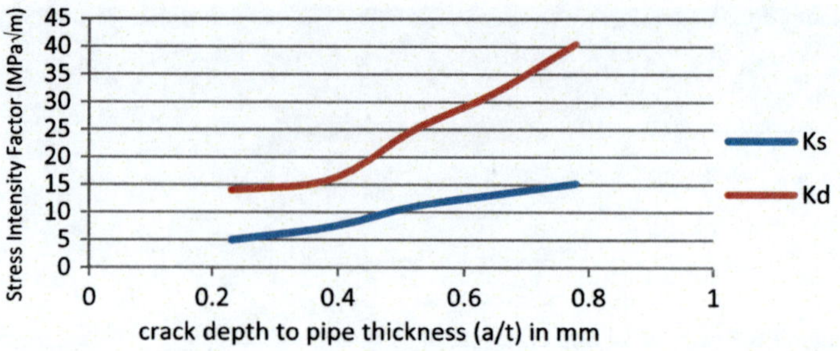

Fig. 16 Variation of K_s (MPa$\sqrt{}$m) stress intensity factor at crack ends at free surface and K_d (MPa$\sqrt{}$m) versus crack depth/ pipe thickness *a*/*t* (mm)

Fig. 17 Variation of crack depth with fatigue cycles for pipe PBSC 8-4

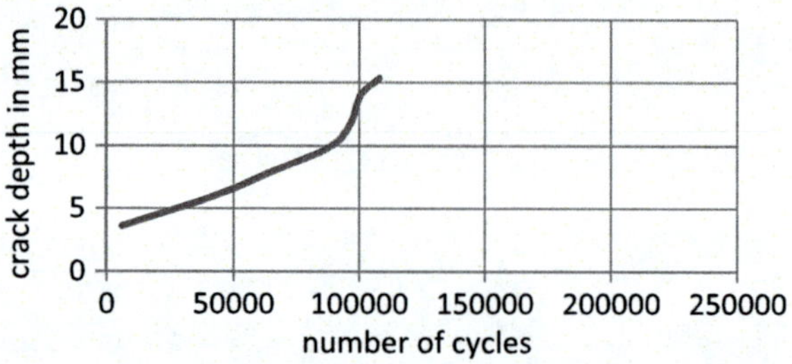

In figures below, stress intensity factor at crack depth or crack tip (K_d) and SIF at crack ends or free surface (K_s) of cracks in MPa$\sqrt{}$m are plotted against crack depth to pipe thickness (*a*/*t*) in mm for pipe PBSC 8-4, as values are obtained after the simulation through ANSYS R19.1 in the present analysis. The values of SIFs K_s and K_d are found to be 15.2 MPa$\sqrt{}$m and 40.5 MPa$\sqrt{}$m at through-wall crack through ANSYS, respectively.

The graph below represents the values of fatigue cycles for pipe PBSC 8-4 obtained at different crack depths in mm.

5 Comparison of FEM Results with ASM Solution and M. Bergman Solution

PBSC 8-4 pipe specimen details of prediction results using ASM solution and M. Bergman solution for SIF values for depth and surface of the crack ends and fatigue life cycles are recorded at different crack depths, while crack propagation process till through-wall crack of the specimen.

Results of ANSYS R19.1 for pipe specimen have been compared graphically with predicted experimental results of ASM and finite element results of M. Bergman solution using ABACUS code available in literature. Data of stress intensity factor at surface (K_s) and stress intensity factor at depth point (K_d) are plotted with crack depth to thickness ratio (*a*/*t*) of the pipe specimen in Figs. 18 and 19 comparing all the three results.

The number of fatigue cycles during propagation of the crack in the pipe specimen of all the three results were plotted against crack depth/thickness (*a*/*t*) for comparison, viz. ANSYS results, ASM experimental results, and M. Bergman results using ABACUS finite element software in Fig. 20.

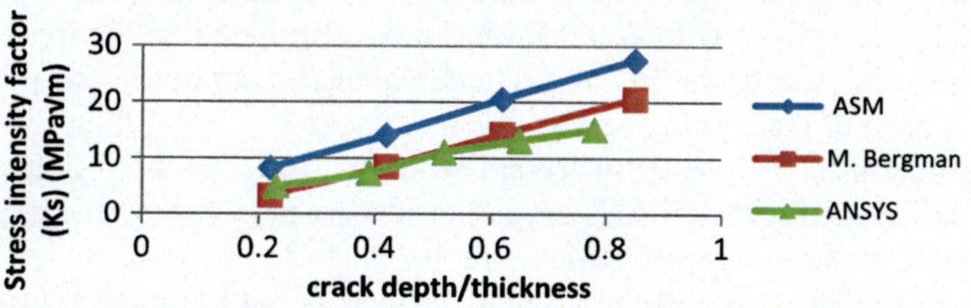

Fig. 18 Variation of SIF values at surface ends (K_s) with *a*/*t*

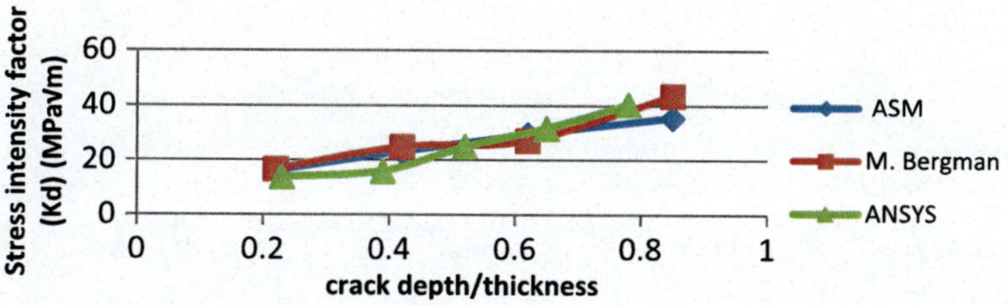

Fig. 19 Variation of SIF values at deepest point (K_d) with *a*/*t*

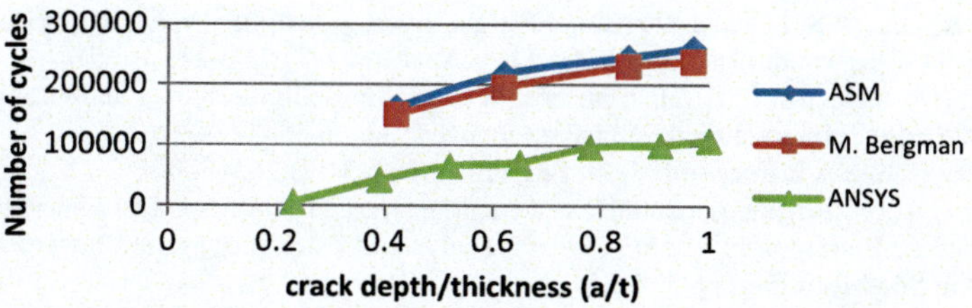

Fig. 20 Number of fatigue cycles vs. crack depth/thickness (*a*/*t*)

6 Conclusions

Piping material selected for the present analysis is ASME SA333Gr6 steel. This is used in the manufacturing of pipes of the Indian Pressurized Heavy Water Reactor (PHWR).

1. The analysis performed through finite element technique-based software ANSYS R19.1 for pressure-based straight piping component of dimension length 2500 mm, outer diameter 219 mm, and thickness of 15.38 mm and loaded vertically downward with 160 KN uniformly distributed on the inner span of 880 mm opposite to the initial circumferential surface crack (2C—113 mm, a—3.5 mm) of the specimen.
2. The solutions shown in the work are accurate due to the used finite element technique-based software ANSYS and results are nearer to other experimental solutions available in open literature [8, 9].
3. As ANSYS saves much time and this method doesn't require a huge infrastructure setup, it saves money and production volume too. This method is even helpful in designing new pipe components by removing flaws of the old designs, which are responsible for less life to the piping components.

References

1. E606–93 (1995) Standard test method for measurement of fatigue crack growth rate- annual book of ASTM standard, vol 03(01)
2. Bennantine JA (1989) Fundamentals of metal fatigue analysis, chapter 2, Prentice Halls, Englewood Cliffs, pp 40–71
3. Farahmand B et al (1997) Fatigue and Fracture mechanics on high risk parts, Chapter 2, ed. 9. Chapman & Hall, page 13
4. Dinesh A, Dinesh S, Dhakad SK (2019) Fatigue crack growth life evaluation of straight pipe component using computer software. Int J Res Anal Rev 6(1):211–215
5. Richard HA, Sander M (2016) Fatigue crack growth. In: Solid mechanics and its applications, vol 227. Springer International Publishing Switzerland, pp 56–57
6. Schutz W (1996) A history of fatigue. Eng Fract Mech 54(2):263–300
7. Basquin OH (1910) The exponential law of endurance test. Proc Am Soc Test Mater 10:625–630
8. Paris Paul C, Endrogen F (1963) A critical analysis of crack propagation laws. J Basic Eng Trans Am Soc Mech Eng, pp 528–534
9. Paris PC, Erdogan F (1963) Trans ASME J Basic Eng 85:528–534
10. Bergman M (1995) stress intensity factors for circumferential surface cracks in pipes. Fatigue Fract Eng Mater Struct 18(10):1155–1172
11. Kelly JM (1991) Solid mechanics, Part—I—an introduction to solid mechanics

Analytical Hierarchy Process in SMEs of Punjab

Satnam Singh, Ankur Bahl, and Guravtar Singh Mann

1 Introduction

AHP approach developed is the combination of two different fields, i.e., maths and interaction of the projected work [1]. AHP is effective in prioritizing the factors to mitigate unforeseen accidents in industries [2, 3]. Implementation of prioritized factors further saves the unallocated funds. Such tools have the capacity to provide growth in industries [4, 5]. In order to get the best parameter selection for occupational safety, it is important that all the responses are optimized simultaneously. AHP is one of the most successful techniques for solving decision-making problems involving the goals, the alternatives for reaching the goals and the criteria for evaluating the alternatives [6, 7]. In this technique, by using the simple mathematical theory of Eigenvector and priority, a number of pair-wise comparison matrices are formed as per their relative importance with respect to the parent element. The advantage of this tool is that it combines both qualitative and quantitative parameters [8, 9]. The detailed steps are as follows:

1. At the topmost level, this comprises a goal or a focus. At the intermediate and lower levels, the approach deals with criteria or sub-criteria and the available alternatives, respectively.
2. Construction of a pair matrix for each level with esteem to higher levels. The relative importance of different alternatives with respect to the immediately above sub-criteria is determined.
3. Application of Eigenvector methods to calculate the relative weight for the pair-wise comparison of options on each criterion.
4. Check the consistency related with matrix. This is achieved using consistency ratio (CR) of consistency index (CI) with the proper range of random index (RI).

S. Singh (✉) · A. Bahl · G. S. Mann
School of Mechanical Engineering, Lovely Professional University, Phagwara, Punjab, India
e-mail: satnam.singh@lpu.co.in

A. Parey et al. (eds.), *Recent Trends in Engineering Design*, Lecture Notes in Mechanical Engineering, https://doi.org/10.1007/978-981-16-1079-0_26

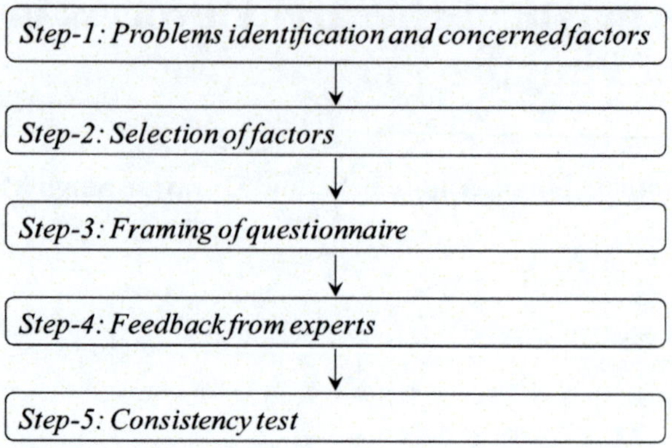

Fig. 1 Methodology for analytic hierarchy process

5. Repeat the above steps.
6. Evaluate the general relative value by linear addition function.

2 Prioritization of Occupational Safety Factors Using AHP

AHP is successfully implemented in different fields like field of manufacturing, design of layout, appraisal of tools, flexible mechanized system, etc. [10, 11]. AHP is based on calculation of an Eigenvalue with different pair-wise comparison [12]. Qualitative and quantitative calculations are carried out by the use of AHP [13, 14].

The objective of this analysis is to utilize the effectiveness of AHP to prioritize the factors associated with the safety of workers in the small and medium size manufacturing industries (SSMIs and MSMIs). A questionnaire is developed for conducting the study using AHP technique. Feedback from knowledgeable and skilled experts is taken on an developed questionnaire for different parameters with pair-wise comparison. Based on the results of **the** earlier chapter, a total of nine main factors are considered with 22 and 40 sub-factors for SSMIs and MSMIs, respectively. The methodology of AHP involves five key steps as shown in Fig. 1. Moreover, all the steps are carried out with full care.

It is ensured after pair-wise comparison, CR is below or nearly 10%, only then the questionnaire is considered good. If CR comes out to be more than 10%, the questionnaire needs to be reviewed if it is not accommodating the possible factors.

3 AHP Hierarchy Model for SSMIs and MSMIs

The hierarchy model for SSMIs is shown in Fig. 2 in the form of nine main factors and 22 sub-factors. The detailed procedure of the AHP method, the pair-wise comparison based on experts' opinion is detailed for the sub-factors f_{41},f_{42},f_{43} of the main factor

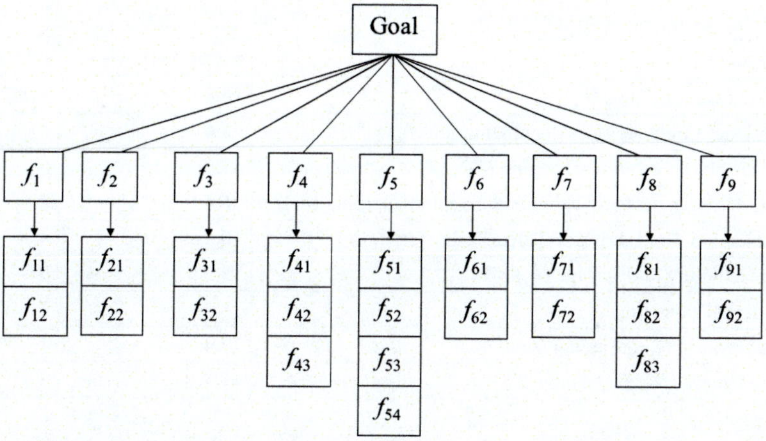

Fig. 2 AHP hierarchy model for SSMIs

f_4 only. These three sub-factors refer to f_{41}: Periodic inspection (PI); f_{42}: Availability of proper machine guards (APMG); f_{43}: Training sessions for tools and equipments use (TPFHT).

The hierarchy model for MSMIs is shown in Fig. 3 in the form of nine main factors and 40 sub-factors. As an example, the pair-wise comparison based on experts' opinion is detailed for the factor f_1 having sub-factors f_{11},f_{12},f_{13}. These three sub-factors refer to f_{11}: Existence of safety policy (ESP); f_{12}: safety department (FSD); f_{13}: Qualified safety and health specialists engineers (QSHS). The factors are prioritized using AHP, and the priority level is tabulated in Table 1. Based on this analysis, the highest priority is found for factor PPEs. Therefore, PPEs are the main and very important parameter, and there is a necessity for its priority to prevent the

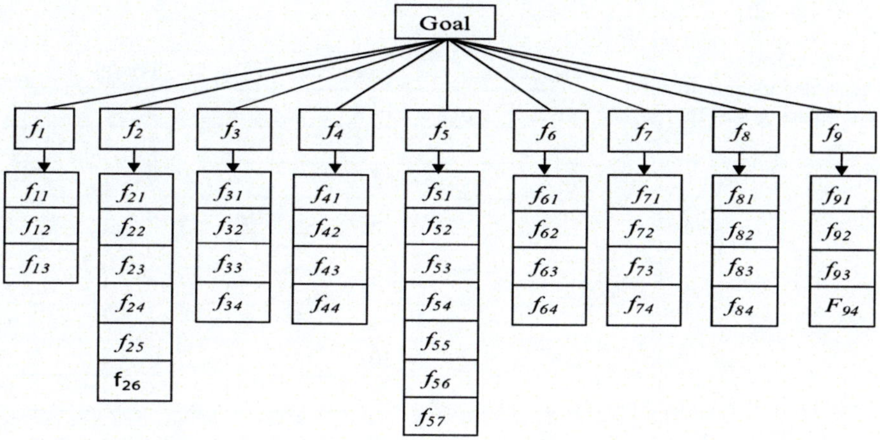

Fig. 3 AHP hierarchy model for MSMIs

Table 1 Values for Eigenvector (SSMIs)

Main factor	Eigen vector (%)	Priority
Organizational attributes (f_1)	16.8	2
Occupational safety services/documentation (f_2)	8.4	5
Workplace layout and housekeeping (f_3)	7.0	6
Equipment and hand tools safety and machine guarding (f_4)	6.4	8
Fire prevention, firefighting and electrical safety (f_5)	10.1	4
Material handling and storage (f_6)	5.9	9
Occupational exposures (f_7)	6.5	7
Personal protective equipment (f_8)	27.4	1
Hygiene factors (f_9)	11.4	3

life of workers and cost at the workplace of SSMIs. Other factors also need to be implemented as per their priority level given in Table 1.

The major advantage of this analysis is that the factors, as well as their sub-factors, are also prioritized. This leads to considerable savings in cost that can be offset for worker welfare. For example, the first main parameter found was PPEs with some sub-factors. The sufficient supply of PPEs is 3.89 and 9.02 times important as compared to maintenance and training of PPEs, respectively. So, necessary actions can be taken accordingly, and SSMIs can take the benefit from this prioritization.

The main factors associated with worker's safety, in MSMIs, are prioritized with AHP expert decision-making technique as tabulated in Table 2. According to expert feedback, the personal protective equipments (PPEs) are again given first priority followed by fire prevention system and organizational attributes. It is suggested that PPEs must be provided to the employee, who are working on the shop floor. This explains and confirms the relevance of AHP methodology for proper prioritization of occupational safety factors.

Table 2 Values of Eigenvector and priority for main factors of MSMIs

Main factor	Eigenvector (%)	Priority
f_1	14.0	3
f_2	7.6	5
f_3	2.5	9
f_4	10.1	4
f_5	22.8	2
f_6	4.6	7
f_7	4.7	6
f_8	30.6	1
f_9	3.0	8

4 Conclusion

The technique, AHP is carried out to prioritize the factors of SSMIs and MSMIs. In case of SSMIs, it is observed that first priority is given to f_8 (personal protective equipment) followed by f_1 (organizational attributes), f_9 (hygiene factors), f_5 (fire-fighting, fire Prevention, and electrical safety), f_2 (occupational safety documentation/services), f_3 (workplace layout and housekeeping, f_7 (occupational exposures), f_4 (machine guarding), and f_6 (storage). It is recommended that SSMIs can follow this hierarchy while implementing the safety measure. On the other hand, factors for MSMIs are also prioritized with weightage given by experts. It is found that first priority is given to f_8 (personal protective equipment) followed by f_5 (fire-fighting, fire prevention, and electrical safety), f_1 (organizational attributes), f_4 (machine guarding and equipment and hand tools safety), f_2 (occupational safety documentation /services), f_7 (occupational exposures), f_9 (hygiene factors), and f_3 (workplace layout and housekeeping). On the basis of AHP analysis, it is suggested that factor f_8 (personal protective equipment) is the prime factor which needs utmost care in both SSMIs and MSMIs. It is suggested to the SMEs that PPE should be given the highest priority and be supported by regular training programs in order to prevent workplace accidents. Further, if the industry is new and wants to implement some of the safety measure, then also industry can follow the above-discussed hierarchy for better and safe working environment.

References

1. Harker PT, Vargas LG (1987) The theory of ratio scale estimation: Saaty's analytic hierarchy process. Manag Sci 33(11):1383–1403
2. Arbel A, Orgler YE (1990) An application of the AHP to bank strategic planning: The mergers and acquisitions process. Eur J Operat Res 48(1):27–37
3. Saaty TL (1985) Decision making for leaders. IEEE Trans Syst Man Cybern 3:450–452
4. Bayazit O (2005) Use of AHP in decision making for flexible manufacturing systems. J Manuf Technol Manag
5. Al-Harbi KMAS (2001) Application of the AHP in project management. Int J Project Manag 19(1):19–27
6. Singh S, Singh LP, Kaur M (2016) Analytical hierarchy process-based methodology for selection of safety parameters in manufacturing industry. In: Advances in safety management and human factors. Springer, Cham, pp 357–366
7. Boucher TO, MacStravic EL (1991) Multiattribute evaluation within a present value framework and its relation to the analytic hierarchy process. Eng Econ 37(1):1–32
8. Cambron KE, Evans GW (1991) Layout design using the analytic hierarchy process. Comput Indus Eng 20(2) 211–229
9. Atsumbe B, Amine J, Umar I, Salawu J (2012) Appraisal of occupational safety in manufacturing industries in Kaduna and Niger states of Nigeria. Int J Eng Res Dev 6:55–62
10. Armacost RL, Componation PJ, Mullens MA, Swart WW (1994) An AHP framework for prioritizing customer requirements in QFD: an industrialized housing application. IIE Trans 26(4):72–79

11. Das MC, Sarkar MCB, Ray S (2012) A framework to measure relative performance of Indian technical institutions using integrated fuzzy AHP and COPRAS methodology. Socio-Econ Plann Sci 46(3):230–241
12. Saaty TL (1990) How to make a decision: the analytic hierarchy process. Eur J Oper Res 48(1):9–26
13. S.Singh, M.Kaur, A.Bahl and R.S. Uppal.: Occupational safety evaluation index-based obstacles prediction for manufacturing industries. Modern Physics Letters B, 34(11), p.2050104(2020).
14. Singh LP, Singh S (2019) Strategic enhancement of workplace safety in small scale manufacturing industries using AHP approach. Int J Anal Hierarchy Process 11(2):269–282

A Study of Virtual Simulation for Five-Axis Machining

Vishaldeep Singh, Hitesh Arora, and Prashant Kumar Pandey

1 Introduction

The multi-axis machining for the manufacturing of the products has been used for long time. The Computer Numeric Control (CNC) machines can move a tool or the part to be manufactured on the different axis of the machine at the same time. In three-axis machines, the part can move in three directions, whereas in five-axis machines the two additional rotations are there. Figure 1 shows the universal head which is mainly used in industries for the manufacturing of parts. These machines are mainly used because of the flexibility in manufacturing and accuracy with which the parts are manufactured. Also, the parts with complex structures are manufactured in less time.

There are many benefits of five-axis machining system. The machining time for the sculptured surfaces is reduced that too with accuracy and better surface finish. Also, this system reduces the multiple setups used during machining of the parts. It also reduces the cost of fixtures which otherwise were used to machine the complex parts. Figure 2 compares the three-axis and five-axis machining system. Figure 2a it shows that if the inclined surface is to be manufactured, in three-axis machining system the tool has to be moved number of times, whereas in five-axis machining system the tool can be tilted and directly the tool can move on the inclined surface. Figure 2b shows that for the contoured surfaces, in three-axis machining the material is uncut, whereas in five-axis machining the material can be easily removed because of the easy tool accessibility. Along with so many advantages of this system, there are

V. Singh · H. Arora (✉) · P. K. Pandey
Department of Mechanical Engineering, Lovely Professional University, Phagwara, Punjab, India
e-mail: hitesh.15774@lpu.co.in

V. Singh
e-mail: Vishaldeep.18537@lpu.co.in

P. K. Pandey
e-mail: prashant.15821@lpu.co.in

A. Parey et al. (eds.), *Recent Trends in Engineering Design*, Lecture Notes in Mechanical Engineering, https://doi.org/10.1007/978-981-16-1079-0_27

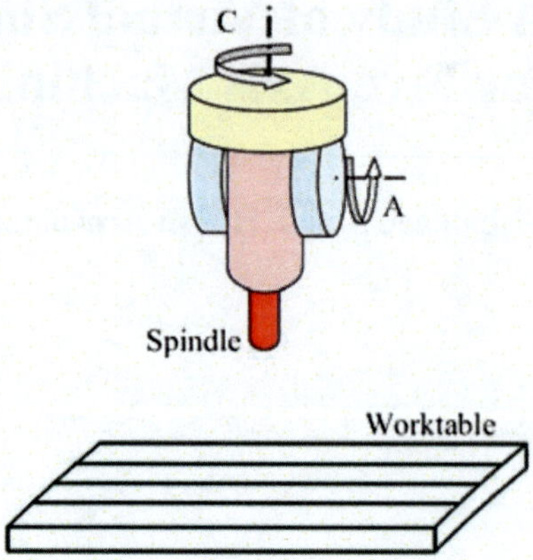

Fig. 1 Interpretation of five-axis machine with a universal head [1]

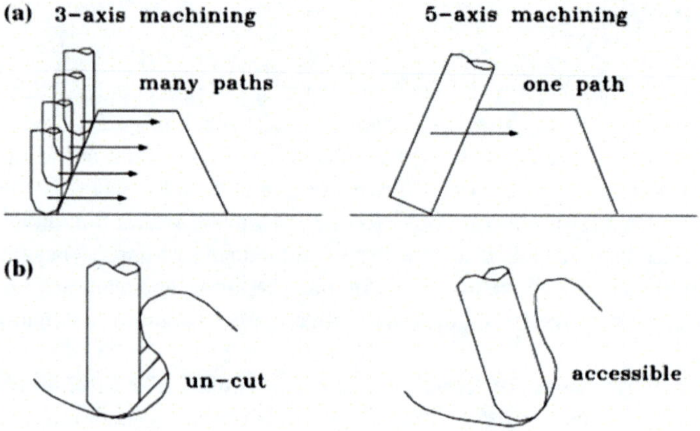

Fig. 2 Comparison three-axis machining versus five-axis machining [2]

certain disadvantages as well. If the tool path is not generated properly, the chances of the collision with the other surfaces could be there. Also, to set the axis needs more care.

There are three linear axes and two rotary axes in five-axis machining centers. Apart from this, there are many configurations of machining centers that are available in the market but three of them are very commonly used in the industry. The first category of the machine head is of the universal head type in which two rotary axes are there. This is also known as table spindle type configuration. In the second configuration, the table is tilted with two of the rotary axis. Whereas the third type is combination of the first two types in which the head rotates and table also rotates [3].

A study has been done with regard to tool path planning, tool path generation, and simulation of the tool path for five-axis machining.

2 Tool Path Planning and Tool Path Generation

Liang et al. [4] proposed a Standard for Exchange of Product Model Data (STEP) model to produce the tool path for the sculptured surface. In his research work, he proposed four components. First of all, the data is extracted, then the slicing of the module is done then the Cutter Location (CL) file is generated and from this, NC code is generated. The tool path is generated from the CAD model directly from his proposed research work. From their proposed work to prepare the tool path, the intermediate data exchange is not required.

Rao et al. [5] implemented the Principal Axis Method (PAM) to machine the surface. In their approach, the curvatures of both design surface and machined surface are compared which are used to find the cutter orientation. The toroidal cutter was aligned to the design surface and so oriented to minimize the scallops. The approach was applied to the five-axis machine configuration and found the scallop height between the movements has considerably reduced. The tool path generated was simulated and verified experimentally and found that the scallop heights were reduced.

Chiou and Lee [6] proposed a Machining Potential Field (MRF) method to produce the tool path for the sculptured surfaces. For the planning of the tool path, the geometry of the surface and cutter is considered along with the orientation of cutter. This method generates the tool path with which the surface finish is improved. They proposed an algorithm which generated the optimal tool path. This tool path generated adjacent tool path and minimizes the overlapping and the scallops. The adjacent tool paths are generated from the initial tool paths. The surface to be machined is divided into two portions that is unfinished and finished area. The tool moves along the cutting direction and produces the sculptured surface. The experimental and simulated results were verified and found that smooth surface finish was produced and with shorter tool length. This approach leads to decrease in the machining time without compromising the quality of the product.

Lin and Ye [7] proposed the Stereolithography or Standard Triangle Language (STL) model to generate the tool path. In the five-axis machining, the cutter contact points make a nonlinear trajectory which leads to undercut or overcut of the part. This error is known as the contour error. In their research work, they worked on the contour error. They controlled the orientation of the tool and step size according to the pre-determined trajectory error.

Duvedi et al. [8] proposed a multi-point method for the planning of tool path. They proposed a drop and tilt method onto the triangulated surfaces to produce the tool path. They proposed the algorithm to generate the tool path in which the tool is dropped on the surface and is allowed to rotate about the pseudo axis till the tool touches the next triangulated surface. This algorithm proposed by the authors will generate the gouge-free tool path in which the tool is assumed to be rotated about the pseudo axis. In this, the author tested the algorithm experimentally and further proposed to optimize the tool path.

Duvedi et al. [9] extended their drop and tilt method to a tensor product Bezier surfaces. In this approach, they dropped the tool so that it touches the first point of contact and then the tool is tilted about tensor product Bezier surfaces. The tool path for Bezier surface was generated and testing was done using toroidal milling cutter. The tool path generated was tested experimentally. This approach can be extended for the generation of tool path at commercial level. The optimization of the tool path and the errors generated like undercut are still to be minimized.

Ezair and Elber [10] presented a technique to generate a smooth and obstacle-free tool path for five-axis machining system. The proposed method can be categorized as C-space method. They proposed an algorithm to generate the tool path and is verified by the simulation. The entire cutter contact curve can be generated for the whole surface and need not generate the cutter contact points. The proposed method in which the cutter contact curve will specify the tool path along with the orientation of the tool. The proposed approach by the authors is verified using the simulation of the machining.

Sharma et al. [11] proposed a new method for the location of tool for five-axis machining. In their research work, they have used the drop and spin method to find the contact points. From the algorithm proposed by them, tool drop locations are generated for the reference line. The toroidal-shaped geometry is formed by the sweep approach in which the dropped tool is rotated about spin axis creating a gouge free surface which was a limitation of the proposed methodology of drop and tilt approach by Duvedi et al. [9]. They tested their approach for convex and concave-shaped surfaces and found that the clean region is formed at the common edge zone.

Zubair and Mansor [12] proposed algorithm which automatically generates tool path for grooved surfaces by using sub-delta volume approach which is extracted from three-dimensional model. This approach fetches the different vertices from the geometry and generates the tool path.

3 Validation of Tool Path Using Machining Simulation

The tool path generated needs to be verified. To validate these tool paths generated, there are different methods. The most common method is dry run method, in which the machine runs without raw material. The limitation of this method is that the worker should be experienced to check the tool path. The other method is to load the soft material like sponge or thermocol to check the tool path. The other method is to measure the coordinates of the machine but it is expensive. The best method to check the tool path is simulation. The tool path and workpiece are generated and the tool is allowed to simulate. Simulation verifies and analyzes the tool path without any physical machining requirement of the workpiece. The idea is to use the simulation results as basis for making decision for the real-life machining process. The input parameters can be varied to study the behavior of the real time process.

Complex sculptured surfaces are widely used in many engineering objects such as automobile components, aerospace parts, turbines blades, and medical implants. The manufacturing of such custom intricate designs is a costly and time-consuming task. Computer-aided design and Computer-Aided Manufacturing technology have been around for decades to assist with design and manufacturing of such components. With CAD/ CAM software, Computer Numeric Controlled machines can machine intricate components with the highest dimensional accuracy. CAD/CAM software simplifies the programming of CNC machines and helps improve their utilization.

Bedi et al. [13] proposed a technique to machine complex parts. In their proposed work, the material removal is increased by aligning the principle axis of machining surface and machined surface. This method will also reduce the height of the scallop. This technique will also reduce the machining time.

Roth et al. [14] in their research work has found the cutting forces using the transformation matrices. To find the cutting force, they found the amount of material removed and the chip generated during the machining process. To find the chip geometry, they used local depth buffer and adaptive depth buffer. The tool was sized in local depth buffer, whereas in adaptive depth buffer, the buffer orientation was matched with the axis of orientation.

Tutunea-Fatan and Feng [1] proposed a kinematic model for the machine tool. This model is used to determine the movement of tool for a particular location and orientation. This model is also helpful for post-processing, design, and analysis of machine error. In their proposed work, coordinate matrix are formed to find the tool length for Spindle Rotating machine and location of workpiece for Rotary Table machine.

Mann et al. [15] in their proposed work have calculated the points on the machined surface. These points are calculated using tool position and the motion of tool. This method is known as imprint point method. They proposed a hierarchical model approach for finding the direction of motion of tool. This model is formed using transformation matrices and can be used on any CNC machine with single tool head.

Sullivan et al. [16] proposed a new technique for the simulation of the tool path for three- and five-axes machines. In composite adaptively sampled distance field (composite ADF), the swept volume of the tool is calculated and the Boolean operation is done on the swept volume by tool and workpiece. The remaining part is the end result which is simulated. This approach increases the simulation speed as very low memory is used in this methodology.

Singh et al. [17] in their research work proposed a simulation of three-axis and five-axis CNC tool path. They used the CAD package to produce the workpiece and tool in the CAD package. The Boolean operation is used to subtract the tool from the workpiece. The verification of the simulation result has been done by ray tracing algorithm. The scallop height was calculated by comparing the actual part and simulated part; it was found that the results are valid with minimum errors.

Shneor et al. [18] proposed a software for the prediction of accuracy of five-axis machined part. It works on the form shaping system theory. The input for this software is geometry of workpiece, tool parameters, and tool path. The results generated from this software are compared with coordinate measuring machine. This simulation result shows the effectiveness of the results.

Tunc [19] in his research work proposed the smart tool path generation for five-axis machining of curved surfaces. The tool path planning, tool path generation, and simulation of generated tool path are discussed. The simulation of the tool path also gives the cutting forces and scallop height for the given tool path. The main aim is to reduce the process time by using simulation in tool path planning.

4 Case Study

In this work, Singh et al. [17] proposed a simulation of five-axis machining tool path for sculptured surfaces. In his study, the macro was created for the machine simulation in which the workpiece can be inserted of defined dimensions in the SolidWorks (student version), and desired tool for the machining purpose is inserted in the software for the defined axis which is generated from the toolpath. This approach used by the author is similar like drop and tilt method of tool location as proposed by Duvedi et al. [9]. The authors have further performed the Boolean subtraction for the particular tool inserted at a particular location. This approach of subtracting the tool from the workpiece leads to perform the infinite calculation to subtract the dropped tool to represent the simulation which is clearly mentioned in their research work. To overcome such issue, the modified macro can be created in which first the tool must be allowed to fall on the workpiece without performing the Boolean operation. Once the tool is dropped at all locations for the desired tool path then the Boolean subtraction operation can be performed. This simulation by drop tool approach for all the cutter contact points has been used by Duvedi et al. [9] which is shown in Fig. 3. Figure 3 shows the dropped location of all the points on the machine surface for the tool path in simulation. This approach will help to give the simulation result in a better way without affecting the performance of the computation of the system during the simulation process.

Figure 4 shows the tool inserted for the various locations along the axis. The tool is allowed to drop on the particular location of the workpiece. The tool dropped on

Fig. 3 Simulation of tool path for machining a sculptured surface by drop tilt method [9]

Fig. 4 Movement of tool over the workpiece

the workpiece is generated with the help of axis. The input of the axis is generated from the two points which are generated from the cutter location of the tool and other points for the axis is the point on the tool along the tilted axis. After inserting the tool for all the points, the Boolean subtraction operation can be performed. The simulation results for the given tool path are shown in Fig. 5 which shows the isometric view and orthographic views. The sculptured surface is produced with the tool path provided.

The similar approach is used to analyze the simulator for the sculptured surface. Figure 6 shows the sculptured surface for generating tool path from STL file. The STL file is generated from the ProE software (student version). The dimensions of

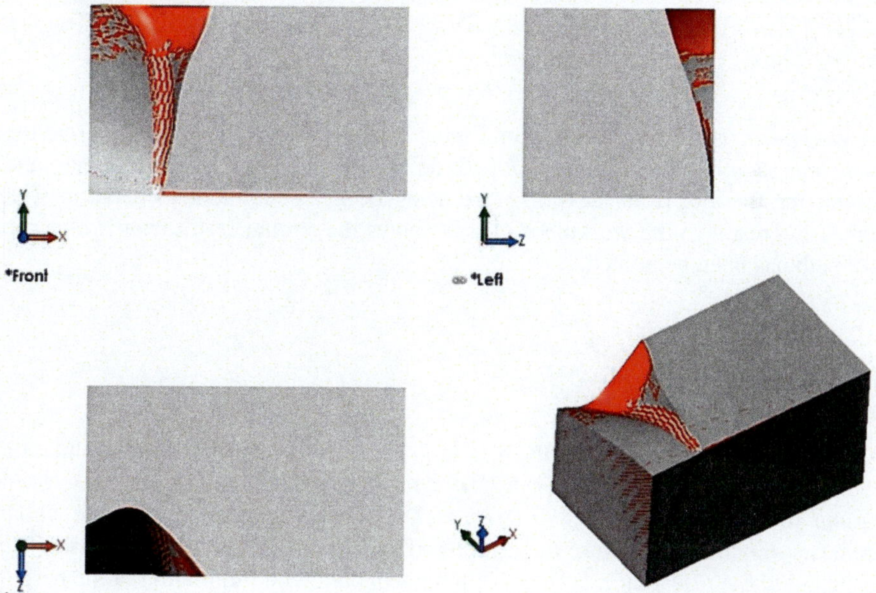

Fig. 5 Simulation results for the sculptured surface

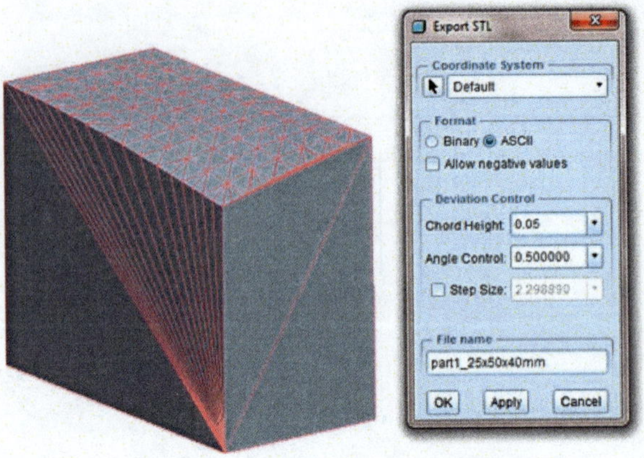

Fig. 6 Sculptured surface for generating tool path

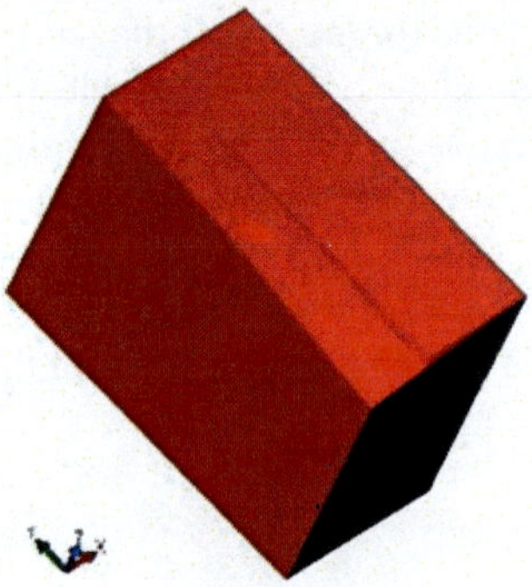

Fig. 7 Simulation result for the sculptured surface

the workpiece are 25 × 50 × 40 mm. Figure 7 shows the simulation result from the same approach by inserting the tool path for various locations of the tool path and subtracting the tool from the workpiece using Boolean subtraction operation. The simulation result of the workpiece clearly shows the similar result from which the tool path has been generated.

5 Summary

A review on the development in the planning, generation of tool path, and simulation of the generated tool path has been discussed in this paper. The authors have discussed various methods and proposed different techniques to generate the tool path which are accurate and manufacture the sculptured surfaces with fine surface finish. The main aim of the researchers is to develop the tool path with minimum scallop height and with minimum errors. The researchers proposed different techniques to simulate these tool paths. A lot of scope is there in this field because most of the researchers have worked to generate the tool path by considering one type of tool and based upon that data the simulation of the tool path has been done. There is no such work was found which can be considered as generalized for all the different tools. To make some general system or algorithm is to be developed, for which the tool path thus generated can be directly be used by the CNC machines. The tool drop method used by Duvedi et al. [9] and Boolean subtraction approach is similar and gives the best results.

References

1. Tutunea-Fatan OR, Feng HY (2004) Configuration analysis of five-axis machine tools using a generic kinematic model. Int J Mach Tools Manuf 44:1235–1243
2. Cho HD, Jun YT, Yang MY (1993) Five-axis CNC milling for effective machining of sculptured surfaces. 31:2559–2573
3. Tsutsumi M, Saito A (2004) Identification of angular and positional deviations inherent to 5-axis machining centers with a tilting-rotary table by simultaneous four-axis control movements. Int J Mach Tools Manuf 44:1333–1342
4. Liang M, Ahamed SS, Van Den Berg B (1996) A STEP based tool path generation system for rough machining of planar surfaces. Comput Ind 32:219–231
5. Rao N, Bedi S, Buchal R (1996) Implementation of the principal-axis method for machining of complex surfaces. Int J Adv Manuf Technol 11:249–257
6. Chiou CJ, Lee YS (2002) A machining potential field approach to tool path generation for multi-axis sculptured surface machining. CAD Comput Aided Des 34:357–371
7. Lin RS, Ye CB (2012) Accurate trajectory control for five-axis tool-path planning. Lect Notes Eng Comput Sci 2196:932–937
8. Duvedi RK, Bedi S, Batish A, Mann S (2015) Numeric implementation of drop and tilt method of 5-axis tool positioning for machining of triangulated surfaces. Int J Adv Manuf Technol 78:1677–1690
9. Duvedi RK, Bedi S, Mann S (2017) Drop and tilt method of five-axis tool positioning for tensor product surfaces. Int J Adv Manuf Technol 93:617–622
10. Ezair B, Elber G (2018) Automatic generation of globally assured collision free orientations for 5-axis ball-end tool-paths. CAD Comput Aided Des 102:171–181
11. Sharma SK, Duvedi RK, Bedi S, Mann S (2019) Five-axis tool positioning of a toroidal end mill tool near common edge between two intersecting surfaces. In: CAD conference, pp 75–79
12. Zubair AF, Abu Mansor MS (2020) Tool-path generation for groove machining feature. IOP Conf Ser Mater Sci Eng 834
13. Bedi S, Gravelle S, Chen YH (1997) Principal curvature alignment technique for machining complex surfaces. J Manuf Sci Eng Trans ASME 119:756–765
14. Roth D, Ismail F, Bedi S (2003) Mechanistic modelling of the milling process using an adaptive depth buffer. CAD Comput Aided Des 35:1287–1303
15. Mann S, Bedi S, Israeli G, Zhou X (2010) (Linda): Machine models and tool motions for simulating five-axis machining. CAD Comput Aided Des 42:231–237
16. Sullivan A, Erdim H, Perry RN, Frisken SF (2012) High accuracy NC milling simulation using composite adaptively sampled distance fields. CAD Comput Aided Des 44:522–536
17. Singh V, Arora H, Pandey PK, Wandra R (2018) A methodology for simulation and verification of tool path data for 3-AXIS and 5-AXIS CNC machining. Int J Mech Eng Technol 9:450–461
18. Shneor Y, Chapsky VS, Shapiro A (2018) Virtual verification of 5-axis machine tools based on workpiece accuracy analysis: software tool instead of expensive machining tests. Procedia Manuf 21:228–235
19. Tunc LT (2019) Smart tool path generation for 5-axis ball-end milling of sculptured surfaces using process models. Robot Comput Integr Manuf 56:212–221

An Investigation of Crack Structural Analysis and Its Propagation of a Taper Roller Bearing Under Different Misalignment Conditions

Rajeev Kumar, Manpreet Singh, and Jujhar Singh

1 Introduction

Bearings are the broadly used elements in rotating machineries and even minor fault in bearing results in abrupt failure that lead to major economic losses. Therefore, the fault diagnosis of bearing has been widely studied in the past few years [1–3]. Any defect in the bearings generates vibration in the system and that can be used as a source for diagnosing fault. Misaligned machinery is more prone to failure due to increased loads on bearings and couplings, alongside misalignments also causes decrease in the overall efficiency of the system. The typical faults of roller bearings are localized defects in the outer-race, inner-race, and on the roller, that defects generate complicated vibration signal, which appears with different complex frequencies. It is important to develop some effective maintenance and diagnosis procedure for feature extraction and computing techniques during dynamic stage to avoid abrupt shutdown of the rotor-bearing system.

Misalignment is basically divided into two subparts, i.e., Angular Misalignment (SAM) in which shaft centerlines intersect and Parallel misalignment in which shaft centerlines are parallel (Fig. 1). In fact, in the most cases, the misalignment is caused by a combination of both.

There are various causes for vibration, but nearly 90% are produced due to misalignment and unbalance of rotating and reciprocating parts. The abrupt failure in the machines can be due to damage in any components like stator, rotor, and

R. Kumar (✉) · M. Singh
Department of Mechanical Engineering, Lovely Professional University, Phagwara, India
e-mail: rajeev.14584@lpu.co.in

R. Kumar
Inder Kumar Gujral Punjab Technical University, Kapurthala, India

J. Singh
Department of Mechanical Engineering, Inder Kumar Gujral Punjab Technical University, Kapurthala, India

A. Parey et al. (eds.), *Recent Trends in Engineering Design*, Lecture Notes in Mechanical Engineering, https://doi.org/10.1007/978-981-16-1079-0_28

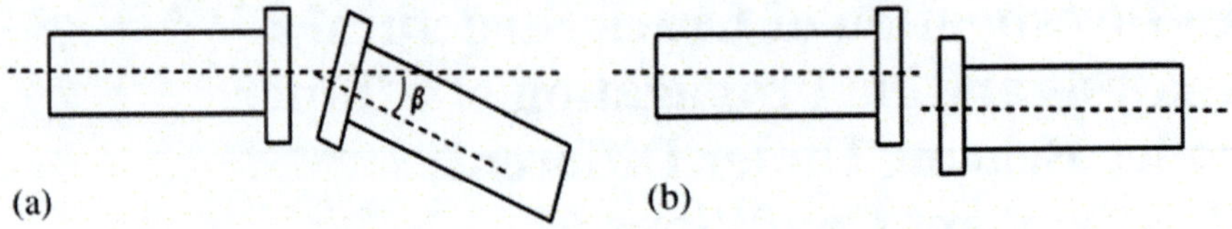

Fig. 1 Graphical explanation of **a** Angular and **b** Parallel misalignment

bearing. These faults further result in excessive vibration, decreases efficiency, extreme heating, decreased torque, etc. To diagnose the root cause, a number of structural conditional monitoring techniques are developed such as Vibration monitoring, Current monitoring, Acoustic Emission Technique, Thermal imaging monitoring, Torque monitoring, Chemical Analysis, Angular Speed, and Power Testing [1]. Earlier the signal processing techniques consist of basically calculating the statistical parameters like RMS, Kurtosis, Standard Deviation, Crest Factor, etc., but that were not too successful and effective for analysis because the signals are highly affected by noise and other related interference like varying speed, load, and other operating conditions. A new feature, i.e., statistical moments that can be able to make the signals insensitive of varying load and speed was proposed. After this, for detecting the faults before the severe stage, the use of normalized skewness and normalized kurtosis is introduced for detecting the fault at early stage. Apart from statistical parameters-based technique, parametric time-domain techniques are also being used for fault diagnosis. Mechefske and Mathew [4] uses Autoregressive models that are used to generate the frequency spectra and interprets the mean (arithmetic and geometric), RMS of spectral difference and found that this technique is more superior in terms of classifying the healthy and faulty bearing than the traditional FFT. Further few more research are done by Mechefske and Mathew [5] in context of autoregressive models that concludes that it works more efficiently for short signals and low-speed bearings.

For analyzing vibration signals, different signal processing techniques are used like Time domain, Frequency domain, and Time–Frequency domain. Time–Frequency domain is the mostly used technique for fault diagnosis in the bearing specifically for non-stationary signals. As time-domain signals do not disclose anything about the frequency components of signals consequently frequency domain and time–frequency domain techniques such as FFT, STFT, Wigner Vile, etc., are been broadly applied for determining the hidden frequency information about the signal. FFT is generally used for converting the time domain into frequency domain, however, the defect frequencies are still not clearly distinguishable due to excessive noise and hindrance, whereas for high SNR, FFT works efficiently. To overcome this problem researchers, work on developing the bi-spectrum, power spectral density, etc.

Meng and Qu [6] explains that the most parameters that possess good diagnosing ability have a close relationship with respect to time. Consequently, techniques like FFT do not provide enough information when used alone because the time domain information is unnoticed. Mori et al. [7] used discrete wavelet transform for prediction of fault and found it more enhanced technique than FFT. For analyzing vibration signals, different signal processing techniques are used like Time domain, Frequency domain, and Time–Frequency domain. Time–Frequency domain is the mostly used technique for fault diagnosis in the bearing specifically for non-stationary signals. Nandi and Ahmed [8] discusses various signal processing techniques used for machine fault diagnosis. Chandra and Sekhar [9] have described the fault detection in rotor-bearing system using time–frequency techniques. Mori et al. [7] uses Discrete Wavelet Transform (DWT) technique to capture information from both time and frequency characteristics and much more helpful for detecting refined time localized changes. Saxena et al. [10] have presented an online fault diagnosis systems procedure for fault diagnosis of ball bearing through Continuous Wavelet Transform (CWT). Castejón et al. [11] have proposed a method for automated diagnosis of rolling bearing using Multiresolution Analysis (MRA) and neural networks a real system technique for the detection of faults at very incipient stage under any conditions. The effectiveness of Artificial Neural Network (ANN) and Support Vector Machine (SVM) for fault diagnosis of ball bearing using machine learning methods has been revealed by Jamadar and Vakharia [12]. Kankar et al. [13] have presented integrating Dimensional Analysis (DA) and Backpropagation Neural Network (BPNN) approach for the detection of localized faults in roller bearings by developed algorithm using experimental data as obtained from developed test rig using defective bearing. Kumar and Singh [14] have described the Motor Current Signature Analysis (MCSA) for bearing fault detection in mechanical systems. Fast Fourier transformation is initially used for the comparison of healthy and faulty bearings. Jalan and Mohanty [15] have investigated the model-based fault diagnosis technique for fault diagnosis of a rotor-based system subject to misalignment and unbalance and it effectively found out the fault condition and fault location using this technique and the concepts of characterization of parallel misalignment in rotating machines was explained by Meroño Pérez [16] using modulated signal of incremental encoders which set the relationship between frequency spectrum of the angular vibration and the modulated signal from the encoders. Simm [17] proposed online monitoring of shaft misalignment using non-contact laser-based measurement technique and it can be used for the monitoring of both offset and angular shaft misalignment in the shaft and can be used where it is difficult to measure misalignment and conventional vibration techniques [18]. Shaft Angular Misalignment (SAM) detection through acoustic emission was analyzed and operated under different operational conditions of load and speed in order to evaluate and impact on the acoustic emission for the detection of SAM under varying operational conditions. This study demonstrated the potential performance of acoustic emission method to analyze the SAM detection and vibration analysis [19]. Wavelets are the short-duration signals that comprises average zero value. It is popular for its capability to present both time and frequency information, but selection of mother wavelet is a challenging task as different wavelets applied on a single signal may produce different results. Wavelet transforms are further sub-divided into DWT, WPT, CWT. Wavelets are characterized by various factors like orthogonality, symmetry, compact support, and vanishing moment. Selection of mother wavelets depends on various factors. Ngui et al. [20] presents and classifies the suitable mother wavelet related to different faults. Flanders [21] states that according to symmetric properties, it is also possible to select the mother wavelet by visual inspection [22].

2 Shannon Entropy

2.1 Shannon Entropy

Entropy is the measure of energy, change in physical performance of rotor-bearing system results to change various parameters and subsequently can be used as a factor for diagnosing faults. Shannon entropy quantifies the randomness of signal, and randomness in vibration spectrum is the result of any sudden change or defect in the bearing. It is generally employed for quality assessment of spackles patterns, spackles pattern with large Shannon entropy called as good quality spackles pattern, which means greater degree of spackles uniqueness the main function of Shannon entropy is suppressing the unwanted noise and capturing the periodic impulse. Shannon entropy represents the behavior of the signal acquired if in any case the randomness is higher it represents the higher negative values. Mathematical representation of Shannon entropy is represented by Eq. (2.1).

$$H(X) = -\sum_{i=1}^{n} p_i \log_2 p_i \tag{1}$$

where X represents the random signal, n represents the number of outcomes with probability function p_i. These statistical features are used for the analysis of fault propagation.

2.2 Orbit Plots

Orbit plots also known as Lissajous pattern is very simple and convenient method to deal with rotating machinery motion. This plot helps us to identify various types of faults associated with rotating components, like misalignment, unbalance, rubbing, lubrication, etc. It basically plots according to the locus of center of rotating shaft.

Orbits plots can be created with the help of connecting proximity probes, eddy current probes, or proximity sensors around the shaft with equal angle from other sensors. Orbit Plot Analysis is one of the basic tools for detecting the fault in machines like misalignment, unbalance, oil whirl, rubbing, and other related faults. The orbit plots are unable to compute the location and severity of fault efficiently. Normally, the orbit plots are used along with another technique to make the results more descriptive and efficient. da Silva Tuckmantel et al. [23] uses the forces and moments developed due to misalignments as it gives rise to vibration, so can be used for analyzing the system signature follows for detection and diagnosis of misalignment and fault, here orbit shapes and displacements spectrum are analyzed by measuring steady-state responses in order to observe harmonic components which leads in detecting misalignment. Patel and Darpe [24] analyses misalignment and its different types like angular, offset, and parallel misalignment. Lot of existing literatures presents work on detecting misalignment but detecting its type is a necessary for better and reliable results. Different types of faults can generate similar frequency spectra, but this tool can be able to separate faults, as well full spectra and orbit plots are used parallelly to analyze the hidden information of signals. Different fault results in different orbit pattern, as [25, 26] shows that banana-shaped, petal-shaped, ellipse-shaped patterns results in misalignment, oil whirl, unbalance, respectively. Qu et al. [27] shows that orbit plots can detect the severity of misalignment as increase in misalignment results in flatness of the resultant orbit plot.

3 Experimentation

For this study, some experimental tests are designed for the capability to detect misalignment from the proposed techniques when applied on different misalignment levels, for this purpose two-coupled shafts are used with taper roller bearing on which angular misalignment is calibrated by shaft alignment kit and measured the angle of misalignment with the help of laser-guided angle measuring tool. A proper experiment setup consists of accelerometer, proximity probe, coupling, and bearing used is shown in Fig. 2.

Angular deviation Shaft is coupled with three-phase motor having voltage 415 V, 2250 rpm and power of 0.75 KW with proper housing, which is fixed, on the bed over

Fig. 2 Experimental test rig

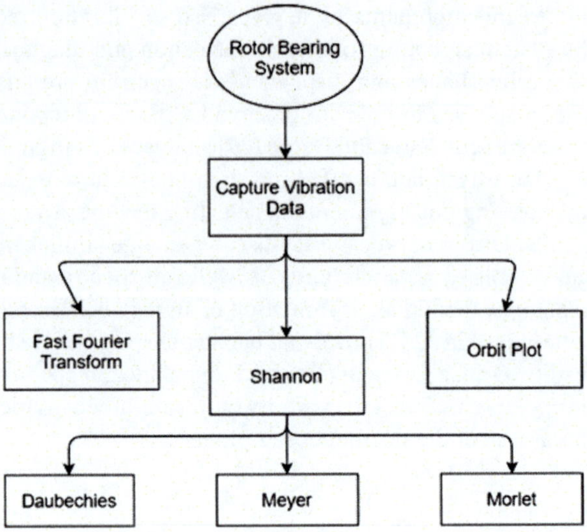

Fig. 3 Research methodology

the main body (Modulus of Elasticity 188.4401 GPa and Hardness 11.02647 GPa) (Fig. 3).

An accelerometer having sensitivity of 100 mV/(m/s^2) and frequency range ($\pm$3 dB) 0.35 to 12,000 Hz is fixed on the top portion of the housing which is further attached with the data acquisition hardware that converts vibration signal into time domain signal. These generated signals are used to analyze FFT graph using MATLAB. For controlling the speed of the motor, Variable Frequency Drive (VFD) used by this can regulate the rpm for different conditions. The bearing is tested at 35 Hz frequency to check the maximum amplitude and approximate natural frequency of the bearing. The proximity probe was used to check the wobbling of the shaft, connected to spectra quest DAQ 501-A demodulator further connected to NI-cDAQ9174 (Card-9234) that converts physical signals to waveforms. Each time the number of samples recorded were 25,600 with sampling rate of 5120. The recorded data is visualized and analyzed using LabVIEW and MATLAB software.

The angular deviation (0.00–1.5 mm) of shaft can be adjusted by adjustable screws from one side. The distance between housings and total shaft length is 380 and 500 mm. It accommodates ½ to 1¼ in. diameter shafts. Table 1 represents the specifications of taper roller bearing (30,205) used in the experiment.

The fundamental principle behind the Fourier Transform is that it decomposes the time domain signal into varying amplitude of sine waves. The resultant spectrum displays the peaks of individual sine waves present in the signals.

The below figures interpret that peak values present at subsequent 1× and 5× frequency are 0.01082 and 0.01175 for Healthy system (Fig. 4), whereas for Faulty system (Fig. 5) the same corresponds to the value of 0.01738 and 0.02235. It is found that the same system under misalignment results in 60% increase in amplitude value at 5× frequency component. For 1× peak, the increase of 90% are seen for faulty

Table 1 Specification of the test bearings

Parameters	Bearing number (30,205)
Inner race diameter (mm)	25
Outer race diameter (mm)	52
Width (mm)	15
Number of rollers	17
Pitch diameter (mm)	38
Roller diameter (mm)	6.39

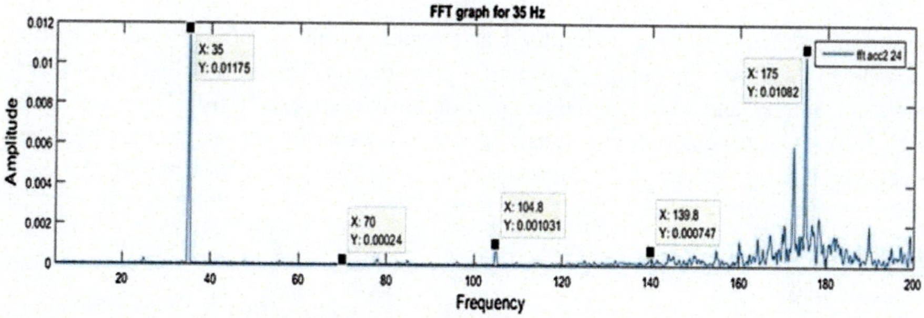

Fig. 4 FFT signatures of healthy bearing at 35 Hz

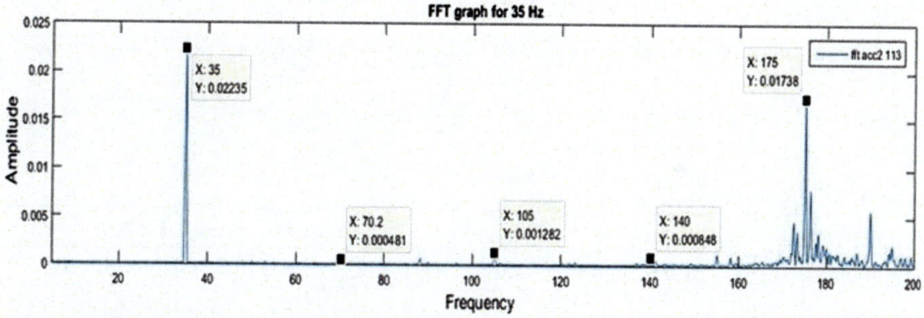

Fig. 5 FFT signatures for faulty bearing at 35 Hz

system. Therefore, it can successfully be able to detect the changes in the system. Different frequency component plays an important role in detection of misalignment using FFT, the analysis at 1×, 2×, 3×, 4×, and 5× are discussed (Table 2).

The peak value at 1× frequency for faulty system is 0.001175 and the peak value at 2× frequency component is 0.00024, or in other words the decrease in the peak value at 2× is seen. It is seen in other related literatures that decrease in 2× peak level signifies the angular misalignment, whereas increase in 2× compared to 1× frequency component results in parallel misalignment For the compilation of result,

Table 2 FFT values for healthy and faulty system

Healthy system (amplitude of FFT)	Faulty system (amplitude of FFT)	Frequency component	Percentage change (%)
0.01175	0.02235	1×	90.2
0.00024	0.000481	2×	100
0.001031	0.001282	3×	24.3
0.000747	0.000848	4×	13.5
0.01082	0.01738	5×	60.6

Shannon entropy has been calculated at frequency rate of 35 Hz with misalignment level as shown in Tables 3, 4, and 5 considering Shannon value for different wavelets, Mayer and Morlet wavelets significantly responds to misalignment under high-frequency zone and continuously increasing with the increase in misalignment level which is not apparently represented in dB2 wavelet as shown in (Fig. 6).

Table 3 Calculated Shannon Value at 35 Hz for range (0.00 mm-1.50 mm) deviation at High-frequency zone

Statistical test	Zone	Level of misalignment (mm)	Wavelet type		
			DB2	Morlet	Meyer
Shannon entropy	High	0.00	1.578×10^4	1.396×10^3	5.032×10^3
		0.50	3.093×10^4	1.963×10^4	2.387×10^4
		1.00	2.427×10^4	2.434×10^4	2.744×10^4
		1.50	3.6337×10^4	3.6793×10^4	3.6983×10^4

Table 4 DB2 wavelet under misalignment

Misalignment level	0 mm	0.5 mm	1.0 mm	1.5 mm
Daubechies	1.578×10^4	3.0938×10^4	2.4279×10^4	3.6337×10^4
Percentage change		+96%	−78.4%	+49.6%

Table 5 Morlet wavelet under misalignment

Misalignment level (mm)	0	0.5	1.0	1.5
Morlet	1.396×10^3	1.9630×10^4	2.4340×10^4	3.6793×10^4
Percentage change (%)		+1300	+23.9	+51.1

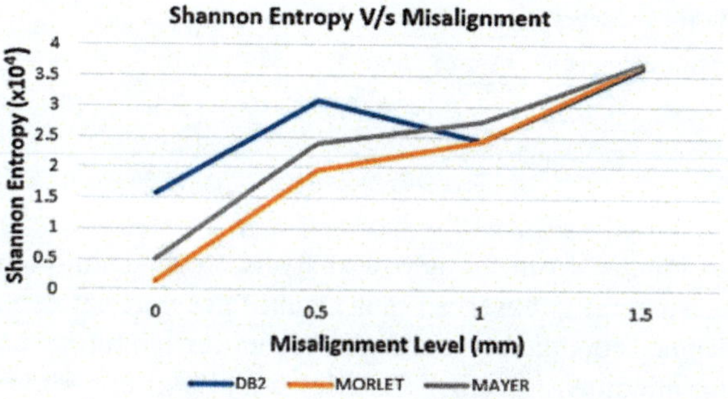

Fig. 6 Shannon entropy with DB2 versus Morlet versus meyer wavelets

3.1 *Daubechies Wavelet*

For db2 wavelet, the value observed at 0 mm is too high when the system is properly aligned, with increase in misalignment that is 0.5 mm a sudden increase of 96% is seen that shows a good response but later at 1.0 mm a abrupt decrease of 78.4% is observed and again increase of 49.6% is observed at 1.5 mm.

Therefore, the abrupt decrease and increase in the magnitude with increase in misalignment level is stated. Hence, db2 wavelet cannot be used as suitable wavelet for the case.

3.2 *Morlet Wavelet*

For Morlet Wavelet, the Shannon entropy under align condition obtained is 1.396×10^3, with increase in misalignment at 0.5 mm or when the system changes its state from aligned condition to unstable condition it shows the increase of 1300%. Till this instance, this wavelet can be used for misalignment detection abrupt increment in Shannon entropy. Further when the misalignment is increased to 1.0 mm the increase of 23.9% is reported and at 1.5 mm the increase of 51.1% is seen, so finally this wavelet is responded successfully and can able to determine the severity of misalignment because with increment in misalignment only the rise in Shannon entropy is seen.

3.3 *Meyer Wavelet*

Now for Meyer Wavelet, when the system changes its state from stable to unstable condition, i.e., misalignment of 0.5 mm increase of 374% is observed and with the increase in misalignment level from 0.5 to 1.0 mm rise of 14.9% and from 1.0 to 1.5 mm rise of 37.4% is seen, therefore this wavelet can also be able to detect the misalignment and its severity (Table 6).

Table 6 Meyer wavelet under misalignment

Misalignment level (mm)	0	0.5	1.0	1.5
Meyer	5.0323×10^3	2.387×10^4	2.7448×10^4	3.6983×10^4
Percentage change (%)		+374	+14.9	+34.7

Finally, from the above analysis on Shannon entropy, after applying different wavelets it is observed that Daubechies wavelet can detect the state change from aligned condition to misaligned condition but further cannot be used for detecting the misalignment severity. The Meyer and Morlet both responded well while changing system state from stable to unstable as well both the wavelets can detect the severity of misalignment. It is concluded that Morlet wavelet shows more deviation than Meyer wavelet but both the Meyer and Morlet can be used for misalignment detection and also detect the severity of misalignment.

3.4 Orbit Plots

Orbit plots trace the center of shaft to plot the orbits and result in identifying the faults; it is seen that the different faults result in plotting different orbit plots. In the experiment performed the testing are performed under various frequency 35 Hz with different magnitude of misalignments that are 0, 0.5, 1, 1.5 mm. It is observed in Fig. 7 that plot with aligned condition (0 mm) is different from all other misaligned cases, i.e., 0.5, 1.0, and 1.5 mm. The experiment conducted is under angular misalignment and the plotted orbit plots result in depicting the banana-shaped pattern [26] and it

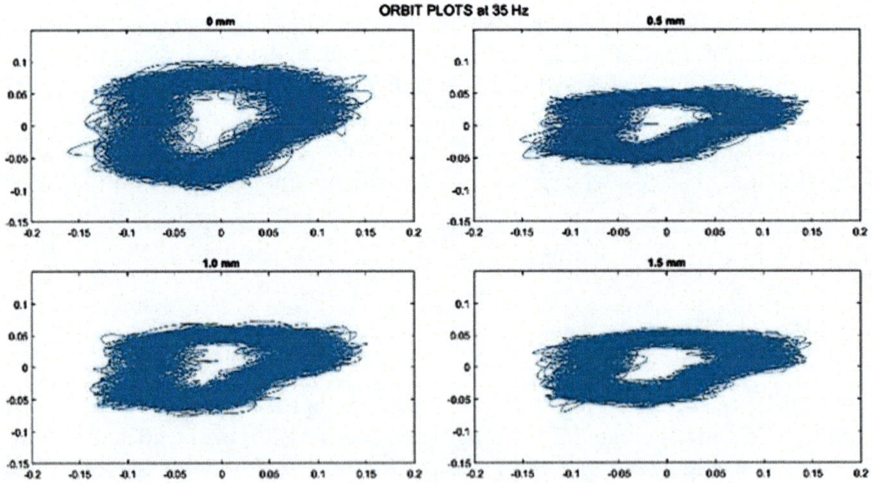

Fig. 7 Orbits plots under different misalignment level

can be also inferred from (Fig. 7) that higher misalignment level results in flatness of the orbit plot [27].

The experiment conducted is under angular misalignment, and the plotted orbit plots result in depicting the banana-shaped pattern. The same type of pattern has been observed in experiment conducted by [28]. It concludes that misalignment results in the formation of banana type pattern are due to misalignment. Orbit plots can successfully be able to detect the faults but in this experiment orbit plot detect the misalignment but unable to detect the type of misalignment.

In Fig. 7, the orbit plots are plotted for different misalignment levels, that is at 0, 1.0, and 1.5 mm. It is clearly seen from the above figure that with the rise in the misalignment value the orbit plot gets more flattened. The same results are seen for the experiment conducted by [27], and concludes the same. So, from the plotted orbit plots against different misalignment levels and available literatures, it can be inferred that higher misalignment level results in flatness of the orbit plots and efficiently leads to determine the severity of faults.

3.5 *Crack Analysis Using SEM*

Scanning Electron Microscope is basically the technique used to capture topography of the material. In this experiment, the intended V-notch is made in the bearing using wire EDM at inner race. Now, this rotor-bearing system starts generating vibrations due to faulty bearing. The SEM images are captured after 124 h of running at 35 Hz under faulty condition. Crack analysis is an important part to study as after minor defect in system, the machines continue to run that further prevents loss in productive time. Crack analysis is an important for analyzing the crack growth rate that further estimates bearing life [29] (Fig. 8).

Finally, after SEM analysis, it is observed that crack propagation firstly starts from outer surface near edges due to sudden and continuous impact of load. Crack

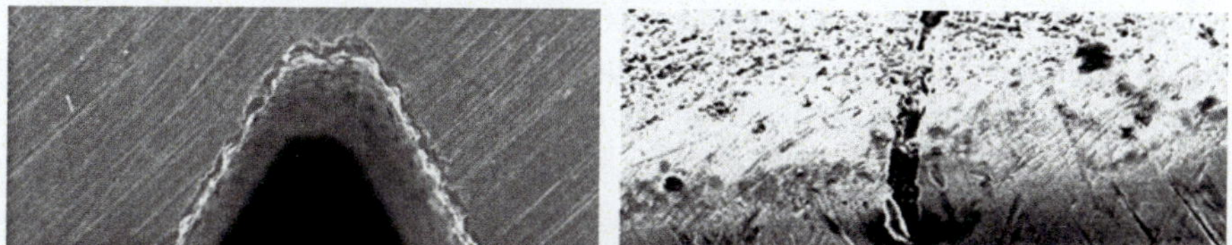

(a) (b)

Fig. 8 SEM image of V-notch at **a** 40× and **b** 450×

analysis can be analyzed in more depth for estimating crack propagation rate and bearing life and that will prevent abrupt failure and sudden breakdown of machines.

4 Conclusions

This research work presents an investigation on detection of misalignment through advanced signal processing techniques. Misalignment is one of the most reported problems in case of rotational machineries that causes sudden breakdown of machineries. Fault detection and condition monitoring is more focused subject nowadays as it led to alarm as even minor change in spectra or characteristics is observed that further result in preventing abrupt failure.

For analyzing the feasibility of this technique, several tests have been performed and change in misalignment level have been significantly identified.

i. From the above analysis on Shannon entropy, using different wavelets, it is observed that Daubechies wavelet fails to detect the misalignment because db2 wavelet shows negative deviation with increase in misalignment. It is observed that with increase in misalignment level from 0.5 mm to 1.0 mm Meyer and Morlet show deviation of +23.9% and +14.9%, respectively, whereas from 1.0 mm to 1.5 mm deviation of +51.1% and +34.7% is observed.
ii. Fast Fourier Transforms are analyzed and found increase of 60% in the amplitude at 5× frequency and increase of 90% in the amplitude at 1× frequency under misalignment compared to healthy system. The peak value at 2× frequency is comparably lower than 1× that depicts the presence of angular misalignment in the system.
iii. Orbit plots are the trajectory traced by the shaft center, using orbit plot it is analyzed and found that it generates the banana pattern that depicts the misalignment and can also predict the severity with increment in flatness of orbit plots.
iv. Trends of statistical parameter are also analyzed by increasing the misalignment level and observed that crest factor is most influenced parameter.
v. It is observed that crack propagation firstly starts from outer surface due to sudden impact load on outer surface.

Finally, it is concluded that Meyer and Morlet wavelets both respond effectively to the case, whereas Daubechies fails to detect the misalignment. Orbit plots can successfully be able to detect the fault and misalignment severity. Crest factor is highly influenced statistical parameter and can be used for detection of misalignment.

This method has an advantage over existing method and is best suited for analyzing misalignment and its severity. Future work could be performed on crack propagation for estimating bearing life and rate of crack growth and for the selection of wavelets which is still a challenging task.

References

1. Rai A, Upadhyay SH (2016) A review on signal processing techniques utilized in the fault diagnosis of rolling element bearings. Tribol Int 96:289–306
2. Singh S, Kumar A, Kumar N (2014) ScienceDirect motor current signature analysis for bearing fault detection in mechanical systems. Proc Mater Sci 6:171–177
3. Vernekar K, Kumar H, Gangadharan KV (2014) Gear fault detection using vibration analysis and continuous wavelet transform. Proc Mater Sci 5:1846–1852
4. Mechefske CK, Mathew J (1992) Fault detection and diagnosis in low speed rolling element bearings part I: the use of parametric spectra. Mech Syst Sig Process 6(4):297–307
5. Mechefske CK, Mathew J (1993) Parametric spectral estimation to detect and diagnose faults in low speed rolling element bearings: Preliminary investigations. Mech Syst Signal Process 7(1):1–12
6. Meng Q, Qu L (1991) Rotating machinery fault diagnosis using Wigner distribution. Mech Syst Signal Process 5(3):155–166
7. Mori K, Kasashima N, Yoshioka T, Ueno Y (1996) Prediction of spalling on a ball bearing by applying the discrete wavelet transform to vibration signals. Wear 195(1–2):162–168
8. Nandi A, Ahmed H (2019) Time-frequency domain analysis. In: Condition monitoring with vibration signals. Wiley, pp 79–114
9. Chandra NH, Sekhar AS (2016) Fault detection in rotor bearing systems using time frequency techniques. Mech Syst Sig Process 72–73:105–133
10. Saxena M, Bannet OO, Gupta M, Rajoria RP (2016) Bearing fault monitoring using CWT based vibration signature. Proc Eng 144:234–241
11. Castejón C, Lara O, García-Prada JC (2010) Automated diagnosis of rolling bearings using MRA and neural networks. Mech Syst Sig Process 24(1):289–299
12. Jamadar IM, Vakharia DP (2016) A novel approach integrating dimensional analysis and neural networks for the detection of localized faults in roller bearings. Meas J Int Meas Confed 94:177–185
13. Kankar PK, Sharma SC, Harsha SP (2011) Fault diagnosis of ball bearings using machine learning methods. Expert Syst Appl 38(3):1876–1886
14. Kumar R, Singh M (2013) Outer race defect width measurement in taper roller bearing using discrete wavelet transform of vibration signal. Meas J Int Meas Confed 46(1):537–545
15. Jalan AK, Mohanty AR (2009) Model based fault diagnosis of a rotor-bearing system for misalignment and unbalance under steady-state condition. J Sound Vib 327(3–5):604–622
16. Meroño Pérez PA, Gómez de León FC, Zaghar L (2014) Characterisation of parallel misalignment in rotating machines by means of the modulated signal of incremental encoders. J Sound Vib 333(21):5229–5243
17. Simm A, Wang Q, Huang S, Zhao W (2016) Laser based measurement for the monitoring of shaft misalignment. Meas J Int Meas Confed 87:104–116
18. Mohanty AR, Fatima S (2015) Shaft misalignment detection by thermal imaging of support bearings. IFAC-Pap Online 28(21):554–559
19. Ferrando Chacon JL, Artigao Andicoberry E, Kappatos V, Asfis G, Gan TH, Balachandran W (2014) Shaft angular misalignment detection using acoustic emission. Appl Acoust 85:12–22
20. Ngui WK, Leong MS, Hee LM, Abdelrhman AM (2013) Wavelet analysis: mother wavelet selection methods [Online]. Available: https://www.researchgate.net/publication/259810533_Wavelet_Analysis_Mother_Wavelet_Selection_Methods. Accessed: 06 Jul 2020
21. Flanders M (2002) Choosing a wavelet for single-trial EMG. J Neurosci Methods 116(2):165–177
22. Safavian LS, Kinsner W, Turanli H (2005) A quantitative comparison of different mother wavelets for characterizing transients in power systems [Online]. Available: https://www.researchgate.net/publication/313604612_A_quantitative_comparison_of_different_mother_wavelets_for_characterizing_transients_in_power_systems. Accessed: 07 Jul 2020

23. da Silva Tuckmantel FW, Cavalca KL (2019) Vibration signatures of a rotor-coupling-bearing system under angular misalignment. Mech Mach Theory 133:559–583
24. Patel TH, Darpe AK (2009) Experimental investigations on vibration response of misaligned rotors. Mech Syst Sig Process 23(7):2236–2252
25. Wu B, Feng S, Sun G, Xu L, Ai C (2018) Identification method of shaft orbit in rotating machines based on accurate Fourier height functions descriptors. Shock Vib
26. Jeong H, Park S, Woo S, Lee S (2016) Rotating machinery diagnostics using deep learning on orbit plot images. Proc Manuf 5:1107–1118
27. Qu L, Lin J, Liao Y, Zhao M (2019) Changes in rotor response characteristics based diagnostic method and its application to identification of misalignment. Meas J Int Meas Confed 138:91–105
28. Konar P, Chattopadhyay P (2011) Shaft orbit pattern recognition for fault diagnosis of induction motor using neural network. In: Proceedings of the 5th Indian international conference on artificial intelligence. IICAI 2011, Dec 2011, pp 288–304
29. Singh G, Kumar R, Singh M, Singh J (2017) Detection of crack initiation in the ball bearing using FFT analysis. Int J Mech Eng Technol 8(7):1376–1382

Development Method, Manufacturing Process of Fibre Reinforced Polymer Composite Type Helical Springs: A Review

Naik Aniket Ashok, Rajeev Kumar, Manpreet Singh, Jaiinderpreet Singh, Piyush Gulati, and Jujhar Singh

1 Introduction

Coil springs or helical springs are the most important and commonly used machine components which aid in vibration reductions, protection against impacts and to certain limits even acts as dampers. Nowadays, vehicular gas emission and fuel efficiency are of major concern [1]. To achieve this, weight reduction of an automobile is being carried out which improves its performance thus saving energy. Also, the current generation is heading towards electric and hybrid vehicles hence reduced weight would be effective in increasing the overall range of vehicles with such powertrains [1]. Combination of two or multiple materials such as a composite can yield improved mechanical behaviour and properties. Essentially a composite made of FRP possesses higher strength to weight ratio and improved elastic strain storage capacity on contrary to alloys.

Initially, FRPC-based technology emerged in the early 1900s. In the mid-1930s, FRPC saw its first application in boat hull components which was basically polyester resin reinforced with fiberglass [2]. The recent research in the field of FRPC has created components which can resist even high speed impacts the example includes bulletproof vest made of aramid fibre which can stop bullets. Even the aircraft carriers make use of FRPC components in the design. A good example is a passenger aircraft carrier Airbus A380 which is one largest of its kind makes use of nearly 20% FRPC parts of which the prime parts are made of carbon fibre reinforced composites [1].

N. A. Ashok
Lovely Professional University, Phagwara, India

R. Kumar (✉) · M. Singh · J. Singh · P. Gulati
Department of Mechanical Engineering, Lovely Professional University, Phagwara, India
e-mail: rajeev.14584@lpu.co.in

J. Singh
Department of Mechanical Engineering, Inder Kumar Gujral Punjab Technical University, Kapurthala, India

A. Parey et al. (eds.), *Recent Trends in Engineering Design*, Lecture Notes in Mechanical Engineering, https://doi.org/10.1007/978-981-16-1079-0_29

FRPCs are not just restricted to aircraft but also finds their applications in spacecraft, marine vehicles such as boats, ships and submarines, along with on land automobiles, sports-related equipment and also in bridges and buildings.

The above-mentioned reasons are justified in the switch of spring materials from conventional stainless steel to FRPC. Compared to different types of mechanical springs, coil springs/helical springs are merged as a part of various mechanical systems, thus incorporating a FRPC coil spring reduces the weight of the suspension system and also does not affect the load-carrying capacity of the system [1]. The leaf spring emerged as a first practical design, the material of which was FRPC in early 1981. Such FRPC leaf springs which were developed by Daugherty then found their application in heavy-duty trucks [3]. Currently, the research work on FRPC-based coil/helical spring is very limited in number yet attracts the interest of scientists in this particular area.

The aim of this paper is on reviewing the gathered open source research articles and analyses the viability of FRPC coil/helical spring. Various methods used by the current research work are studied and accordingly their competency is analysed.

2 Literature Survey

The practical approach with FRPC-based helical spring is not much popular in the technical field. Research paper that is available based on such application is much minimal. Short review on available research papers has been presented below.

T. S. Manjunatha and D. Abdul Budan fabricated FRPC-based spring and performed experiments for the spring's strength. For the study, three different types of springs are fabricated using glass fibre, carbon fibre and glass/carbon fibre in + 45° orientation and tests are conducted to check the mechanical behaviour of the springs. The helical spring is fabricated using Filament Winding Technique (FTW), for which a mandrel made of cast iron having the profile of the spring is fixated between the centres of the lathe. Silicone gel is applied throughout the mandrel body which acts as a mould releasing agent. Due to shear load acting on the spring, the fibres are cut at 45° angle. These fibre tapes are wound on the mandrel by dipping them in the bath of epoxy resin. This is done till the desired thickness is achieved. Tests are performed on the spring after curing process which took 24 h. Mechanical properties such as stiffness, shear stress and failure load maximum compression are determined by the researchers. The findings of the experiments showed that the spring rate of carbon fibre is 24% more compared to the glass fibre and 10% more than that of glass/carbon fibre (Table 1).

To conclude with the results of the experiment, the weight of spring in test is lesser than that of steel spring and comparatively provides good stiffness [4] (Fig. 1).

Mehdi Bakhshesh and Majid Bakhshesh performed the experiment for the steel spring with that of composite helical spring of three types namely, E-glass/Epoxy, Carbon/Epoxy and Kevlar/Epoxy. The researchers have been modelled using SOLIDWORKS software followed by analysis on ANSYS software. The results achieved,

Table 1 Mechanical properties of the FRPC spring

Properties	Glass fibre	Carbon fibre	Glass/carbon fibre
Spring constant (N/mm)	4.83	6.36	5.75
Shear stress (N/mm^2)	83.00	79.67	95.49
Failure load (N)	1000	1500	1200
Maximum compression (mm)	83	80	77

Stainless Steel Mandrel

Springs after curing

Fig. 1 Manufacturing of FRPC helical spring [4]

i.e., the numerical ones and the theoretical ones were found to be on par with each other. The stresses acting on the composite helical spring are lesser compared to the steel spring also the value is more when the fibre strands are aligned to the direction of the applied load. Though the spring weight has been reduced, changing the percentage composition of fibres in carbon/epoxy composite does not affect the overall weight of the spring. The safety factor in consideration for fibre alignment must be perpendicular to the loading as mentioned by the researchers and the choice of material would be Carbon/Epoxy for the spring. The manufacturing of the spring has been carried out by Resin Transfer Moulding (RTM) process.

In this study, steel spring has been replaced by FRPC-based composite spring including g E-glass/Epoxy, Carbon/Epoxy and Kevlar/Epoxy. On these springs, FEM analysis were performed to obtain the shear stress. The loading conditions are taken to be static. SOLID45 was the element selected for the analysis which has three degrees of freedom at each node allowing up to 250 multiple types material layers. The results from analysis showed that the shear stresses acting within the spring is more, with the fibres in the direction of loading. If fibre directions are perpendicular to the load acting, shear stresses acting would be zero [5]. Ekanthappa et al. manufactured composite cylindrical helical spring in order to replace conventional steel spring used in vehicles by means of Spring winding technique. For the manufacturing purpose, a steel-based reusable mandrel which is of the required spring shape is used. This mandrel is fixated between the lathe and a mould releasing agent such as the silicone gel is applied on the mandrel. As the mandrel is rotated, E-glass fibres are passed through the polymer bath and are laid down on the mandrel till the required spring dimension is attained. The spring after curing for 48 h is carefully removed from the mandrel.

The various spring parameters are noted by testing it on UTM by following all the test procedure according to ASTM standards. The test parameters are as follows (Table 2).

In this study, the average values of the test results indicate a linear curve for load versus deflection analysis. And the spring stiffness is found to be 9.95 N/mm. The study states that the fabrication of reusable mandrel which is made of mild steel is rather easy and therefore makes the fabrication of FRPC spring simpler. The Spring winding technique is cost effective and simple to perform [6] (Fig. 2).

Chang-Hsuan Chiu et al. fabricated and studied four different types of composite helical springs, namely unidirectional laminates (AU), rubber core unidirectional laminates (UR), unidirectional laminates with braided outer layer (BU) and rubber core unidirectional laminates with braided outer layer (BUR). The study tries to understand the significance of the rubber core and braided outer layer on the spring's

Table 2 Test parameters of FRPC spring

Properties	Values
Spring constant	9.95 N/mm
Max compression	56 mm
Load at Max. compression	550.00 N
Shear stress	81.76 N/mm^2
Fibre volume fraction (%)	60
Weight of the spring	424 g

Reusable Mandrel

End products: Fibre Reinforced Plastic Springs

Fig. 2 Fabrication of FRPC helical spring using a reusable mandrel [6]

Table 3 Cross sections of four different types of FRPC Springs indicating different structures [7]

Structure name	AU	UR	BU	BUR
Structure materials	All unidirection	Unidirection + Rubber core	Outer braided + Unidirection	Outer braid + Unidirection + Rubber core
Cross-sections	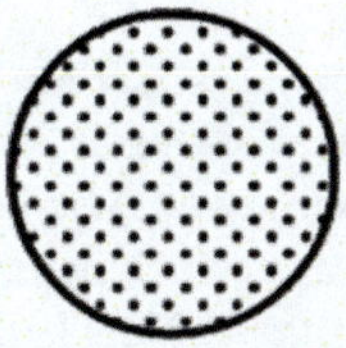	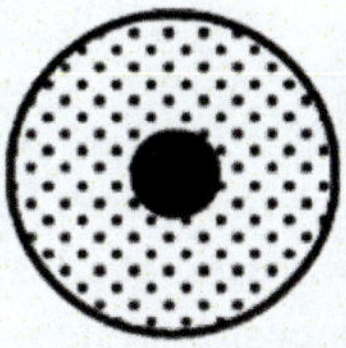	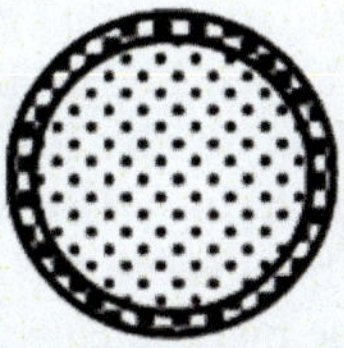	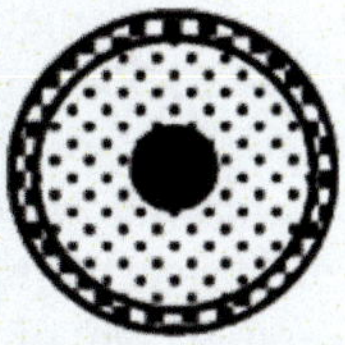
Dimen-sions	76.53×41.5 cm^2	76.53×37.5 cm^2 3 mm rub-ber core	76.53×37.5 cm^2 3 K carbon fibre outer braid (one layer)	76.53×37.5 cm^2 3 K carbon fibre outer braid (one layer)

mechanical properties. Overall conclusion drawn after the studies indicates that all four types of springs are lighter as well as stiffer compared to the conventional steel springs. And amongst all the four springs, BUR spring structure shows failure load due to compression, i.e., 3297 N with spring constant almost equates to 160 N/mm. The study also shows that with rubber core, the failure load in compression is increased to about 12%, whereas the spring with braided outer layer improves the spring constant by 16% and increases the failure load in compression by 18%. The following Table 3 gives the illustration of spring structure:

Suresh G. et al. performed analysis on FRPC coil spring which had reinforcement of woven roving fibre (WRF) and polymer used was thermoset (epoxy resin) with four different level additions of nanoclay (Garamite). The varying amount of nanoclay composite was of order 0, 1, 2, 3%. The aim of the study was to identify the load sustaining capability, the spring stiffness and the amount of weight savings compared to that of a steel spring. The researchers used similar dimensions of an existing light commercial vehicle's suspension spring and fabricated a composite spring.

According to the study, the developed composite springs were lighter than the steel counterparts but the stiffness achieved was lesser than the steel. In order to increase the stiffness, the dimension of the FRPC springs needs to be increased but this leads to increase in the weight of the spring. This fact limits the application of composite springs to mostly lightweight vehicles or EV's for which higher spring stiffness isn't a requirement. Also, the study states that manufacturing such helical spring is challenging and consumes time compared to the steel springs. But for mass production, use of CNC winding machines and automated processes might aid in cost reduction [8]. Anil Antony Sequeira et al. performed a comparative analysis on helical steel springs and composite springs by finite element method. Stiffness to weight ratio of the two springs is of prime importance in this current research. The behaviour of carbon and Kevlar composite springs and helical steel spring is investigated under static analysis. The objective of the research is to collate the various mechanical parameters of the composite springs to that of steel spring such as stiffness, maximum load-bearing capacity, weight reduction in FRPC spring. Here, all the necessary design parameters were kept same for all three springs. ANSYS software was used to study the mechanical behaviour of the springs. The study showed that specific modulus which is basically a ratio of Young's modulus to mass density of the material of Carbon FRPC spring is highest and the Kevlar FRPC spring is lowest. Mass of Kevlar FRPC spring is lower compared to Carbon FRPC spring yet the load & deflection characteristics of steel helical spring is better than that of composite helical springs. Use of Carbon FRPC spring would be beneficial over helical steel spring because of the lower weight advantage which in turn reduces the overall weight of the mechanical suspension system of the vehicle [9].

Sancaktar E. and Gowrishankar S. studied the natural frequency of composite-based helical springs. The fabrication method utilised is filament winding process and they were able to compare the effect of number of turns, ratio of cylinder diameter to wire diameter and material being used on the natural frequency of the helical spring. This experiment was performed on three different types of composite springs which had varying number of turns from 7 to 6 along with coil/wire diameters to understand the effect of these variables on the spring's natural frequency. For the springs, three different types of PVC tubbing are selected which decide on different wire diameters, followed by passing of fibre strands through these tubes which were dipped into the resin bath previously. The researchers have broken down the fabrication process into three stages. In the first stage, amount of glass/carbon fibres which can accommodate the PVC tube space is realised. The second stage helps in understanding the resin bath to fibre stand requirement and finally the third stage, mounting of the fiber-filled PVC pipe on the winder. The last stage determines the required cylinder diameter and fixes the desired shape for the spring to be fabricated. The natural frequency of the spring is recorded with the help of a digital camera. Results show that the natural frequency of these springs can be decreased by increasing the diameter of the coil and also providing for more number of turns. This novel approach of fabricating composite helical spring is more versatile and cost saving [10].

B. S. Azzam performed optimisation study for the design of composite helical spring in which, the aim is about minimisation of spring weight and also maximisation of spring stiffness. Mathematical formulations are used to state the objective functions and constraints and made into multiple-criterion optimisation problem in a non-linear and non-equality constraint forms. In the research, parameters affecting the above design aims are divided into three groups. Manipulation of functional properties of the spring consisting of the induced axial deflection and maximum stress in the spring belonging to the first group. The second group including the materials shear strength and shear modulus, here the properties depend on the reinforcement fibre materials, the fibre volume, angles to which fibres are braided in the spring wire. The third group is about the geometrical shape group of the spring. This optimisation study shows that spring index is directly affected by the helix angle also including mechanical properties of the composite material and the minimum helix angle should be restricted to 20°. The optimisation technique used in this particular research has the ability to give general optimum values for the spring parameters for multiple types of composite material for varying load [11].

Renugadevi K et al. made an attempt to fabricate a composite helical spring using natural fibres extracted out of Calatropis Gigantea (CG). A sub-tropical plant CG has high seasonality with high yield. The fibres are extracted from the stem of the plant. In the current study, the fibres are twisted and treated with different concentrations Level of sodium hydroxide aqueous solution comprising concentration of 1, 2, 5 and 6 wt% for approximately 1 h. After the process of cleaning with water of pH7 and drying at 70° in an oven, the fibres are then sealed in a polythene bag. Due to alkali treatment, the diameter of the fibre decreases from 1 to 0.7 mm. Araldite LY556 (Epoxy) is selected as a matrix material due to excellent adhesion to cleaner surfaces, greater dimensional stability, high resistance to chemical atmospheric corrosion and to this a 10% hardener (HY951) is added which improves the physical and mechanical properties and also aids in providing sufficient gel duration, viscous nature and also adds up to re-mould time.

The fabrication of spring involves the use of two-wheeler spring as a die element. A mild steel shaft is held on a lathe with either of its end fixated to the lathe. The spring die which has its surface covered with petroleum jelly is then inserted on the shaft with a 1 mm clearance. The petroleum jelly ensures easy removal of the product in the end. The alkali-treated fibres are then dipped in the resin and fed over the spring die. The chuck is rotated in clockwise direction until the fibres are fed completely over the die. After the curing process of the product, tests were performed according to the ASTM standards A125-96 of compression. The results for varying levels of NaOH have been listed in the following Table 4.

FEA analysis was performed, and it was found that 4 wt% NaOH-treated FRC spring does produce stiffness to amount of 1 N/mm with 1.3 GPa of shear modulus. Energy-dispersive X-ray analysis shows that the external surface has an O/C ratio of 0.07 thus indicating that the epoxy matrix behaves as a protective layer for the spring [12] (Fig. 3).

Choi B. L. and Choi B. H. introduced the carbon fibre as reinforcement for a composite coil spring taking into considerations the static spring rate. Resin Transfer Moulding (RTM) process was used to manufacture the spring. In this process, the fibres are filled with the epoxy resin and cross-linking hardening takes place; this happens in vacuum conditions. Displacement tests are performed, and the static spring rate of the spring is determined once the curing process is completed. The research paper also puts forward a deterministic approach in developing the helical spring with the use of computational and experimental methodologies. In this particular, study all the necessary design parameters in case of a steel springs that have been taken into consideration and based on these the composite spring has been designed. The fibre volume and void volume for the manufactured CFRP Coil spring are 64.4% and 3.5%, respectively. In the study, the shear modulus in case of composite of 45° ply angle is determined to be 16.8% of steel. The estimated shear modulus is found to be on par with the experimental results for the current research. A weight reduction of 55% compared to conventional steel springs has been achieved by the researchers [13] (Fig. 4).

Table 4 Compression tests over different levels of NaOH concentration [12]

Displacement (mm)	Load				
	1% NaOH	2% NaOH	4% NaOH	5% NaOH	6% NaOH
2	1.75	2	2	0	2
4	3	4	4	2	4
6	3.75	5	6	2	5
8	5	6	8	2	6
10	6	8	10	4	8
12	8.25	9	12	4	8
14	8.25	10	14	4	10
16	8.3	12	15	6	10
18	9.5	13	16	6	11.5
20	12	14	20	8	13
22	12	12	24	8	14
24	14.25	14	26	10	15
26	15	14	28	10	15.5
28	14.5	14	–	12	16
30	15	16	–	14	17.5
32	16	18	–	14	18
34	18	18	–	16	22
36	19.25	–	–	28	24
38	24	–	–	20	30
40	–	–	–	22	44
42	–	–	–	24	–
44	–	–	–	26	–
46	–	–	–	34	–
48	–	–	–	52	–

3 Conclusion

Currently, ever-changing innovation in technology demands for more energy savings, cost effectiveness, improved machine performance. To make this possible, composite material which is in fact lightweighted, high quality is being used. Hence, a composite helical spring certainly is an option to look for in place of conventional steel springs. Researchers have shown interest in this particular area, thereby giving various value-added benefits with the use of FRPC Helical springs.

FRPC Helical/coil springs are basically lighter than their steel counterparts. Their use can be categorised for light-duty vehicle which includes EVs or vehicles with smaller engine capacities. The issue with lower stiffness in composite springs can be overcome by increasing the size to reasonable amount.

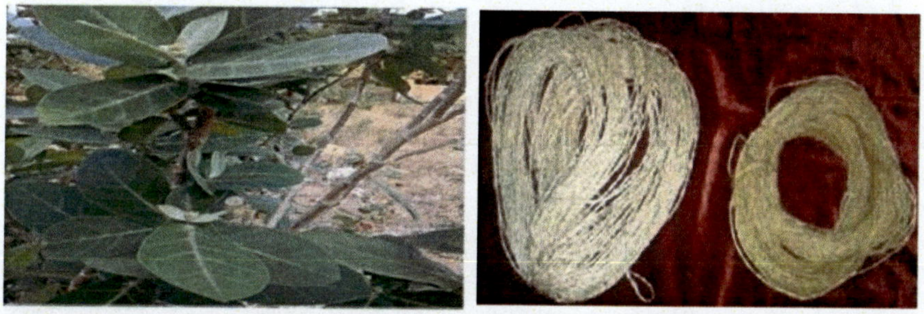

Plant fibres, before and after NaOH treatment

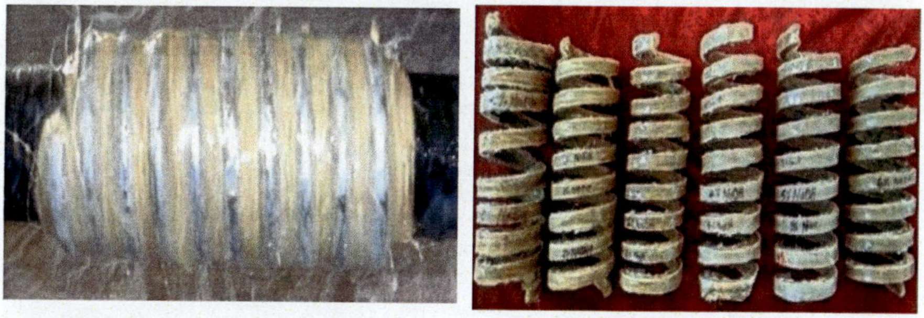

Moulded fibres on spring die after applying the epoxy resin

Fabricated spring of varying NaOH treatment level

Fig. 3 Calatropis gigantea fibre reinforced composite helical spring [12]

Fig. 4 RTM process being utilised for manufacturing of CFRP spring [13]

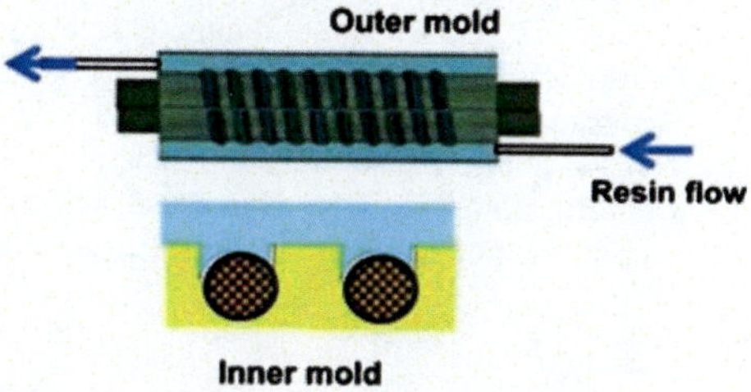

Manufacturing of FRPC helical spring is challenging as compared to the conventional steel springs. Manufacturing time required to make composite spring is more. Also, proper resin impregnation through the fibre strands is still a difficult process to achieve.

Most of the above researches use Filament Winding Technique (FWT) to manufacture FRPC springs with different mandrel setups. Improvement in such FWT systems have been carried out by some researches.

Majority cases show that FRPC helical springs have better mechanical properties against the steel springs. The challenge is with manufacturing of FRPC springs as these need special design process considering the dimensional optimisation and preciseness which is critical for the proper functioning of such springs. Various researches have used ANSYS software for proper analysis of FRPC springs.

References

1. Senthil Kumar M, Vijayarangan S (2006) Static analysis and fatigue life prediction of steel and composite leaf spring for light passenger vehicle. J Sci Indus Res 66:128
2. Hensher DA (2016) Fiber-reinforced-plastic (FRP) reinforcement for concrete structures: properties and applications, vol 42, 1st edn. Elsevier
3. McGeehin P (1982) Composites in transportation: design and current developments. Mate Desi 3:378–387
4. Manjunatha TS, Budan DA (2012) Manufacturing and experimentation of composite helical springs for automotive suspension. Int J Mech Eng Robot Res 1:229–241
5. Bakhshesh M, Bakhshesh M (2012) Optimization of steel helical spring by composite spring. Int J Multidiscip Sci Eng 3(6):47–51
6. Ekanthappa J, Shankar GS, Amith BM, Gagan M (2016) Fabrication and experimentation of FRP helical spring. IOP Conf Ser Mater Sci Eng 149(1):012098
7. Chiu CH, Hwan CL, Tsai HS, Lee WP (2007) An experimental investigation into the mechanical behaviors of helical composite springs. Compos Struct 77:331–340
8. Suresh G, Vignesh R, Aravinth B, Padmanabhan K, Thiagararajan A (2014) Fabrication and analysis of nano composite cylindrical helical spring. Int J Innov Res Sci Eng Technol 3(1):1208–1213
9. Sequeira AA, Singh RK, Shetti GK (2016) Comparative analysis of helical steel springs with composite springs using finite element method. J Mech Engi Auto 6:63–70
10. Sancaktar E, Gowrishankar S (2009) Natural frequencies of composite cylindrical helical springs manufactured using filament winding. In: ASME 2009 international design engineering technical conferences and computers and information in engineering conference. American Society of Mechanical Engineers, pp 785–790
11. Azzam BS (2009) An optimum design for composite helical springs. J Automob Eng 2010(224):347
12. Renugadevi K, Devan PK, Thomas T (2019) Fabrication of calotropis gigantea fibre reinforced compression spring for light weight applications. Compos Part B Eng 172:281–289
13. Choi BL, Choi BH (2015) Numerical method for optimizing design variables of carbon-fiber-reinforced epoxy composite coil springs. Composites 82:42–49

Condition Monitoring of Worm Gearbox Through Oil Analysis

J. S. Bhat and B. U. Sonawane

Abbreviations

AGMA	American Gear Manufacturers Association
TAN	Total Acid Number

1 Introduction

Tribology is the craft of utilizing operational assessment of issues related to uncommon financial consequences especially maintenance, reliability, and wear of equipment parts starting from the rocket to family apparatus [1]. Lubricating oil ability will be reducing surface wear and tear, by avoiding direct metal contact between the rubbing surfaces. Lubricants play a crucial role in almost every type of machinery, and their presence is frequently basic to the speed, reliability, and life of the machines [2].

As part of the product development technique, oil manufacturers are required to monitor and evaluate the condition of lubricating oil for various applications [3]. Moreover, the equipment manufacturer needs to check the performance of lubricating oil to decide the strategy that will empower their equipment to convey the best execution and life which consumes much of the time [4–6]. Lubricating oil

J. S. Bhat (✉)
Department of Mechanical Engineering, K.I.T.'s College of Engineering, Kolhapur, Maharashtra, India

B. U. Sonawane
Department of Manufacturing Engineering and Industrial Management, College of Engineering, Pune, Maharashtra, India
e-mail: bus.prod@coep.ac.in

A. Parey et al. (eds.), *Recent Trends in Engineering Design*, Lecture Notes in Mechanical Engineering, https://doi.org/10.1007/978-981-16-1079-0_30

plays a key role in providing conditional knowledge of the machine through tribological and distinctive aging rules point of view [7, 8]. Worm gears are normally utilized in numerous applications, for example, escalators, packaging equipment, small machinery, conveyors, etc., and are expected to work in all shifts every day on the shop floor [9, 10].

2 Literature Survey

The following are some research papers published related to Condition monitoring of lubricating oil:

Kumar et al. [1] presented that any deterioration of lubricating oil will lead to the wear and further decreased performance of mechanical components. The deterioration of lubricating oil is decided based on physical, chemical, and tribological properties. Such changes are directly dependent on the state of the mechanical system of all tribo elements, and based on its properties. The understanding of the study of used oils is very complicated, because the individual evaluations are interconnected. Trend analysis is important to reach a certain conclusion.

Narwariya et al. [2] explained that transmission lubricant tests are conducted on an axle performance test system. Oil viscosity, type of additive, and quality strongly affect typical gear failures. Concentration of wear metals obtained from the oil used provides valuable data about the cause and degree of the degradation that occurred and progressed in the service of the mechanical part.

Rathore et al. [3] explained a policy of diagnosis and prognosis procedures for the advancement of effective preventive maintenance that can support in decision-making by understanding the process of deterioration of any equipment elements by picking up and investigating information to comprehend its situation. The condition monitoring of the oil gives information about the working of a worm gearbox. The data picked up from this information can be utilized to decide the status of equipment fitness and oil degradation. The approach toward deciding the health of equipment is called a diagnostics of equipment. It plays a crucial role in predictive maintenance for checking the equipment's health. It will help to recognize faults from equipment and is useful to take corrective action.

Ghodake et al. [4] reported that Oil Analysis, Temperature analysis, Vibration analysis, and Acoustic Emission Analysis are used to investigate the fault in worm gear. The operating condition of the gearbox is leading to a deterioration of the gearbox lubricant used in industry applications. Investigation facility of oil analyzes the diverse condition parameters to screen the degree of deterioration and estimating life expectancy.

Bhat et al. [5] described that lubricating oil is necessary to perform specific functions to keep up the gearbox operation in the industrial application. Lubricating oil breakdown is the most influencing aspect which causes the breakdown of any industrial equipment. Lubricating oil breakdown at the sliding pair encourages rapid heating and rubbing at contact surfaces and ultimately loss of the components. This paper provides the wear particle ferrographic analysis used for extreme pressure gear oil use for worm gearboxes.

Jain et al. [6] detailed that machine lubrication is an extremely critical part of engine operations. Lubricating oil has the ability to minimize friction, transfer heat, transmit energy, secure from wear, avoid corrosion, and gas sealing. So tracking the quality of the lubricating oil at regular intervals is becoming very relevant. The used oil can be checked and analyzed for its wear debris to predict the failure of the engine parts in advance.

Wakiru et al. [7] outlined that a prognostic manages to anticipate the advancement of imperfections of any machine elements and engaged with an early diagnostics of imperfection with an adequate lead time for empowering monitoring and actions. Condition-based preventive maintenance of any mechanical equipment preserves the worth of such a equipment and the correct operation of the equipment for a longer period of time can be reflected, security of the plant and its administrators.

Bandyopadhyay et al. [8] stated that small wear particles are normally produced during machine operation. Strange wear conditions, bigger particles begin to be produced, these conditions, in the long run, cause worm gearbox breakdown. Wear of sliding parts brings about diminished mechanical productivity and an unrecoverable loss of material as wear debris.

Karanovi et al. [9] proposed that any deterioration of lubricating oil will lead to the wear and further decreased performance of mechanical components. The deterioration of lubricating oil is decided based on physical, chemical, and tribological properties. These properties are based on viscosity, Total Acid Number (TAN), water content, wear particles, Total Base Number (TBN), insolubles, etc. The various oil analysis techniques like Viscometers, DR Ferrography, and Microscopic analysis are used to determine these properties.

Bhat et al. [10] explained that condition monitoring of lubricating oil quality has become a significant issue in the present manufacturing and automobile company. Assessment of the gear oil, given a precisely depicted application, demonstrates an extremely amazing method for monitoring the condition of the worm gearbox, introducing early warning of likely issues that may bring about the breakdown of the worm gearbox. Lubricating oil change interval can be adjusted based on manufacturer guidelines and besides that a condition monitoring technique that recognizes changes in the lubricating oil based on parameters like viscosity, oxidation, water concentration, contamination, percentage of sludge, and change in lubricant chemistry.

Zhu et al. [11] demonstrated that condition monitoring of the lubricant is an effective technique behind the breakdown to figure out the root cause analysis. The factor dependable for lubricating oil breakdown is viscosity, contamination, temperature, moisture content, foreign material. Maintenance professional can predict, prevent, and eliminate breakdown of worm gearbox through oil analysis.

Solving worm gear failure is challenging and needs more attention based on the concerned study and reported discrepancies. The key function of lubricant is to curtail the adverse effects of friction-related tribological strategies and also maintain properties due to change in temperature for sustainability and secure-mechanical structures; hence all maintenance types consist of lubrication as a crucial role of the entire maintenance process. Consequently, gear oil assessment is based entirely on an accurately defined procedure. The assessment presents symptoms of early caution of possible complications that may lead to the breakdown of power equipment. Like the mechanical components, the lubricant itself could therefore be affected resulting in changes in lubricating oil properties along with increased or decreased viscosity, corrosion, degradation, load-bearing power, and many more. There is opportunity for research in the area of lubricating oil conditioning and monitoring for lubricant quality evaluation and remaining life forecast of deteriorated lubricating oil. Oil analysis is used to evaluate physical, chemical, wear debris analysis, and tribological properties of lubricating oil. Previously research activities have been focused on either physical or chemical properties as well as wear debris analysis assessment of oil. Now, there is scope to explore the physical, chemical properties, and wear debris analysis along with tribological properties. This new method can give better insight on performance evaluation of gear oil used in worm gearbox.

In the following section, lubricating oil's physical and chemical properties affecting the performance of worm gearbox are addressed.

3 Methodology

The purpose of the experimental setup is to gather quantitative data for condition monitoring of EP lubricants (Servo gear HP 140) used for worm gearbox and to receive a better knowledge and understanding of the equipment's lifetime and performance. Oil analysis deals with an evaluation of physical–chemical properties in worm gearbox needed for condition monitoring and fault assessment [11]. Lubricating oil has to be changed after 500 h or one month when it is new and 2500 h or six months regularly according to AGMA standard [12]. Viscosity, Total Acid Number (TAN), water content, wear particles, Total Base Number (TBN), insolubles, etc., are critical properties of lubricating oils. Viscosities, Total Acid Number (TAN), wear particles, and water content are required to check for performance evaluation of EP 140 Gear oil.

Figure 1 illustrates the steps involved in the methodology. The goal of the above

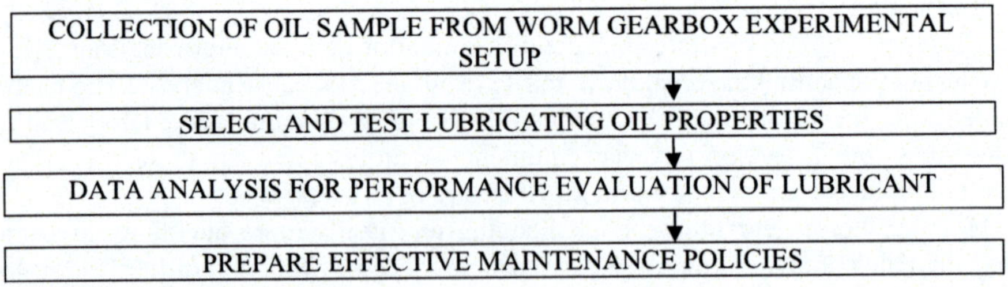

Fig. 1 Steps involved in the methodology

Table 1 Gearbox working condition

Parameter	Value
Load	10 kg
Input speed	1440 rpm
Output speed	72 rpm
Oil capacity	650 ml

Table 2 Standard used for oil testing

Parameter	Standard
Kinematic viscosity	ASTM D 445
Total acid number	ASTM D 664

procedure is to gather quantitative information for condition monitoring of lubricating oil and acquire a better insight and understanding of the status and working of the worm gearbox. Oil viscosity and Total Acid Number (TAN) have been checked according to worm gearbox industrial application.

The oil samples were collected at predefined running durations of 100 h while the system is in operation, and from the same place. The known working conditions are given in Table 1.

Speed reduction worm gearbox is aimed for constant ratio (1:20) with the right angle between output shaft and input shaft. The selection of critical properties and the testing facility was decided according to oil manufacture guidelines and followed the international standards. Standards set out in Table 2 were used for the parameters being examined and evaluated.

This data analysis will be useful to get information about lubricating oil condition and will be beneficial for maintenance decision as well as industry for effective maintenance policies which will reduce lead time and cost of breakdown.

4 Result and Discussion

Table 3 indicates the result of oil properties which is discussed in this section

Viscosity and Total Acid Number (TAN) were selected for condition monitoring of worm gearbox lubricant. The measurement of the lubricating oil viscosity is an important test, because it shows the basic physical property. Viscosity is the critical oil parameter, for example, viscosity influences oil film thickness, friction between gear

Table 3 Result of oil properties

Time (h)	0	100	200	300	400	500	600	700
Kinematic viscosity @ 40 °C (cSt)	333.6	308.5	305.1	305.6	309.4	304.8	308.6	305.7
Total acid number TAN (mg KOH/g)	0.2	0.3	0.5	0.7	0.7	0.8	0.9	1

Fig. 2 Tendency of viscosity value variation

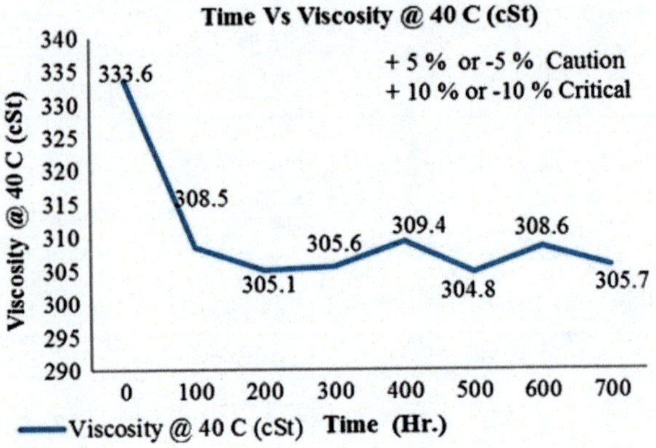

pair, internal or external leakage, and generation of heat. The tendency of viscosity over time is shown in Fig. 2.

The highest deviation of the oil viscosity value was found to be – 8.5 % between used and unused oil sample. The viscosity decreased due to considerable rise in temperature leading to oil degradation. This further leads to friction and wear between gear pair. If viscosity crosses cautionary limit, then one needs to investigate the root cause and make a preventive action. As a result of these consequences, the process becomes expensive. Oil viscosity usually decreases with aging process.

Total Acid Number (TAN) is a significant parameter for the condition monitoring of lubricating oils. The sum of acidic compounds inside the gear oil is called TAN that is indicated as the number of milligrams of potassium hydroxide like the acidic mixture. A change in oil's acid concentration may arise from numerous sources like acid contaminants, wrong oil, depletion of the alkaline reserve, by-products of oxidation, etc., which can lead to an increase in acid concentration. The tendency of total acid number over time is shown in Fig. 3.

Fig. 3 Tendency of TAN value variation

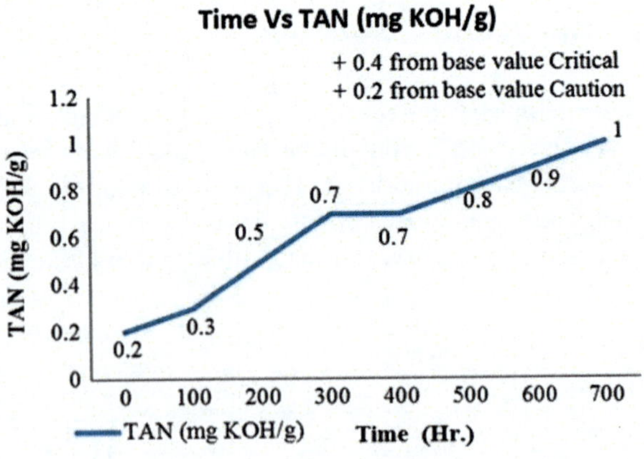

Total Acid Number (TAN) value is increased continuously during the time of operation; compounds are generated which increases the acidity. If a sudden increase in the TAN value is observed, the oil must be thoroughly checked. According to previous experience, oil becomes acidic because of the oxidation process due to aging. When TAN value increases to 0.7 mg KOH/g from the initial value because of oil oxidation then the oil passes the critical limit. Because of this oil becomes more acidic as well as corrosive to lubricate the parts.

5 Conclusion

Following conclusions can be drawn based on the Condition Monitoring of Worm Gearbox through Oil Analysis:

- Based on the literature review, it can be concluded that the parameters like viscosity, Total Acid Number (TAN), wear particles, water content, Total Base Number (TBN), insolubles, etc., become very critical.
- The experimental results reveal that there is 8.0% decrease in viscosity due to degradation oil.
- There is increase in Total Acid Number (TAN) due to oil oxidation which leads to corrosion of lubricated parts.
- It becomes evident from the results that the need for condition monitoring of lubricating oil of worm gearbox is necessary as preventive measure.
- Based on condition monitoring of lubricating oil as a preventive measure, one can take decision regarding replacement of oil.
- Condition monitoring of oil helps to diagnose the quality of lubricant without having to dismantle the complete worm gearbox.
- Lubricating oil analysis presents the earliest information about the worm gearbox failure which can save cost and time on reactive maintenance.
- Maintenance managers are progressively autonomous in arriving at the suitable decision based on acquired qualitative data about the machine's condition and state.

Acknowledgements The authors are thankful to Chougule Engineering Works, Kolhapur, for the financial and technical support for the present research work. Special thanks to Department of Manufacturing Engineering and Industrial Management, College of Engineering Pune and Department of Mechanical Engineering, K.I.T.'s College of Engineering, Kolhapur for providing Laboratory and Library facilities.

References

1. Kumar A, Dewangan SK, Rakesh GL (2016) A review on comparison of lubricants and development of optimization model of rolling mill. Int J Innov Res Multidisc Field 2(7):189–194
2. Narwariya PS, Sharma P, Sharma N, Chauhan P, Sharma P, Jain A (2018) A review paper on condition monitoring of gear oil. J Adv Res Appl Sci 5(4):422–424
3. Rathore SS et al (2019) An overview of diagnostics and prognostics of rotating machines for timely maintenance intervention. IOP Conf Ser Mater Sci Eng 691:012054, 1–9
4. Ghodake SB, Mishra AK, Deokar AV (2016) A review paper on fault detection of worm gearbox. Int Adv Res J Sci Eng Technol 3(1):161–164
5. Bhat JS, Sonawane BU (2019) Analysis of wear particles in lubricating oil of worm gearbox using ferrography. Manufact Technol Today 18(5):34–40
6. Jain A et al (2016) A review on "condition monitoring of used lubricating oil". Int J Innov Res Sci Eng Technol 6(9):18300–18305
7. Wakiru JM et al (2018) Analysis of lubrication oil towards maintenance grouping for multiple equipment using fuzzy cluster analysis. IOP Conf Ser Mater Sci Eng 393:012011, 1–9
8. Bandyopadhyay PK et al (2017) Application of oil condition monitoring techniques for improving critical equipment availability in steel plant. Tribol Online 12(2):37–41
9. Karanovi VV et al (2018) Benefits of lubricant oil analysis for maintenance decision support: a case study. IOP Conf Ser Mater Sci Eng 393:012013, 1–7
10. Bhat JS, Sonawane BU (2018) Evaluation of performance deterioration of the used equipment oil in sliding pair. Int J Res Adv Technol 6(8):2061–2066
11. Zhu J et al (2013) Survey of lubrication oil condition monitoring, diagnostics, and prognostics techniques and systems. J Chem Sci Technol 2(3):100–115
12. AGMA 9005 Standard

Mathematical Formulation of Load Frequency Control for Multi-area Multi-source Realistic Power System Considering Energy Storage Devices

Ardhala Bala Krishna, Sobhit Saxena, and Vikram Kumar Kamboj

Nomenclature

CES	Capacitive energy storage
TCPS	Thyristor controlled phase shifter
ISO	Independent system operators
TRANSCO	Transmission companies
GENCO	Generation companies
DISCO	Distribution companies
GRC	Generation rate constant
Cpf	Contract participation factor
DPM	Disco participation matrix
ACE	Area control error
B_1, B_2	Frequency bias constant of area-1 and 2
δ_1^0, δ_2^0	Power angles of equivalent machines of the two areas
$\Delta\phi$	Angle of the phase shifter
k_ϕ	Stabilization gain
T_{gh}	Electric governor gain
δ_1^0, δ_2^0	Power angles of equivalent machines of the two areas
K_i, K_p	Integral and proportional gains of the system

A. B. Krishna · S. Saxena (✉) · V. K. Kamboj
School of Electronics and Electrical Engineering, Lovely Professional University, Phagwara, Punjab, India
e-mail: sobhit.23364@lpu.co.in

A. B. Krishna
Department of Electrical and Electronics Engineering, KKR & KSR Institute of Technology and Sciences, Guntur, Andhra Pradesh, India

V. K. Kamboj
Department of Electrical and Computer Engineering, Schulich School of Engineering, University of Calgary, Alberta, Canada

A. Parey et al. (eds.), *Recent Trends in Engineering Design*, Lecture Notes in Mechanical Engineering, https://doi.org/10.1007/978-981-16-1079-0_31

Δp_{D2}, Δp_{D1}	Local area-2 and area-1 loads
Δp_{l4}, Δp_{l3}, Δp_{l2}, Δp_{l1}	Per unit (pu) loads of DISCOs
Δp_{Gci}	Produced steady-state set-out limit power
k_r	Gain constant
T_r	Turbine time constant
T_{sg}	Governor time constant
T_{rs}	Hydro turbine speed governor reset time
T_{rh}	Hydro turbine speed governor transient droop time constant
T_{cd}	Compressor discharge system time
X_g, Y_g	Governor servo time and droop time constants
B_g, C_g	Valve positioner time constants
T_f	Fuel dynamics of turbine
T_{cr}	Turbine compressor discharge
$\Delta F_i(s)$	ith area frequency fluctuation
T_w	Time constant for water
T_{TCPS}	Time constant of TCPS device
K_{ces}	CES gain constant
T_{ces}	CES time constant
T_{12}	Tie line synchronizing coefficient
Δp_{Gi}	ith GENCO generated power change
Δp_{Li}^{uc}	Unagreed DISCO of the ith area power demand
Δp_{lj}	Demand load of jth DISCO
λ, Y, X, Z	Uncontracted, state, control, and disturbance load vectors

1 Introduction

Load frequency regulation (LFR) in a multi-area realistic electric power system (multi source) has been the hot theme in the modern era and in the last few decades, several researchers have introduced many innovations to achieve requisition and supply loads. After deregulation, the dynamic electricity sector has engaged with new industry players such as ISO, GENCO, DISCO, and TRANSCO. The individual contribution of these organizations increases the complexity in the power system. Such newly formed organizations, under very modest and distributed control systems, have their own responsibility to maintain unity, protection, and reliability. In this new electricity market environment, frequency regulation is becoming more important to provide greater reliability and protection for power. Furthermore, the number of DISCO rises regularly for reaching domestic and commercial applications considerably. The current power systems facing challenges in terms of frequency regulation, which raises concerns for power system operators and regulatory agencies [1, 2].

Here, the size of the electric power system raises regularly due to huge requirements in all sectors (for example, commercial, agricultural, and industrial). Due to

Table 1 Configuration parameters for the proposed model

Area	GENCO	DISCO	CES	TCPS
1	Hydro, thermal, gas	2	1	1
2	Hydro, thermal, gas	2	1	

these kinds of issues, there may be a possibility to the failure of each individual generating unit which leads to complete or practical collapse of the electric power supply that is common to be encountered. Therefore in the modern era, individual power system frequency regulation is difficult to maintain. It leads to a multi-area electric power system with tie line error control methodologies in the open market. This research paper presents the complete mathematical modeling of the two areas system which includes Thermal, Hydro and gas generating sources with PI controller.

Recently, several traditional control methods have been employed to resolve current load frequency regulation issues throughout the electric power system. Table 1 displays the configuration of the proposed model. Table 2 provides a literature survey and history of recent approaching methodologies focused on various multi-area realistic multi-source FACTS devices and specific types of controllers introduced by different researchers in order to preserve the LFR of the electric power system. Figure 1 displays the representation of the proposed mathematical modeling. Figure 2 illustrates the block diagram representation of the proposed mathematical model with EV/BEV charging/discharging, along with renewable energy sources and energy storage devices (ESD).

2 Mathematical Modeling

Here, two areas with three GENCO (thermal, gas, and hydro) sources and power exchanged CES in each area are taken into consideration. A TCPS is integrated with a tie line, to control the sudden changes in the tie-line power. To control the entire power system, an optimization PI controller is used. The most commonly used PI controller is for load frequency control strategies. The benefit of this regulator is that the steady-state error is reduced to zero.

From Fig. 1, area-1 net power is the amount of the energy provided with a thermal ($\text{PF}_{\text{thermal1}}$), hydro ($\text{PF}_{\text{hydro1}}$), and gas ($\text{PF}_{\text{gas1}}$) generating sources.

The total outcome generation of area-1 is

$$\text{POWER1} = \text{PF}_{\text{thermal1}} + \text{PF}_{\text{hydro1}} + \text{PF}_{\text{gas1}} \pm \text{PF}_{\text{ces1}} \tag{1a}$$

Here, P_{thermal1} is the thermal energy generation, P_{hydro1} is the hydro energy generation, P_{diesel1} is the diesel energy generation, and P_{ces} is the exchanged capacitive energy storage bank. The net amount of area-2 generating power is

Table 2 Literature review on recently proposed models focused on various multi-area realistic deregulated multi-source FACTS devices and specific types of controllers

Author name/year	Generating sources	Control areas/system type	Deregulated/traditional	Controller type/structure	FACTS/other devices	References
Dhundhara et al. (2018)	Thermal, hydro, and gas	2-area, multi-source	Deregulated	PI controller	CES and TCPS	[3]
Ankush et al. (2020)	Thermal, hydro, PV, diesel, fuel, and biomass	3-area, multi-source	Traditional	PID controller	CES and TCPS	[4]
Farahari et al. (2013)	Thermal	2-area, single-source	Traditional	PID controller	SMES	[5]
Kangarlu et al. (2014)	Thermal	Single area, infinite bus system	Traditional	CHB cascade H bridge level	SMES	[6]
Chidambaram et al. (2013)	Thermal	2-area, single-source	Deregulated	Integral and fuzzy logic controller	RFB	[7]
Pappachen et al. (2016)	Hydro and thermal	2-area, multi-source	Deregulated	ANFIS controller	SMES and TCPS	[1]
Shankar et al. (2016)	Thermal, hydro, and gas	2-area, multi-source	Deregulated	Integral controller	RFB	[8]
Arya (2017)	Thermal, hydro, and gas	2-area, multi-source	Traditional	Fractional order fuzzy PID controller	CES and TCPS	[9]
Ponnusamy et al. (2015)	Hydro and thermal	2-area, multi-source	Deregulated	Integral controller	SSSC and CES	[10]
Bhatt et al. (2010)	Thermal, hydro, and diesel	2-area, multi-source	Traditional	Integral controller	CES and TCPS	[11]
Mukherjee et al. (2009)	Thermal and hydro	SMIB—single machine infinite bus system	Traditional	Integral controller	CES	[12]

(continued)

Table 2 (continued)

Author name/year	Generating sources	Control areas/system type	Deregulated/traditional	Controller type/structure	FACTS/other devices	References
Mahto et al. (2017)	(IHPS) isolated wind-diesel hybrid power system	2-area, multi-source	Traditional	PID with scaling factor-based fuzzy logic controller	CES	[13]
Chaine et al. (2015)	Thermal	2-area, single-source	Traditional	Integral controller	SMES	[14]
Selvaraju et al. (2016)	Thermal	2-area, multi-source	Deregulated	PI controller	SMES and RFB	[15]

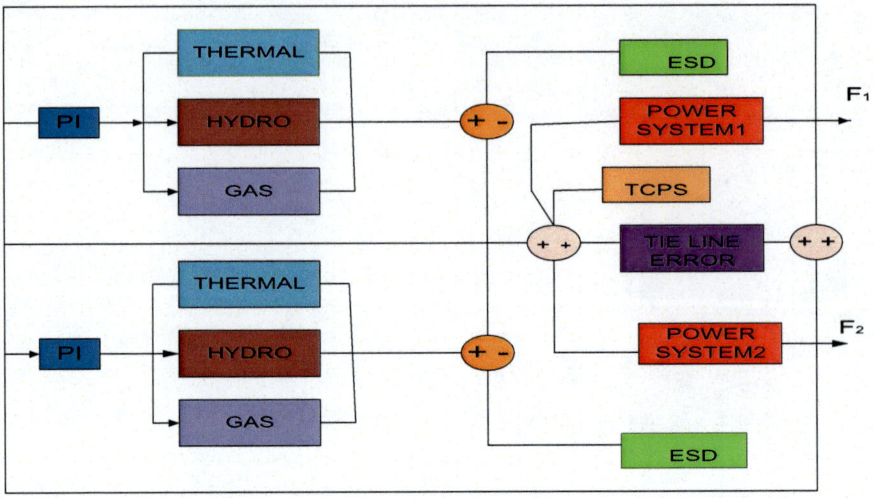

Fig. 1 Block diagram representation of the proposed mathematical model

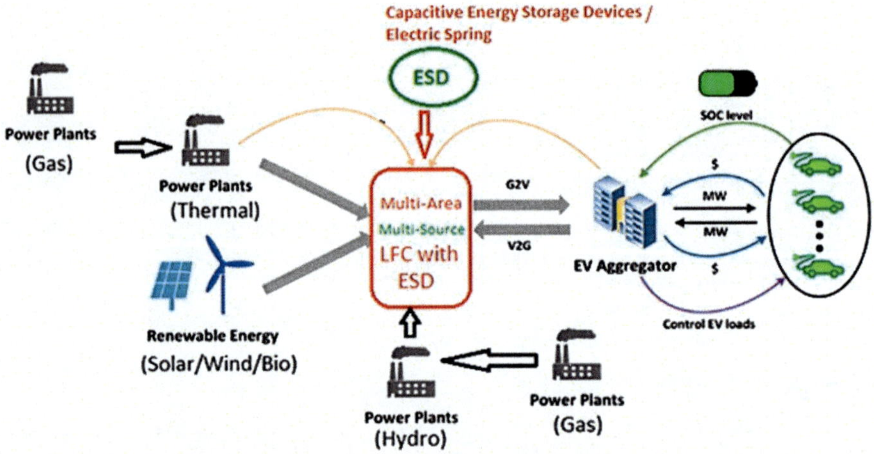

Fig. 2 Basic model with including EV/BEV and renewable sources

$$\text{POWER2} = \text{PF}_{\text{thermal2}} + \text{PF}_{\text{hydro2}} + \text{PF}_{\text{gas2}} \pm \text{PF}_{\text{ces2}} \tag{1b}$$

The total generating net amount of power is as in Eq. (1c):

$$\text{Total POWER} = \text{POWER1} + \text{POWER2} \tag{1c}$$

A thermal generator consists majorly of two parameters, namely governor and reheat turbine including GRC.

$$\text{thermal governor} = \frac{1}{sT_{sg} + s}, \quad \text{reheat turbine} = \frac{sT_r K_r + 1}{sT_r + 1} \tag{2a}$$

A hydro generator has majorly three parameters namely, hydro governor, hydro turbine, and transient droop compensator.

Transfer functions for the major elements are

$$\text{hydro governor} = \frac{1}{sT_{gh} + s}, \quad \text{hydro turbine} = \frac{-sT_w + 1}{0.5sT_w + 1},$$
$$\text{transient droop compensator} = \frac{sT_{rs} + 1}{sT_{rh} + 1} \tag{2b}$$

The gas turbine control design consists of four control circuits: governor delay, valve positioner, fuel dynamics, and compressor discharge system.

$$\text{Compressor discharge system} = \frac{1}{sT_{cd} + 1}, \quad \text{Gas governor} = \frac{1 + X_g s}{sY_g + 1},$$
$$\text{Valve positioner} = \frac{1}{sB_g + C_g}, \quad \text{Fuel dynamics} = \frac{1 - T_{cr}s}{sT_f + 1} \tag{3}$$

2.1 *CES Model*

A capacitive energy storage (CES) performs with a greater efficacy than the existing energy storage devices (ESD). Similar major features, such as quick discharge and charge rate, prevent the loss of efficacy. CES presented in equation (4) having quick response time for high cycle numbers, high power density, long serviceable lifespan and sufficient capability to distribute more and daily power requisition to grid makes it suitable for ESD [13].

$$\Delta p_{\text{ces}} = \left[\frac{k_{\text{ces}}}{1 + sT_{\text{ces}}}\right] + \left[\frac{1 + sT_{c1}}{1 + sT_{c2}}\right] + \left[\frac{1 + sT_{c3}}{1 + sT_{c4}}\right] \Delta F_i(s) \tag{4}$$

2.2 *TCPS Model*

Among the FACTs controller devices, thyristor controlled phase shifter (TCPS) has widespread range of uses for the power transmission line in series with the applications by a degree of compensation in realistic novel power systems. TCPS maintains the power grid's efficiency and stability limit by offering flexible power scheduling under many (having to change) operating environments. It can be used to increase a transmission line's power-transfer capability by damping the swings in power through regional and cross-area oscillations [16]. Under challenging conditions, the TCPS supports the real flow of power through transmission lines (tie lines), slows down high frequency, and regulates the electric power system voltage by adjusting the difference between the phase angles.

As shown below, the exponential voltage regulation between area-1 and area-2 of the tie-line connection is indicated by Eq. (5).

$$\Delta p^{o}_{\text{tieline}}(s) = \frac{2\pi T^{o}_{12}}{S}(\Delta F_1(s) - \Delta F_2(s)) \tag{5}$$

The actual power flow exchange between area-1 and area-2 can be defined as shown below after integrating a TCPS device into the proposed model:

$$\Delta p^{\text{real}}_{\text{tieline},12} = \frac{|V_1||V_2|}{X_{12}}\sin(\delta_1 + \delta_2 - \phi) \tag{6}$$

Here, $\delta_1, \delta_2,$ and ϕ are from their original values, $\delta_1^0, \delta_2^0,$ and ϕ^0, respectively. The voltage control oscillation between the two area tie line power as shown by Eq. (7) follows a limited signal approximation approach.

$$\Delta p^{\text{real}}_{\text{tieline},12} = T_{12}(\Delta\delta_1 + \Delta\delta_2) + T_{12}\Delta\phi) \tag{7}$$

Here,

$$T_{12} = \frac{|V_1||V_2|}{X_{12}}\cos\left(\delta_1^0 + \delta_2^0 - \phi\right) \tag{8}$$

In addition, the angular deviation can be written as

$$\Delta\delta_1 = 2\pi\Delta F_1 \mathrm{d}t \quad \text{and} \quad \Delta\delta_2 = 2\pi\Delta F_2 \mathrm{d}t \tag{9}$$

Here, applying the Laplace transform of Eq. (7),

$$\Delta p^{\text{real}}_{\text{tieline},12}(s) = \frac{2\pi T^{o}_{12}}{S}(\Delta F_1(s) - \Delta F_2(s)) + T_{12}\Delta\phi(s) \tag{10}$$

Equation (10) demonstrates that the angle of the phase shifter ($\Delta\phi$) controls the exchange of power flow in the tie line. The mathematical expression represents that the angle(s) of the phase shifter is represented as

$$\phi(s) = \Delta F_1(s)\frac{k_\phi}{1 + sT_{\text{TCPS}}}\Delta\text{Error} \tag{11}$$

$\Delta F_1(s)$ is the frequency fluctuation in area-1, which is taken as the error signal depicted in (22). And hence Eq. (10) can be written as

$$\Delta p_{\text{tieline},12}^{\text{real}}(s) = \frac{2\pi T_{12}^{o}}{S}(\Delta F_1(s) - \Delta F_2(s)) + \Delta F_1(s)\frac{k_{\phi}}{1 + sT_{\text{TCPS}}}T_{12} \tag{12}$$

2.3 PI Controller Design

Despite their simple control structure, PI controllers are now the type of controller most commonly used in technological innovation. A PI controller is a device that regulates the pressure, temperature, flow, speed, and many industrial management flow variables. PI controller as an application of temperature control is virtually universal and occurs in various chemical and scientific processes along with automation.

$$G(s) = k_p + \frac{k_i}{s}$$

The sources of power generation listed above from the different GENCOs of the novel hybrid power system are discussed here. For power demand, DISCO is free to contract any GENCO. DPM helps envision contracts under GENCO-DISCO.

In general, the representation for the DISCO participation matrix, for two areas, three sources, and two DISCOS in each area is

$$\text{DPM} = \begin{bmatrix} cpf_{v11} & cpf_{v12} & cpf_{v13} & cpf_{v14} \\ cpf_{v21} & cpf_{v22} & cpf_{v23} & cpf_{v24} \\ cpf_{v31} & cpf_{v32} & cpf_{v33} & cpf_{v34} \\ cpf_{v41} & cpf_{v42} & cpf_{v43} & cpf_{v44} \\ cpf_{v51} & cpf_{v52} & cpf_{v53} & cpf_{v54} \\ cpf_{v61} & cpf_{v62} & cpf_{v63} & cpf_{v64} \end{bmatrix} \tag{13}$$

Each dimension of the cpf-matrix measures the fraction of the overall amount load power contracted by DISCOs, and the corresponding GENCO involved in the deal satisfies this demand. Here, cpf_{vij} replicates the contract participation matrix factor, $\sum_{i=1}^{n} cpf_{vij} = 1$, here $j = 1, 2, \ldots, k$, n replicates the sum of GENCOs, and k replicates the sum of DISCOs [17].

The mathematical expression for the local load in a realistic deregulated open market is illustrated as

$$\Delta p_{l2} + \Delta p_{l1} = \Delta p_{D1} \tag{14}$$

$$\Delta p_{l4} + \Delta p_{l3} = \Delta p_{D2} \tag{15}$$

Every GENCO supply produced or contracted by the power system as proposed under the study can be illustrated as follows:

$$\Delta p_{Gci} = \sum_{j=1}^{\text{GENCO}_{\text{total}}} cpf_{vij} * \Delta p_{lj} \tag{16}$$

Likewise, when any DISCO violates the agreement by requiring more than predefined power or contractual limitations, then the unidentified contractual demand is expressed as unagreed more load, and this demand is only met by the GENCO which belongs to the very same region as the DISCO [16]. GENCOs are assigned the unagreed power load demands premised on area participation factors (apfs). Besides, area-1 apfs were taken into account as $\text{PF}_{\text{gas}_{13}}$, $\text{PF}_{\text{ther}_{11}}$, and $\text{PF}_{\text{hydr}_{12}}$. For area-2, apfs were taken into account as $\text{PF}_{\text{ther}_{21}}$, $\text{PF}_{\text{hydr}_{22}}$, and $\text{PF}_{\text{gas}_{23}}$. According to their apfs and predefined contract, each GENCO except for the output power change can be formulated as

$$\Delta p_{Gi} = cpf_{i1} + \Delta p_{l1} + cpf_{i2} + \Delta p_{l2} + cpf_{i3} + \Delta p_{l3} + cpf_{i4} + \Delta p_{l4} + apf_i \Delta p_{Li}^{uc} \tag{17}$$

And hence, the local unagreed power demand is

$$\Delta p_{L1}^{uc} + \Delta p_{l2} + \Delta p_{l1} = \Delta p_{D1} \tag{18}$$

$$\Delta p_{L2}^{uc} + \Delta p_{l4} + \Delta p_{l3} = \Delta p_{D2} \tag{19}$$

Here, Δp_{L2}^{uc} and Δp_{L1}^{uc} are the unagreed load requisition of the respective area-2 and area-1. The unexpected demand or change in the allocated power causes variations in the tie-line power flow and is the area control error (ACE) [8]. You can measure the set-out tie-line power exchange among the areas rates as

$$\Delta p_{\text{tieline},12}^{\text{allocated}} = \sum_{n=1}^{3} \sum_{m=3}^{4} cpf_{nm} \Delta p_{lm} - \sum_{n=4}^{6} \sum_{m=1}^{2} cpf_{nm} \Delta p_{lm} \tag{20}$$

The real value that passes through the tie line may be represented as given by Eq. (21)

$$\Delta p_{\text{tieline},12}^{\text{real}} = \frac{2\pi T_{12}}{s} (\Delta F_1 - \Delta F_2) \tag{21}$$

The error which occurs in the tie-line power due to the set-out limit in power flow can be described as in Eq. (22),

$$\Delta p_{\text{tieline},12}^{\text{error}} = \Delta p_{\text{tie},12}^{\text{real}} - \Delta p_{\text{tieline},12}^{\text{allocated}} \tag{22}$$

If the real tie-line power flow exceeds the error limits in power flow, according to the current flow, set value ($\Delta p_{\text{tieline},12}^{\text{error}}$) is zero, and called the condition of constant state. An area defined by Eq. (23) represents a linear combination of the power flow error in tie line, and the weighted frequency fluctuation, ACE, can be written as

$$\mathrm{ACE}_1 = \Delta F_1 * B_1 + \Delta p^{\mathrm{error}}_{\mathrm{tieline},12} \quad \text{and as well as,} \quad \mathrm{ACE}_2 = \Delta F_2 * B_2 + \Delta p^{\mathrm{error}}_{\mathrm{tieline},21} \tag{23}$$

The linear dynamics nature of current mathematical model contemplated in a competitive electricity market for frequency control may be indicated using the static variable differential mathematical equation as defined illustrates in (24)

$$A = PX + QY + RZ + S\lambda \tag{24}$$

S, P, Q, and R are constant Compatible Matrix measurements tied up with them [16].

Equations (25)–(28) will display the matrices for the proposed scheme under study in summary form:

$$X = [\Delta F_1 \Delta F_2 \Delta p^{\mathrm{real}}_{\mathrm{tie},12} \Delta p_{Gc1} \Delta p_{Gc2} \Delta p_{Gc3} \Delta p_{Gc4} \Delta p_{Gc5} \Delta p_{Gc6} \Delta p_{\mathrm{ces1}} \Delta p_{\mathrm{ces2}} \Delta p_{\mathrm{TCPS}}]^T \tag{25}$$

$$U_S = [U_{S1} U_{S2}]^T \tag{26}$$

$$p = [\Delta p_{l1} + \Delta p_{l2} + \Delta p_{l3} + \Delta p_{l4}]^T \quad \text{or} \quad p = [\Delta p_{D1} \Delta p_{D2}]^T \tag{27}$$

$$\text{and} \quad p' = [\Delta p^{uc}_{l1} \Delta p^{uc}_{l2}]^T \tag{28}$$

In any of the several electrical power markets—a concept known as vehicle-to-grid power or V2 G power—the BEVs could sell both power and storage capacity back to the grid. Given that EVs are operating on electricity, the high market penetration of EVs is likely to boost electricity demand. Power organizations are confronted that such an excess supply could result in an increase in the peak load which will also allow them to bring extra power back to the grid. Moreover, certain EVs are stationed over 90 percent of the time; when not powered, such EVs could be used as large-scale distributed batteries and can supply the energy grid with power storage and ancillary services. This idea is also known as the Vehicle-To-Grid (V2G) control system.

3 Conclusion

A full mathematical model for multi-area hydrothermal gas sources based on a deregulated practical power system model in conjunction with traditional controller, energy storage device, and TCPS design is modeled in the paper. Such framework will be used in the future for multi-area multi-source realistic deregulated electric power systems including different energy storage devices and renewable energy sources with an EV/BEV grid power system, where load frequency regulations with various optimization methodologies and traditional controllers are applicable.

References

1. Pappachen A, Peer Fathima A (2016) Load frequency control in deregulated power system integrated with SMES-TCPS combination using ANFIS controller. Int J Electr Power Energy Syst 82:519–534. https://doi.org/10.1016/j.ijepes.2016.04.032
2. Chandra Sekhar GT, Sahu RK, Baliarsingh AK, Panda S (2016) Load frequency control of power system under deregulated environment using optimal firefly algorithm. Int J Electr Power Energy Syst 74:195–211. https://doi.org/10.1016/j.ijepes.2015.07.025
3. Dhundhara S, Verma YP (2018) Capacitive energy storage with optimized controller for frequency regulation in realistic multisource deregulated power system. Energy 147:1108–1128. https://doi.org/10.1016/j.energy.2018.01.076
4. Dutta A, Prakash S (2020) Load frequency control of multi-area hybrid power system integrated with renewable energy sources utilizing FACTS & energy storage system. Environ Prog Sustain Energy 39(2):1–13. https://doi.org/10.1002/ep.13329
5. Farahani M, Ganjefar S (2013) Solving LFC problem in an interconnected power system using superconducting magnetic energy storage. Phys C Supercond Appl 487:60–66. https://doi.org/10.1016/j.physc.2013.02.005
6. Kangarlu MF, Pahlavani MRA (2014) Cascaded multilevel converter based superconducting magnetic energy storage system for frequency control. Energy 70:504–513. https://doi.org/10.1016/j.energy.2014.04.025
7. Chidambaram IA, Paramasivam B (2013) Optimized load-frequency simulation in restructured power system with redox flow batteries and interline power flow controller. Int J Electr Power Energy Syst 50(1):9–24. https://doi.org/10.1016/j.ijepes.2013.02.004
8. Shankar R, Chatterjee K, Bhushan R (2016) Impact of energy storage system on load frequency control for diverse sources of interconnected power system in deregulated power environment. Int J Electr Power Energy Syst 79:11–26. https://doi.org/10.1016/j.ijepes.2015.12.029
9. Arya Y (2017) AGC performance enrichment of multi-source hydrothermal gas power systems using new optimized FOFPID controller and redox flow batteries. Energy 127:704–715. https://doi.org/10.1016/j.energy.2017.03.129
10. Ponnusamy M, Banakara B, Dash SS, Veerasamy M (2015) Design of integral controller for load frequency control of static synchronous series compensator and capacitive energy source based multi area system consisting of diverse sources of generation employing Imperialistic Competition Algorithm. Int J Electr Power Energy Syst 73:863–871. https://doi.org/10.1016/j.ijepes.2015.06.019
11. Bhatt P, Roy R, Ghoshal SP (2010) GA/particle swarm intelligence based optimization of two specific varieties of controller devices applied to two-area multi-units automatic generation control. Int J Electr Power Energy Syst 32(4):299–310. https://doi.org/10.1016/j.ijepes.2009.09.004
12. Mukherjee V, Ghoshal SP (2009) Application of capacitive energy storage for transient performance improvement of power system. Electr Power Syst Res 79(2):282–294. https://doi.org/10.1016/j.epsr.2008.06.013
13. Mahto T, Mukherjee V (2017) A novel scaling factor based fuzzy logic controller for frequency control of an isolated hybrid power system. Energy 130:339–350. https://doi.org/10.1016/j.energy.2017.04.155
14. Chaine S, Tripathy M (2015) Design of an optimal SMES for automatic generation control of two-area thermal power system using Cuckoo search algorithm. J Electr Syst Inf Technol 2(1):1–13. https://doi.org/10.1016/j.jesit.2015.03.001

15. Selvaraju RK, Somaskandan G (2016) Impact of energy storage units on load frequency control of deregulated power systems. Energy 97:214–228. https://doi.org/10.1016/j.energy.2015.12.121
16. Abraham RJ, Das D, Patra A (2015) AGC system after deregulation considering TCPS in series with the tie-line. Int J Emerg Electr Power Syst 16(3):281–295. https://doi.org/10.1515/ijeeps-2013-0165
17. Donde V, Pai MA, Hiskens IA, Member S (2001) Simulation and optimization in an AGC system after deregulation. IEEE Trans Power Syst 16(3):481–489

Optimization of Process Parameters for a Better Surface Finish of ABS Parts Prepared by Fused Deposition Modelling (FDM): A Comprehensive Review

Harpinder Singh Sandhu, T. Rohit Raj, L. Venkatasubramaniam, and Prateek Tiwari

1 Introduction

Due to the aid of rapid prototyping techniques, today, it feasible to construct relatively complex parts depending upon the computer 3D geometries. A majority of the rapid prototyping processes are able to create parts from a diverse family of common and special materials. Material selection is done depending on the type of rapid prototyping technology used [1].

Over the last few years, FDM technology has attained a huge general public interest due to its use and ease of machinery. Today, several FDM machines are designed and manufactured by hobbyists as a do-it-yourself (DIY) machine for personal use [2]. Although 3D printing technology has been around since the late 1980s, it is now recognized as one of the most important technological developments of the twenty-first century. A plethora of different 3D printing techniques has been invented over the years. Fused Deposition Modelling (FDM) was first invented and is still reckoned as well-liked and favourable today [3].

Fused Deposition Modelling (FDM) is a 3D printing technology (Fig. 1) that works on an "additive manufacturing" principle that lays down material layer by layer. A heating chamber is installed in which the filament is inserted where it is melted down and pushed out of the nozzle at a specific maintained rate based on the relative rate of movement of the three axes [4]. Scott Crump developed this technology first in the late 1980s. FDM technology is processed by software known as the STL file (stereolithography file format). Thereafter, we have to slice the model with another program for the build process. Support structures may be generated if necessary [5]. The thermoplastic material is extruded to form layers as the material hardens after extrusion from the nozzle and hence the final model is formed [6]. An extrusion nozzle turns the flow on and off and a plastic filament is unwound from a coil. A worm-drive is used to push the filament into the nozzle at a controlled

H. S. Sandhu (✉) · T. R. Raj · L. Venkatasubramaniam · P. Tiwari
School of Mechanical Engineering, Lovely Professional University, Phagwara, Punjab, India

A. Parey et al. (eds.), *Recent Trends in Engineering Design*, Lecture Notes in Mechanical Engineering, https://doi.org/10.1007/978-981-16-1079-0_32

Fig. 1 Schematic illustration of FDM technology

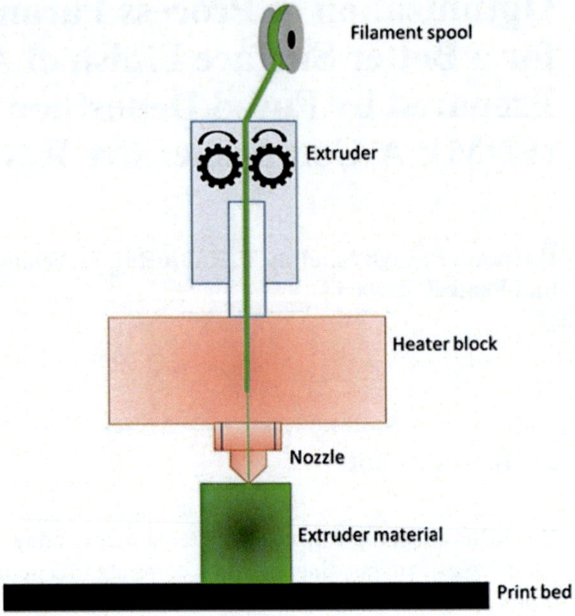

rate [7]. The glass transition temperature, the point at which it melts, and its variety of filaments determine the liquefier block temperature. Then, the high-temperature nozzle is used to extrude the molten filament and after that, it hardens on the base film on the build plate [8]. After this stage, the material is melted by heating the nozzle. A numerically controlled mechanism is used to move it in both vertical and horizontal directions. A computer-aided manufacturing (CAM) software package controls the nozzle, and the bottom-to-top approach is applied to build the part layer-wise. The extrusion head is moved by stepper motors. The mechanism uses a three-axis *X*–*Y*–*Z* rectilinear movement [9].

1.1 Empirical Contemplations in FDM Printing

The mechanical, as well as the physical, attributes of the 3D printed designs are impressive [10]. Heat transfer traits and physics of materials used in the filament are one of the primary criteria in the selection of materials [11]. In the 3D printer software, layer thickness, angle of raster, and inter-layer road gap width are pliant variables [8, 12]. Some other variables are also included such as the speed of extrusion, the layer height, infill density, the nozzle temperature, and the height of the build platform [13–15]. Fill density introduces the level to which the empty space can be filled within the printer with resin, from 0, 100 to empty, solid [16]. The standard of these features results in the staging outcomes in required perforation, interrelation, and object framework [8]. The raster (matrix with printed rows) interval and raster angle can be controlled during the FDM printing process, resulting in a porous membrane and porous distribution in the matrix [12]. Build plate features also significantly affect the affluent printing of prototypes. For example, the support material is necessary so that the base does not move so as to follow the extruded material in the build plate. According to the build, the platform should be large enough such that during printing, it has a proper hold on the print part, however, it should not be too large as it may control the detachment of the material from the build plate as soon as the printing process is complete [17].

The strength of the column is an important property of a filament. The strength of the lower column prevents the melting of the filament over liquefaction by bending or buckling. The strength of the filament column is due to the diameter, tensility, and ductility of the filament [18]. Moreover, to the strength of the column, the drop of pressure in the nozzle is responsible for the force that is necessary for melting through the nozzle, which is dependent on the viscosity and geometry of the print head of the melt. The melting index and pressure drop are inversely proportional to each other in relation that the lower the melting point of the filament, the higher the pressure. Conversely, as the melting point gets higher, it smoothens the filament consumption but induces a steady flow [18].

1.2 Material Properties of ABS Material

ABS material is one of the most abundantly used polymers in the world and touches all parts of our lives [19]. Acrylonitrile butadiene styrene (ABS) is a thermoplastic that resists heat and impact. It is a copolymer attained by the polymerization of acrylonitrile and styrene in the presence of polybutadiene, each based on a ratio of three, combining three monomers to form plastics of different grades. ABS polymers for FDM machines today offer more impact, flexural and tensile strengths [19]. Table 1 shows the various physical, thermal, and mechanical properties of ABS material when it is subject to different loads and limits under subjective environments.

2 Results and Discussion

The parameters in focus here are layer thickness, speed of deposition, air gap, raster angle, raster width, and fill density. Each parameter at some value influences the surface finish of the part, which opens the door for multiple combinations of parameter values used together. It forms a scale for the product ranging from a bad surface finish to a good one. We are interested in the combination which will provide us with the best surface finish (Fig. 2) [23].

Table 1 Properties of ABS materials [20–22]

Property	Density (kg/m^3)	Young's modulus (MPa)	Yield strength (MPa)	Poisson's ratio	Specific heat capacity (J/kg-K)	Thermal Conductivity (W/mK)	Tensile Strength (MPa)	Elongation at break (%)	Izod impact strength (J/m)
Value	1200	2200	31	0.4	1386	0.2256	40	50	34

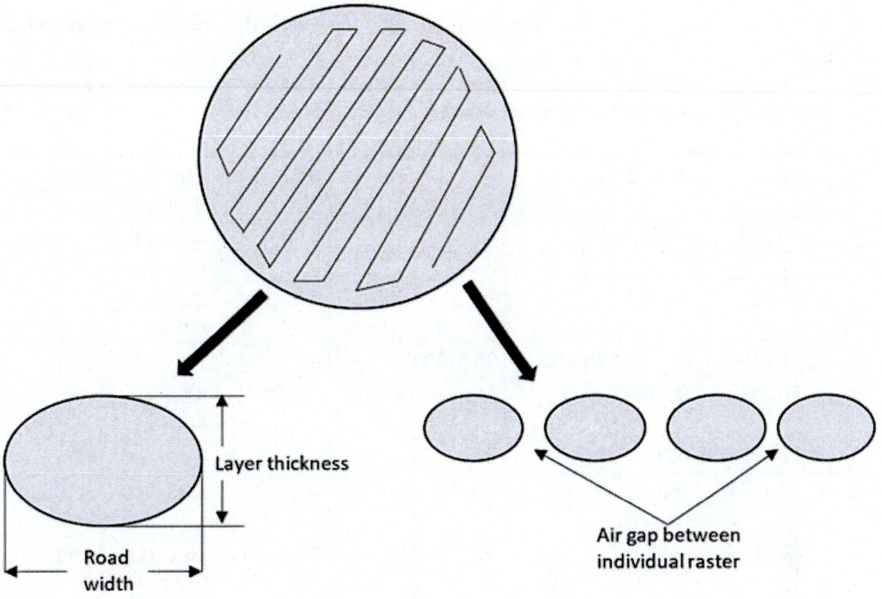

Fig. 2 Process parameters

2.1 *Layer Thickness*

It is the layer height (in mm) of the material deposited when added in succession [24]. A higher value of layer thickness is optimum for good dimensional accuracy but lowers surface finish as the layers take a long time to solidify and during this course, the next layer is added. So, a lower value (refer Table 2) is chosen to enhance the product quality. In addition to that, it improves the tensile strength of the part [25–32].

2.2 *Build Orientation*

It defines the optimal direction/orientation of how the part should be positioned so that it does not topple during the process and require less support material [24]. Build orientation plays an important role as it helps to keep the part from toppling during the FDM process and uses less support material so that there are less scrap and filling of the surface. Ali et al. [33] experimented with various angles in X, Y, Z directions by manually changing them and concluded that the FDM process for parts with orientations $X = 0°$, $Y = 45°$, $Z = 0°$, and $X = 45°$, $Y = 0°$, $Z = 0°$ cannot be performed. The layer thickness is to be kept low in the range of less than 0.300 mm.

Table 2 The process parameters have proven to be minimizing the surface roughness in the previous researches

S. No.	Layer thickness (mm)	Raster width (mm)	Speed of deposition (mm/s)	Fill density (%)	Raster angle (°)	Air gap (mm)	Remarks	References
1.	0.254	0.2832	–	–	10°	0.02	Flat surface	[25]
	0.254	0.6213	–	–	8°	0.02	Inclined surface	
	0.254	0.5102	–	–	37°	0.05	Curved surface	
2.	0.1152	–	36.3324	26.9730	–	–	–	[26]
3.	0.254	0.508	–	–	45°/90°	– 0.01	ABS M–30i	[27]
4.	0.178	0.305	–	–	–	–	–	[28]
5.	0.1778	(Fine rasters)	–	–	–	–	ABS–400	[29]
6.	0.254	–	–	–	90°	0.02	Stratasys P–400 ABS	[30]
7.	0.007	–	–	–	–	–	–	[31]
8.	0.127	0.407	–	–	–	0.0000004	ABS P400	[32]
9.	0.254	0.4564	–	–	0°	0	Ra = 2.10	[37]
10.	0.1788	–	–	–	–	–	Flat surface	[33]
11.	0.1778	–	100	–	–	–	Ra = 3.82	[35]
	0.254		150				Ra = 2.79	
	0.3556		200				Ra = 2.88	
12.	0.3556	0.537	200	50	45°	–	ABS400	[38]
13.	0.25	–	–	100	45°	0.02	Flat surface	[39]
14.	0.3556	0.537	150	90	45°, 90°	0.01	CNT–ABS and CFR–ABS	[40]
15.	0.2	–	200	80	90°	0.02	ABS	[41]
16.	0.35	–	100	–	30°, 60°	0.02	ABS cylindrical parts	[42]
17.	0.2	–	80–100	40	–	–	50% ABS and 50% PETG	[43]

(continued)

Table 2 (continued)

S. No.	Layer thickness (mm)	Raster width (mm)	Speed of deposition (mm/s)	Fill density (%)	Raster angle (°)	Air gap (mm)	Remarks	References
18.	0.1		100				(60% ABS + 40% PETG)	[44]
19.	0.174	–	–	100	–	–	ABS plus P430	[45]
20.	0.2	–	45	15	–	–	–	[46]
21.	0.1778	–	–	–	–	–	Investment casting ABS pattern (Solid part fill pattern)	[47]
22.	0.1778	0.7258			45°	– 0.001	–	[36]
23.	0.127	0.2032	–	–	45°	0.5588	Ra = 4.01	[24]

2.3 *Air Gap*

It is the space in between the consecutive beads of deposited FDM material (here ABS) [24]. Air gap values can be positive, negative, or zero. A zero or positive air gap has less build time compared to the negative one. The reason is that the toolpaths in the negative gaps are closed and require more time to fill while in a positive air gap, the spacing is filled by air thus reducing the time to fill the material. There is a decrease in the volume of the part when it is toggled from negative to positive [34].

2.4 *Raster Angle*

It refers to the angle made by the path of the nozzle with the *X*-axis of the bed. [24] It is seen that the surface roughness value is lowered when the raster angle nears 90°. The reason is at higher values, servo motors of both *X*- and *Y*-axes are required for deposition and this causes fluctuation resulting in voids and bad surface finish. Hence, values lower than 90° are to be used; refer to Table 2 for values [34].

2.5 *Raster Width*

Raster or road width is defined as the width at which beads are deposited. Extrusion nozzle diameter plays an important role in determining it [24]. A higher value of raster width will increase the perimeter of the bead, which will require a stylus to travel longer distances giving a poor surface finish. Nozzle extrusion diameter determines the width so a low raster width combined with a low speed of deposition will help improve the surface finish of the part [34].

2.6 Speed of Deposition

Speed of deposition refers to the rate (mm/s) at which the nozzle deposits the material [24]. A high rate will reduce the time taken to complete the FDM process but the surface finish will deter. It is generally used for prototypes where surface finish is not that much of a concern. In our study, a lower speed deposition value (refer to Table 2) is optimum for obtaining the quality we intend the 3D printed part to process [26, 35, 36].

3 Conclusion

This study provides a detailed understanding of the FDM process, parameters, and streamlining of process parameters to ameliorate the surface finish of ABS polymer models. Fused Deposition Modelling is proven to be a highly flexible process to fabricate High–Fidelity prototypes. A review of the various process parameters demonstrated to lessen the surface roughness of ABS models in past researches has been focused. Values of six process parameters recognized to be exacerbating the finish namely, layer thickness, raster width, deposition speed, fill density, raster angle, and air gap from existing researches have been compared. Various optimum surface finish parameters have been discussed for symmetric as well as non-symmetric parts such as cylinders, flat surfaces, curved and also complex 3D parts. We suggest that the steady parameters like layer thickness (0.350 mm, 0.254 mm), raster width [0.2832 mm (flat surface), 0.5102 mm (curved surface), and 0.6213 mm (inclined surface)], and nozzle diameter (T14 for part material and T16 for base material) should be lower for a better surface finish in symmetric parts and variable parameters such as the speed of deposition must be elevated in complex unsymmetric 3D printed parts for finer surface finish.

References

1. Dawoud M, Taha I, Ebeid SJ (2016) Mechanical behavior of ABS: an experimental study using FDM and injection moulding techniques. J Manuf Process 21:39–45
2. Dudek P (2013) FDM 3D printing technology in manufacturing composite elements. Arch Metall Mater 58(4):1415–1418
3. Oyvind Kallevik Grutle: Designing a 5–axis 3D printer (2015)

4. Evans B (2012) Practical 3D printers: the science and art of 3D printing. Apress, New York
5. Kun K, Miskolczi I, Fodor A (2015) 3D nyomtató építése és fejlesztése. Gradus 2(2):155–159
6. Kovács JG (2002) Gyors prototípus eljárások II. Gyakorlati megvalósítások, pp 103–107
7. Morovic L (2004) Rapid technológie: rapid technologies. In: Automation and CA systems in technology planning and manufacturings. Pozna University of Technology, pp 177–183
8. Chia HN, Wu BM (2015) Recent advances in 3D printing of biomaterials. J Biol Eng 9:1
9. Kun K (2016) Reconstruction and development of a 3D printer using FDM technology. Proc Eng 149:203–211
10. Polak R, Sedlacek F, Raz K (2017) Determination of FDM printer settings with regard to geometrical accuracy. In: Katalinic B (ed) Proceedings of the 28th DAAAM international symposium, pp 0561–0566
11. Zein I, Hutmacher DW, Tan KC, Teoh SH (2002) Fused deposition modeling of novel scaffold architectures for tissue engineering applications. Biomaterials 23:1169–1185
12. Chen H, Fuhlbrigge T, Zhang G, Masood S (2007) Application of fused deposition modelling in controlled drug delivery devices. Assemb Automat 27:215–221
13. Skowyra J, Pietrzak K, Alhnan MA (2015) Fabrication of extended–release patient–tailored prednisolone tablets via fused deposition modelling (FDM) 3D printing. Eur J Pharm Sci 68:11–17
14. Goyanes A, Chang H, Sedough D (2015) Fabrication of controlled release budesonide tablets via desktop (FDM) 3D printing. Int J Pharm 496:414–420
15. Pietrzak K, Isreb A, Alhnan MA (2015) A flexible–dose dispenser for immediate and extended release 3D printed tablets. Eur J Pharm Biopharm 96:380–387
16. Goyanes A, Buanz ABM, Basit AW, Gaisford S (2014) Fused–filament 3D printing (3DP) for fabrication of tablets. Int J Pharm B 476:88–92
17. Long J, Gholizadeh H, Lu J, Bunt C, Seyfoddin A (2017) Application of fused deposition modelling (FDM) method of 3D printing in drug delivery. Curr Pharm Des 23(3):433–439
18. Genina N, Holländer J, Jukarainen H, Mäkilä E, Salonen J, Sandler N (2016) Ethylene vinyl acetate (EVA) as a new drug carrier for 3D printed medical drug delivery devices. Eur J Pharm Sci 90:53–63
19. Han S, Xiao Y, Qi T, Li Z, Zeng Q (2017) Design and analysis of fused deposition modeling 3D printer nozzle for color mixing. Adv Mater Sci Eng 2:1–12
20. Yilmaz N, Bal H (2014) Comparison on hydrodynamic behavior of shape–adaptive and rigid marine propellers
21. Tasdemir M, Babat V, Yerlesen U (2014) Effect of friction and wear parameters on acrylonitrile butadiene styrene/aluminum–boron carbide–glass spheres polymer composites. Mechanics 20(4):407–413
22. Ferreira R, Quelho de Macedo R (2017) Residual thermal stress in fused deposition modelling. In: Associacao Brasileira de Engenharia e Ciencias Mecanicas - ABCM
23. Rosen GD, Stucker B (2015) Introduction and basic principles in additive manufacturing technologies. Springer, New York, pp 1–18
24. Onwubolu G, Rayegani F (2014) Characterization and optimization of mechanical properties of ABS parts manufactured by the fused deposition modelling process. Int J Manuf Eng 2014:1–13
25. Khan M, Mishra S (2020) Minimizing surface roughness of ABS–FDM build parts: an experimental approach. Mater Today Proc 26:1557–1566
26. Pramanik D, Mandal A, Kuar AS (2020) An experimental investigation on improvement of surface roughness of ABS on fused deposition modelling process. Mater Today Proc 26:860–863
27. Dinesh Kumar S, Nirmal Kannan V, Sankaranarayanan G (2014) Parameter optimization of ABS–M30i parts produced by fused deposition modeling for minimum surface roughness. Int J Curr Eng Technol. ISSN 2277-4106
28. Galantucci LM, Lavecchia F, Percoco G (2009) Experimental study aiming to enhance the surface finish of fused deposition modeled parts. CIRP Ann Manuf Technol 58:189–192
29. Horvath D, Noorani R, Mendelson M (2007) Improvement of surface roughness on abs 400 polymer materials using design of experiments (DOE). Mater Sci Forum 561–565:2389–2392

30. Raju M, Gupta M, Bhanot N, Sharma V (2018) A hybrid PSO–BFO evolutionary algorithm for optimization of fused deposition modelling process parameters. J Intell Manuf 30:2743–2758
31. Sood AK, Ohdar RK, Siba M (2012) Experimental investigation and empirical modelling of FDM process for compressive strength improvement. J Adv Res 3(1):81–90
32. Khan MS, Dash JP (2019) Enhancing surface finish of fused deposition modelling parts. In: Kumar L, Pandey P, Wimpenny D (eds) 3D Printing and Additive Manufacturing Technologies. Springer, Singapore, pp 45–57
33. Ali F, Chowdary BV, Maharaj J (2014) Influence of some process parameters on build time, material consumption, and surface roughness of FDM processed parts: inferences based on the taguchi design of experiments
34. Anitha R, Arunachalam S, Radhakrishnan P (2001) Critical parameters influencing the quality of prototypes in fused deposition modelling. J Mater Process Technol 118(1–3):385–388
35. Nancharaiah T, Raju D, Raju V (2010) An experimental investigation on surface quality and dimensional accuracy of FDM components. Int J Emerg Technol 1(2):106–111
36. Venkata Subba Reddy Y, Siddikal P, Saleem SM (2016) Improving the dimensional accuracy and surface roughness of FDM parts using optimization techniques. IOSR J Mech Civ Eng 16:18–22
37. Dong G, Wijaya G, Tang Y, Zhao YF (2018) Optimizing process parameters of fused deposition modeling by Taguchi method for the fabrication of lattice structures, vol 19, pp 62–72
38. Sheoran AJ, Kumar H (2020) Fused deposition modeling process parameters optimization and effect on mechanical properties and part quality: review and reflection on present research. Mater Today Proc 21(Part 3):1659–1672
39. Abeykoon C, Sri-Amphorn P, Fernando A (2020) Optimization of fused deposition modeling parameters for improved PLA and ABS 3D printed structures. Int J Lightweight Mater Manuf 3(3):284–297
40. Dev S, Srivastava R (2020) Experimental investigation and optimization of FDM process parameters for material and mechanical strength. Mater Today Proc 26(Part 2):1995–1999
41. Mora SM, Gil JC, Lopez AMC (2019) Influence of manufacturing parameters in the dimensional characteristics of ABS parts obtained by FDM using reverse engineering techniques. Proc Manuf 41:968–975
42. Saini M (2019) Optimization the process parameter of FDM 3D printer using Taguchi method for improving the tensile strength. Int J All Res Educ Sci Methods (IJARESM) 7. ISSN: 2455-6211
43. Yadav D, Chhabra D, Gupta RK, Phogat A, Ahlawat A (2020) Modeling and analysis of significant process parameters of FDM 3D printer using ANFIS. Mater Today Proc 21(Part 3):1592–1604
44. Nuñez PJ, Rivasa A, García-Plazaa E, Beamudb E, Sanz-Loberac A (2015) Dimensional and surface texture characterization in fused deposition modelling (FDM) with ABS plus. Proc Eng 132:856–863
45. Rahman H, John TD, Sivadasan M, Singh NK (2018) Investigation on the scale factor applicable to ABS based FDM additive manufacturing. Mater Today Proc 5:1640–1648
46. Hafsa MN, Ibrahim M, Wahab MdS, Zahid MS (2013) Evaluation of FDM pattern with ABS and PLA material. Appl Mech Mater 465–466:55–59
47. Wang CC, Lin T-W, Hu S-S (2007) Optimizing the rapid prototyping process by integrating the Taguchi method with the Gray relational analysis. Rapid Prototyp J 13(5):304–315